T0177491

ENERGY STORAGE SYSTEMS

Energy Storage Systems

System Design and Storage Technologies

Armin U. Schmiegel

University of Applied Science Reutlingen
REFUdrive GmbH, Pfullingen

OXFORD
UNIVERSITY PRESS

Great Clarendon Street, Oxford, OX2 6DP,
United Kingdom

Oxford University Press is a department of the University of Oxford.
It furthers the University's objective of excellence in research, scholarship,
and education by publishing worldwide. Oxford is a registered trade mark of
Oxford University Press in the UK and in certain other countries

Published in the United States of America by Oxford University Press
198 Madison Avenue, New York, NY 10016, United States of America

British Library Cataloguing in Publication Data

Data available

Library of Congress Control Number: 2023931812

ISBN 978–0–19–285800–9

DOI: 10.1093/oso/9780192858009.001.0001

Printed and bound by
CPI Group (UK) Ltd, Croydon, CR0 4YY

This book is dedicated to all my dear friends, colleagues, and competitors who have made it their mission to contribute to decarbonization. And to my family.

Contents

1
Introduction

The idea for this book derived from my lectures 'Energy Storage Systems' and 'Electrical Storage Systems'. These lectures are part of the master programme 'Decentralized Energy Generation and Energy Management' at the University of Applied Science, Reutlingen. Both lectures and this book are influenced by my many years of professional experience, first as project manager and development engineer for solar home storage and then as development manager at REFU Elektronik, later known as REFUdrive. Here we develop battery systems, solar, battery, and drive inverters. Applications range from stationary battery power plants for the balancing energy market, to the electrification of mobile working machines.

When I started working on this book, I was initially only interested in presenting the different storage technologies. However, it turned out that it was not enough to explain what distinguishes a lead acid battery from a lithium ion battery. My professional experience had shown that the design of electrical storage systems is a complex task. Many design contradictions have to be identified and resolved. Thus, in elaborating my lecture, I developed a toolbox that allows the design task to be solved in a structured way. Some readers may object that cooking recipes are not beneficial to creativity. I disagree. A good cook who has learned his craft works through the individual steps of a recipe in a structured way and is still creative in the design of the meal. Structure creates free space for creativity.

There are a number of very good textbooks on energy storage. Most of them focus on the underlying technology. Depending on their preference, the authors put different emphasis on the discussion of the technology and its physical, chemical operating principles. These textbooks are very important, and I have deep respect for the many details that open up when studying these works. However, what is unfortunately usually treated in these books as a marginal topic is the challenge of integrating these technologies into systems. The simple question 'What do I need to do to integrate storage technology X into my system Y to make it work?' is rarely answered. Countless very enjoyable discussions with students, researchers, collaborators, and customers show that the challenges often lie in these questions. Therefore, in this book, I try to give the reader tools to find answers to these questions.

The recipe for designing electrical storage systems consists of three steps. First, gather the requirements for the energy storage system. In this book, I formulate requirements as user stories. This form, developed out of SCRUM, is not familiar to everyone, but has proven itself in my professional practice.

After the requirements are known, I deal with the question of how the power is distributed in the energy storage system. For this purpose, the power flows of the energy

Energy Storage Systems. Armin U. Schmiegel, Oxford University Press. © Armin U. Schmiegel (2023).
DOI: 10.1093/oso/9780192858009.003.0001

storage system are described in a power flow diagram. This simple representation is independent of the technical realization and can be adapted to include any technical realization. This abstract mathematical method has proven especially useful in determining optimal system configurations. Engineers—myself included—sometimes tend to commit to a technical realization too early. It is perhaps in our personality, or in our education, that once we have committed ourselves, we defend that idea. But the road to hell is paved with good intentions. Often I had to realize that, in retrospect, a different realization would have been better. Therefore, I developed this technique that allows me to make a decision only after the facts have been established.

The last step is the actual system design, which is then also harmonized with the requirements. In system architecture, there is usually no 'wrong'—there is always only a 'different'.

This recipe can be found in every chapter about storage technologies. I have tried to include very many application examples. The experience in my lectures shows that the students and I enjoy the joint design very much. It is a compensation for the tedious learning of the new tools. I hope the readers of this book will also enjoy these application examples.

I would like to conclude the introduction with some notes for readers. The two most important chapters are 3 and 4. Here the tools are introduced. If you as a reader already have a solid knowledge of power electronics, Chapter 5 may not give you any significant new insights. However, one should familiarize oneself with the description of power electronics components from a systems engineer's point of view in Section 5.3. After the toolbox is assembled, we start with mechanical storage in Chapter 6 and then learn about other storage technologies step by step.

I decided to include exercises directly in the text and to show the solutions immediately. Those who wish, and who possess the necessary self-discipline, can cover the solution after reading the task and solve the task themselves. The exercises can also be seen and understood as examples. Both ways are possible.

I wish all readers much joy and many new insights. This book is a common journey, and so we should now go on the way. Wonder and learn.

2

The purpose of energy storage systems

2.1 Introduction

Before we take a deeper look at the mathematical description and modelling of storage systems, we first want to investigate how and where storage is used. There are a large variety of applications for energy storage systems. In fact, there is hardly any area in our lives where energy storage is not used.

We start in Section 2.2 with a discussion about the reason storage is needed at all, and describe the basic applications of storage. Since energy storage is closely related to the concept of energy and power, we address this aspect. We also describe how storage systems can be characterized by two quantities: power and energy demand.

In Section 2.3 we investigate the different applications of storage systems. We start with very small systems, which are used mainly in everyday mobile objects. We then take a look at mobility applications such as electrified or hybridized vehicles and mobile working machines.

In mobile applications, energy is usually stored before the journey begins so that this energy can be used during the trip. The energy storage is used as a food ration for the journey. In Section 2.3.3, we look at another large class of applications: stationary storage systems. Their use is closely linked to our electricity grid and its characteristics. Therefore, at this point we explain the structure of the power grid and the various ways in which energy storage systems can be used in it.

2.2 What storage is used for

Storing things is a common principle in nature. Plants store water to survive dry periods. Squirrels gather food to fall back on in winter. Other animals do not store their provisions in hidden places, but carry them as winter fat directly in their bodies. This ensures that their bodies can survive during times of seasonal food shortage.

We humans 'store' money that we don't need now, to spend later. We fill up the tank of our car before we start a long trip and, unfortunately, we also put on some body fat when we eat too much, because our metabolism wants to make sure that we still have enough energy reserves in bad times. What all these examples have in common is that the process of storing is always linked to the goal of using a surplus

Energy Storage Systems. Armin U. Schmiegel, Oxford University Press. © Armin U. Schmiegel (2023).
DOI: 10.1093/oso/9780192858009.003.0002

of something in the present to cover a shortage in the future. With storage, therefore, one can separate production and consumption in time, but spatial separation is also possible: a healthy breakfast at home covers the energy demand while we work. The car is filled up at the gas station and consumes the energy of the fuel while driving across the country.

There is also a third application. If the chosen method of transport does not allow the total amount of something to be transported immediately and completely, then the difference can first be stored temporarily so that the actual transport process can take place in several steps.

2.2.1   Energy and power

In this book we look at storage systems that are designed to store energy. First of all, energy is a physical quantity that provides information about how much power can be extracted or added into a physical system. Consider a system which has the energy content E_0 at $t = 0$ s. During a period from $t = 0$ s until $t = T$ s, power is added and extracted from this system. The time series describing this process is $P(t)$. Adding time $t = T$ s, the amount of energy of the system can be calculated by:

$$E(T) = E_0 + \int_{t=0}^{t=T} P(s)ds. \tag{2.1}$$

Exercise 2.1 Consider a mobile phone which already has a state of charge of 24 Wh. The mobile is charged during the night (8 h) with 15 W and discharged over the working day (12 h) with 10 W. What is the energy content of the mobile battery?

Solution: Using eqn (2.1) the energy content equals:

$$E(T) = 24 \text{ Wh} + 8 \text{ h} \cdot 15 \text{ W} - 12 \text{ h} \cdot 10 \text{ W} = 24 \text{ Wh} + 120 \text{ Wh} - 120 \text{ Wh} = 24 \text{ Wh}.$$

There are three groups of energy that we will encounter again and again: electrical energy, mechanical energy, and thermal energy. In Tab. 2.1 below we have listed descriptions of each of these.

However, the definition of energy presented in eqn (2.1) shows that in principle every physical system to which an energy can be assigned is also a kind of energy storage. For this to be the case, we just need to be able to add energy to the system at some time and subtract energy at a different time. In fact, not all physical systems can also serve as an energy storage. This is because it is also necessary for the energy to be stored for a sufficiently long period, as the following example illustrates.

Table 2.1 Different types of energy discussed and used in this book, including their definitions.

Potential energy	$E_{pot} = mg\Delta h$	The energy needed to lift a mass m from height h_1 to h_2. g is the gravitational acceleration, which depends on the location. $\Delta h = h_2 - h_1$ is the height difference.
Kinetic energy	$E_{kin} = \frac{1}{2}mv^2$	The energy stored in a mass m, having a velocity of v.
Rotational energy	$E_{rot} = \frac{1}{2}J\omega^2$	The energy stored in a rotating body. J is the moment of inertia, which reflects the mass distribution around the rotating axis. ω is the angular velocity of the rotating body.
Electrical energy	$E_{el} = UIt$	The energy needed to provide a current flow I on a given voltage level U over a time period t.
Energy of capacitor	$E_C = \frac{1}{2}CU^2$	The electrical energy stored in a capacitor, with the capacitance C and the voltage level of U.
Energy of an inductance	$E_L = \frac{1}{2}LI^2$	The total electrical energy stored in an inductance with the inductivity of L, while a current of I flows through the inductance.

Exercise 2.2 We want to store 3 Wh in the form of kinetic energy in an object having a mass of 12 kg. We want to extract the stored energy after one day. What distance has the object travelled in this time?

Solution: The kinetic energy is given by

$$E_{kin} = \frac{1}{2}mv^2 = 3 \text{ Wh} = 3 \cdot 3.6 \text{ kJ} = 10.8 \text{ kJ}.$$

The object therefore has a velocity of

$$v = \sqrt{\frac{2 \cdot E_{kin}}{m}} = \sqrt{\frac{2 \cdot 10,8 \text{ kJ}}{12 \text{ kg}}} = 42 \frac{m}{s} = 151.2 \frac{km}{h}.$$

The longer this energy is stored in the object, the longer the object will move at $151.2 \frac{km}{h}$. If the energy is to be retrieved after one day, the object has already travelled a distance of 3,628.8 km. Thus, translational energy is more suitable for storing energy over a short time.

2.2.2 Efficiency: The cost of transformation

As Exercise 2.2 shows, not every form of energy is equally suitable for every storage application. Hence, there may be a need to transfer energy from one type of energy into another, more suitable one. In this book the main types of energy are: mechanical energy, be it in the form of kinetic energy, rotation energy, or potential energy; electrical energy—that is, current and voltage; and thermal energy. These different forms of energy can be converted into each other (Fig. 2.1).

Not every type of energy can be transformed well into another type of energy. Electrical and mechanical energy, for example, can be transformed into each other

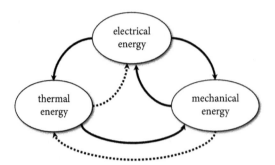

Fig. 2.1 Storage systems deal with the storage of mechanical, electrical, and thermal energy. These energies can be transformed into each other. Some transformations are quite efficient, i.e. only a small part of the transferred energy is lost. The solid lines mark these efficient transformations. The dotted lines mark transformations which show greater energy losses; these are not commonly used in energy storage system applications.

very well, because the movement of a magnet can generate an electrical current via the Lorentz force. With the help of an electric motor, movement can be generated from an electric current and electric current from a movement, if a magnetic field is available. In Section 5.2.6 we will look at this in more detail and describe the necessary technical components.

Electrical energy can also be easily transformed into thermal energy. Thermal energy arises by itself, because every electric current generates heat via the resistance of a conductor.

For the conversion of thermal energy into electrical energy, the thermoelectric effect can be used, in which the heating of a conductor causes current to flow. However, this effect is very small, so the conversion usually takes place via the intermediate step of converting thermal energy into mechanical energy. An example of this is the combustion of coal or gas to generate electrical energy, In this case, the heat causes water to evaporate and this steam is used to drive a turbine. The rotation of the turbine drives a generator, which produces electrical power.

The quality of these various conversions is described by the efficiency. The efficiency is the ratio between the energy E_{in}, which we want to convert or transfer, and the energy that is actually available afterwards, E_{out}:

$$\eta = \frac{E_{out}}{E_{in}}. \tag{2.2}$$

In Section 3.2.1 we will look at the efficiency η and its mathematical definition in more detail. The efficiency η is an important element that we need for describing and designing energy storage systems.

In order to describe an energy storage system, we must therefore be clear about which types of energy occur and how they are transferred into each other. The efficiency is an important criterion for deciding which types are the right choice. In practice, the application already determines which types of energy are present. If we look at a vehicle with an electric or hybrid drive, the types of energy used are already determined: the combustion engine is used to convert heat into mechanical energy. The mechanical energy is used to rotate the axle and thus set the car in motion, which corresponds to a transformation of thermal energy into mechanical energy.

However, the internal combustion engine can also drive the axle of a generator, which uses the movement to produce electricity that charges a battery or drives an electric motor. In this case, thermal energy is thus transformed into kinetic energy, then into electrical energy, and then into kinetic energy.

If the combustion engine is not used, but only the battery and the electric motor, stored electrical energy is converted into kinetic energy.

For the design of the vehicle's system components, the question now arises as to which losses occur during these conversions. These are influenced by the technical realization.

Exercise 2.3 We want to design a power train of an hybrid vehicle. Two different sets of components, A, B, are available for the combustion engine, the electric motor, and the battery. The components have different efficiencies.

In Tab. 2.2 the efficiencies are shown. The most efficient system configuration is now to be chosen for the design. It should be taken into account that the share of utilization of the combustion engine and the electric drive can be different. For this purpose, the weighting factor α shall be used, whose value range lies between 0 and 1. $\alpha = 1$ corresponds to the sole use of the combustion engine. If $\alpha = 0$, the vehicle is driven exclusively electrically.

Solution: The total efficiency is the weighted sum of the individual efficiencies:

$$\eta_i = \alpha\,\eta_{\mathrm{comb}_i} + (1-\alpha)\,\eta_{\mathrm{drive}_i}\eta_{\mathrm{bat}_i},$$

where $i = (A, B)$ is the respective configuration. In Fig. 2.2 the efficiency curves for the two configurations with different weights is shown. Configuration A is better than configuration B at $\alpha = 1$. This is not surprising, because at high α the use of the combustion engine predominates and the combustion engine A is more efficient than engine B. At $\alpha = 0$, however, B is better: here the use of the electric drive train dominates, and this is more efficient at B than A.

Table 2.2 Efficiencies of the available vehicle components.

System component	Efficiency
Combustion engine A	$\eta_{\mathrm{comb}} = 43\%$
Combustion engine B	$\eta_{\mathrm{comb}} = 39\%$
Electric drive A	$\eta_{\mathrm{drive}} = 74\%$
Electric drive B	$\eta_{\mathrm{drive}} = 81\%$

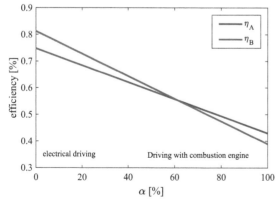

Fig. 2.2 Efficiency of the configuration A and B as a function of α.

2.2.3   The influence of charging and discharging power

We learn from Exercise 2.3 that the usage of the vehicle has an important influence on its design consideration. If it is primarily driven by the combustion engine, a different configuration is better than for the case where the vehicle can be operated completely electrically. In a vehicle that has integrated both system components, for example an electric power train and an internal combustion engine, two variables determine how much driving is done electrically and how much with an internal combustion engine: the energy content of the vehicle battery and its charging and discharging capacity.

The energy content deals with the question of how much energy is needed between two charging processes. This is, of course, followed by the questions of volume, weight, and cost. Since storage always brings additional effort and costs, we always try to make the capacity only as large as is really needed. In our example, the size of the battery influences the proportion of journeys made with the electric motor or the combustion engine. If the battery is empty, the driver has to rely on driving with the combustion engine. A large battery is therefore associated with a small value for α. A small battery means that α becomes one.

The charging and discharging power indicates how much energy can be added to or removed from the storage in relation to a particular amount of time. This parameter also has an influence on the question of how large the share of use of the combustion engine and the electric drive is: if the battery is designed in such a way that it can only be operated with low discharging power, the combustion engine must be served as a support when higher accelerations are required. Whenever the driver 'steps on the gas', the combustion engine will then provide additional power to support the electric motor. This of course has an effect on α, whose value would shift towards one. But there is another effect. An electric motor works both as a motor and as a generator— that is, it is able to charge the battery by converting kinetic energy into electrical energy. But if the battery's maximum charging power is limited, not all of the power that could be served to charge the battery during a braking manoeuvre can be used. Depending on the design of the vehicle, the portion that cannot be stored is converted into friction and heat by a mechanical brake or converted into heat with the help of a brake resistor. The charging power therefore also has an influence on the value of α. If the charging power is small, less braking energy is transferred back into the battery and has to be burned. This reduces the amount of pure electric driving and shifts the value of α towards one.

Storage systems and their applications can therefore be described by three aspects: which forms of energy occur and are stored? How large is the energy content that the storage system must provide to meet the application's requirement? What charging and discharging power is required for the application? In the following section, we will use these three questions to characterize different applications of energy storage systems.

2.3   Applications of energy storage systems

Energy storage systems have a very wide range of applications. In this section we look at some of these and describe them using the three criteria described in Section 2.2: energy content, charging and discharging power, and which types of energy are used. We start with mobile applications, then look at the electrification of vehicles and mobile machinery. Both fields of application are characterized by the fact that they are mobile applications—that is, the storage system is moved. Then we look at two stationary applications. In these cases, the storage system is not moved and can therefore be significantly larger. The first area we deal with is buildings, both in connection with a power grid and as a component of a self-sufficient energy supply. The last area deals with the possibility of using energy storage to support the power grid.

2.3.1 Mobile applications

In this section we look at the use of storage for mobile applications, and more specifically consumer electronics, tools, and gardening equipment. At the end of the twentieth century, the energy and power densities of primary and secondary energy storage devices became so high that it was possible to equip devices that previously had to be connected to a power grid with batteries as well. Tools that could previously be served with a cable have now become cordless. Cordless drills and cordless screwdrivers are examples of this.

At the same time, the market for mobile consumer electronics developed. Laptops and mobiles were constantly developed further and became technology drivers for the increase of energy and power density of energy storage devices. The use case is always a spatial and temporal shift in the usage of energy. We charge our mobiles overnight at home to be reachable anywhere, any time, during the day via email, phone, online messenger, to be able to watch kitty videos anytime and anywhere, or to be able to listen to our favourite song from our record collection or playlist. The storage in the laptop is charged in the library or docking station to be served in a lecture or transit from one work place to another. Thanks to cordless technology, it is now possible to serve the electric screwdriver in places without a power socket. And anyone who has to mow a lawn with lots of trees and shrubs plus tables and paddling pools appreciates not having to fuss with a cord or constantly refill the fuel.

If we want to develop an energy storage system for a mobile device, we have to find a solution for the conflicting product requirements of portability, price and operating time, and durability. The smaller the energy storage system, the higher the portability of the device. Volume and weight increase if the energy content increases. An increase in the available power also increases the volume and weight. Although the technical reasons are different, the same applies to the price. The more energy or power is to be available, the higher the price.

Operating time and service life are influenced by the choice of storage technology. But with the same storage technology, the influencing variables are essentially reduced to the capacity of the energy storage system. The larger the capacity, the longer the unit can be operated for before it needs to be recharged. This also extends the service life, as it takes longer for a user to become aware of an age-related reduction in storage capacity. But the increase in storage capacity goes hand in hand with an increase in cost and volume. This means that we have two contradictory requirements that must be weighed against each other appropriately.

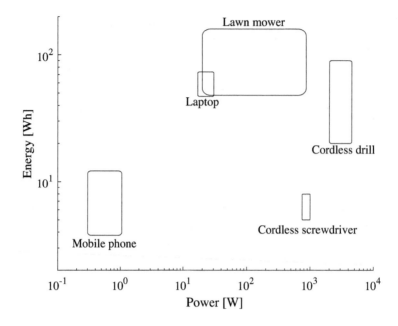

Fig. 2.3 Energy and power requirements of various mobile applications.

The diversity of use cases is also reflected in the requirements for energy content and power demand. In Fig. 2.3, these are shown for the applications mentioned above. In this figure the power and the energy demand are shown for the different applications. Note that the axis is on a logarithmic scale: this allows a better comparison between applications where the difference in energy and power demand is more than one order of magnitude.

The applications with the highest power requirements are the mobile tools. Cordless screwdrivers and drills require power ranging from 750 W to 2000 W. However, this power is not applied permanently. A screwdriver only needs the power for a short time, during the insertion or removal of the screw. This process can be between 2 and 30 s depending on the length of the screw, the nature of the part being assembled, and the skill of the user.

Exercise 2.4 A cordless screwdriver has a storage of 5.4 Wh. For screwing or unscrewing screws, a power of 100 W is required for 2 s. In order to assemble a wall shelf, 25 screws need to be screwed. How many wall shelves can be assembled with the screwdriver if no mistakes are made?

Solution: The energy content of a screw driving operation is

$$100 \text{ W} \cdot 2 \text{ s} = 200 \text{ Ws} = \frac{200 \text{ Ws } 1 \text{ h}}{3{,}600 \text{ s}} = 0.05 \text{ Wh}.$$

Thus, without recharging, the fully charged screwdriver can drive approximately

$$N = \frac{5.4 \text{ Wh}}{0.05 \text{ Wh}} = 108$$

screws. This would be enough for about four wall shelves in a row.

In the case of the drill and the screwdriver, the power requirement naturally depends very much on the specific application. It takes different amounts of time to screw in a screw that is 10 cm long, depending on whether the material is soft or hard. Hence, the power requirement is different. However, it is essential in this application that the load is not permanently applied. Both devices are designed for work which is regularly interrupted, allowing the electronics and batteries to cool down.

For the battery-powered lawnmower, the power range of the energy storage system is between 20 and 1,600 W and the available storage is 20–160 Wh. The lawnmower is an application having a continuous operation with a steady discharge. The power requirement depends on the environmental conditions. Wet, tall grass demands more power from the lawnmower than short, dry grass. This is also a reason why robotic lawnmowers mow the lawn every day and not just once a week. The peak performance of a lownmower is very rarely requested.

The energy and power range covered by laptop batteries overlaps with the lower range of the lawnmower. This is not surprising, because often not only the same storage technology is used, but also the same components: lithium ion batteries using round cells of the 16850 type (see Section 8.4). The electronics market relies heavily on the use of components that are installed in high quantities in different devices, which means that the energy storage systems for different types of laptops are very similar to each other in their technical properties. The same can be observed with mobile phones. The data sheets of laptop batteries show the same voltage range and the same electro chemistry. The advantage is that the costs for an energy storage system are reduced due to the higher number of units and lower product diversity.

For the laptop, the energy content is between 45 Wh and 75 Wh. Here, too, we have to find a compromise between volume, weight, operating time, and price. The operating time depends on the use of the laptop. Playing a CD or continuously downloading files requires about 18 W. If the laptop is used for 3D gaming or numerical simulations, the power demand varies from 21 W to 30 W (Mahesri and Vardhan, 2004).

The picture for mobile phones is similar to that for laptops. However, the energy content and power demand is considerably lower. The available capacity is 4 Wh to 13 Wh. The power demand is 0.4 W to 1 W (Carroll *et al.*, 2010; Ardito *et al.*, 2013; Tawalbeh *et al.*, 2016).

Mobile phones are at least on standby all of the time, to assure that they can receive calls and the user can be informed about the latest information from the internet at in any time and in any place. This means that the energy storage system is continuously discharged. Since the size and volume of a mobile phone is limited, a lot of development work has been put into reducing the power demand and increasing the energy density of the battery.

Energy storage applications can be divided into two categories: energy applications and power applications. The mobile phone and the laptop are energy applications—that is, their storage is served to provide energy over a long period of time. The screwdriver is a power application: its storage is used to call up high power for a short time. Therefore, it does not continuously make low-power requests, but instead has intermittent high power demands.

The 2D division of applications into energy demand and power demand in Fig. 2.3 unfortunately does not allow these two applications to be easily distinguished between. Characterization is not so simple at first. This is where the E-Rate helps us by allowing us to differentiate between energy applications and power applications.

The E-Rate is the ratio of the power demand and the available storage capacity of an application. These rates are shown in Fig. 2.4 below. E-Rates above one are referred to as power applications; for example, here the energy storage system serves to provide power. E-Rates below one, on the other hand, are referred to as energy applications; for example, here the storage serves to provide as much energy as possible over a long period of time.

The E-Rates shown in Fig. 2.4 confirm the previous observations. Mobile phones and laptops are basically energy applications. The power demand is reduced so that the energy storage system can provide its energy over a long period of time. Cordless screwdrivers and cordless drills are power applications. High power is required over a short period of time in relation to the energy content. Lawnmowers represent an application that lies between these two areas. There are lawnmowers that are designed for low power over a long period of time and there are products that can also provide peak power intermittently. What is surprising here is the range of values of the E-Rates. Should one wish to purchase a lawnmower, Fig. 2.4 teaches us that we must look very carefully at which E-Rate the lawnmower provides. Otherwise, we may buy it for a power application, but find that instead we have an energy application. In this case, we will need to plan for many charging breaks while mowing the lawn.

2.3.2   E-Mobility and mobile machinery

In this section we look at the application of energy storage systems for vehicles and mobile machines. Similar to mobile devices, in these systems energy is charged into storage to be consumed at another location. Mobility is focused on using the energy to get from one place to another. With mobile machines, there is also the aspect of work. Mobile machines not only serve to get from one place to another, but also need to do a job on the spot.

We start our investigations with vehicles whose application is to transport people or goods.

There are various possibilities for powering a vehicle. These are basically derived from the forms of energy and their transmission that we have presented in Fig. 2.1: to drive a vehicle, electrical energy or thermal energy is transformed into kinetic energy. If we want to use electrical energy, we use an electric motor that draws its energy from a battery or a fuel cell, for example. If we want to use thermal energy, we use a combustion engine. In both cases, the energy is transferred to a powertrain.

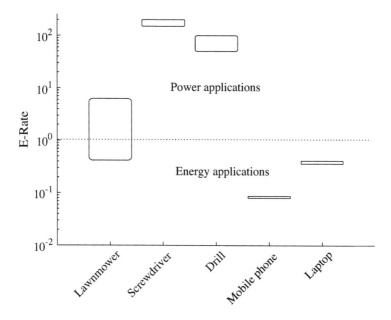

Fig. 2.4 E-Rates for different mobile applications.

There are four ways to combine these technologies. In Fig. 2.5, we have illustrated these. Historically, A), the conversion of thermal energy into kinetic energy was the first form of mobility. Originally, this was realized by steam engines. Today, internal combustion engines powered by fossil fuel are a common occurrence. The disadvantage of this technology is that kinetic energy cannot be converted back into thermal energy. The energy flow always goes in one direction only. If the speed is to be reduced, then brakes must be applied. This also releases thermal energy, but this energy is lost and is not returned to the fuel tank.

In approach B), an electric powertrain is used. This consists of an electrical energy source, an energy storage system, and an electric motor. The advantage of this approach is that mechanical energy can be converted back into electrical energy. If a storage unit is available, it is discharged during the acceleration phase and charged during the braking phase. This process is called recuperation.

Unfortunately, electrical energy storage systems have not yet reached the energy density of fossil fuels. Petrol has an energy density of 8.8 kWh/l, whereas a lithium ion battery has an energy density of 0.4 kWh/l. However, the efficiencies are also different. The conversion of petrol into kinetic energy via an internal combustion engine has an averaged efficiency of about $\eta = 30\%$. A purely electric powertrain, on the other hand, has an averaged efficiency of $\eta = 85\%$.

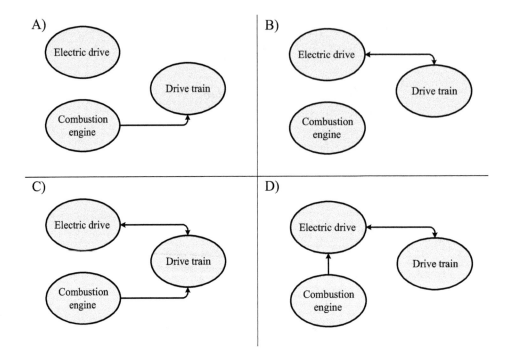

Fig. 2.5 Overview of different powertrain concepts of vehicles. A) corresponds to the drive with the aid of an internal combustion engine. B) is a full electric drive. C) and D) represent hybrid drive concepts. In C) both drives can power the vehicle separately. In D) the combustion engine generates electrical energy, which is served by an electric motor for the drive.

Exercise 2.5 The storage of a vehicle is to be designed so that it can convert $\kappa = 400$ kWh kinetic energy. How much volume is required for the energy storage system if petrol with an energy density of $\rho_V = 8.8$ kWh/l or a lithium ion battery with an energy density of $\rho_V = 0.4$ kWh/l is used? The efficiency of the combustion engine is $\eta = 30\%$, and that of the electric drive is $\eta = 85\%$. Recuperation, i.e. the possibility of storing kinetic energy again, is not to be considered in this task.

   Solution: The volume is given by the formula

$$V = \frac{\kappa}{\eta \cdot \rho_V}.$$

The target capacity κ is corrected by the efficiency. As a result, if the drive is less efficient, the storage must be larger.

   If petrol is served as the energy source, we need a volume of:

$$V = \frac{400 \text{ kWh}}{30\% \cdot 8.8 \text{ kWh/l}} = 151 \text{ l.}$$

In the case of a lithium ion battery, the volume is:

$$V = \frac{400 \text{ kWh}}{85\% \cdot 0.4 \text{ kWh/l}} = 1{,}176 \text{ l.}$$

The lithium ion battery needs 10 times as much volume for the same energy content.

But be careful! In contrast to the internal combustion engine, a lithium ion battery can store braking energy, i.e. the range of the vehicle would be considerably greater, as the storage unit would only have to compensate for the braking and friction losses that occur while driving.

The calculation in Exercise 2.5 shows that the energy storage system with fossil fuels takes up significantly less volume than an electrical energy storage system. For this reason, mixed forms have also become established in practice. These want to ensure a long range by using the combustion engine, but at the same time offer the advantage of recuperation by integrating an electric drive. Concept C) in Fig. 2.5 represents a hybrid drive in which both electrical energy and thermal energy are served for the drive.

In concept D), in contrast, the internal combustion engine serves to generate electrical energy. The basic idea is that the combustion engine is operated at the optimal operating point so that the efficiency is improved. Diesel-powered locomotives work with this concept. Recuperation is not provided for in this concept, unless there is still a battery on the electric powertrain.

When considering the energy requirements of a vehicle, it must be taken into account that the discharge is needed for the acceleration phase. During travel, on the other hand, only friction needs to be compensated. If it is possible to store kinetic energy during a braking process, this can lead to a considerable increase in range.

Exercise 2.6 One vehicle has an internal combustion engine and another has a full electric drive. The electric vehicle has a storage capacity of $\kappa = 40$ kWh. The vehicle with the combustion engine has a tank with a capacity of 30 l. The fuel used is petrol with an energy density of $\rho_V = 8.8$ kWh/l. Both vehicles have the same mass, $m = 1{,}500$ kg.

The vehicle is to accelerate to 50 $\frac{km}{h}$ within one minute and then drive straight ahead for 10 min, requiring an average power of 500 W to compensate for friction losses. This is followed by full braking, which lasts half a minute.

How many of these driving cycles can be driven with the combustion engine and how many with the electric drive? For simplicity, we assume that the mass of the vehicle is independent of the filling level of the tank. Furthermore, we assume that recuperation has an efficiency of $\eta_{EV} = 90\%$. The combustion engine has an efficiency of $\eta_{ICE} = 30\%$.

Solution: At a speed of $v = 50$ $\frac{km}{h}$, the kinetic energy is:

$$E_{\text{kin.}} = \frac{1}{2}mv^2$$

$$= \frac{1}{2}1{,}500 \text{ kg} \left(50 \ \frac{km}{h}\right)^2$$

$$= \frac{1}{2}1{,}500 \text{ kg} \left(\frac{50{,}000 \text{ m}}{3{,}600 \ h}\right)^2$$

$$= 144{,}699.1 \text{ J} = 40.18 \text{ Wh} = 0.0402 \text{ kWh}.$$

During travel, friction losses of $P_{\text{loss}} = 500$ W must be compensated. The energy required for this is given by:

$$E_{\text{loss}} = \int_{t=0}^{t=T} P_{\text{loss}} dt$$

$$= P_{\text{loss}} \cdot T$$
$$= 500 \text{ W} \cdot 10 \text{ min}$$
$$= 5{,}000 \text{ Wmin} = 0.083 \text{ kWh}.$$

Since the internal combustion engine vehicle cannot recover energy during braking, we can determine the number of possible driving cycles based on this data. The energy content of the tank is:

$$\kappa_{\text{ICE}} = V \cdot \rho_{\text{V}} \cdot \eta_{\text{ICE}} = 30 \text{ l} \cdot 8.8 \text{ kWh/l} \cdot 30\% = 79 \text{ kWh}.$$

The number of driving cycles then results in:

$$N = \frac{\kappa_{\text{ICE}}}{E_{\text{kin.}} + E_{\text{loss}}} = \frac{79 \text{ kWh}}{0.0402 \text{ kWh} + 0.083 \text{ kWh}} = 641.23 \approx 641.$$

In the case of the electric vehicle, energy is recovered during the braking process. This energy is:

$$E_{\text{rekub}} = E_{\text{kin}} \cdot \eta_{\text{EV}} = 0.0402 \text{ kWh} \cdot 90\% = 0.036 \text{ kWh}.$$

The energy demand for N cycles is

$$E_N = E_{\text{kin.}} + (N - 1) E_{\text{kin}} (1 - \eta_{\text{EV}}) + N E_{\text{loss}}.$$

The number of cycles is obtained (with $E_N = \kappa_{\text{EV}}$) from:

$$N = \frac{\kappa_{\text{EV}} - \eta_{\text{EV}} E_{\text{kin.}}}{(E_{\text{kin.}} \cdot (1 - \eta EV) + E_{\text{loss.}})}$$

$$= \frac{20 \text{ kWh} - 90\% \cdot 0.0402 \text{ kWh}}{(0.0402 \text{ kWh} \cdot 10\% + 0.083 \text{ kWh})}$$

$$= 459.249 \approx 459.$$

Although the electric vehicle has about 50% less storage capacity, its range is about 71% of the range of the internal combustion engine.

The size of the energy storage and the charging and discharging capacity varies greatly depending on the type of vehicle. In Fig. 2.6 the energy content and power level are shown for different vehicles. The peak power was used for the power demand. Peak power refers to the maximum power draw. Note, that the peak power cannot be drawn permanently. This is because the components of the storage unit or the drive train are not designed for a continuous load of this power rate.

As can be seen in Fig. 2.6, the diagram breaks down into two areas. Below a power of 5,000 W are applications with small or light vehicles: roller, scooters, and bicycles with electric auxiliary drive. Above this limit are cars.

The power and energy range of eScooters, eBikes, and eRollers is in the same range as that of the lawnmower from Fig. 2.3. This is because the technical components

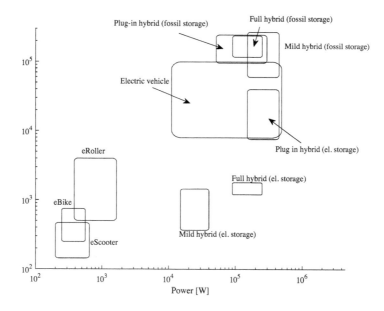

Fig. 2.6 Energy contents and power level of vehicle applications. In cases where both fossil and electrical energy storage were used, the electrical energy storage was also entered separately.

are similar. For example, the same type of battery cells are used for the storage technology. The reason for this is that the requirements in terms of volume, price, and weight are comparable.

eScooters and eBikes are typically not used for long-distance travel. The driving time—if no tracking tours are undertaken—is in the range of minutes or hours. Strictly speaking, both vehicles are hybrid drive systems. While they do not use a combustion engine that serves fossil fuels, they do use human muscle power.

The highest power demand occurs when a vehicle is starting up. Here, in addition to rolling friction, static friction must also be overcome and the mass of the vehicle and load must be accelerated to a target speed. With eScooters and eBikes, the rider helps with their muscle power. During the ride, the stored energy only serves to compensate for friction losses. The discharge rate is considerably lower and is additionally compensated by the assistance of the rider. Therefore, eScooters and eBikes are more of an energy application than a power application.

With an eScooter, the second energy source is omitted. eScooters are not hybrid drive applications. Here, the entire power is drawn from the electric storage unit. As can be seen in Fig. 2.6, the energy content but also the power demand of these vehicles is greater than that of eBikes.

Above a power of 10 kW are the passenger cars. Apart from vehicles powered by an internal combustion engine alone, there are four categories into which cars with alternative drives are divided: electric vehicle, mild hybrid, plug-in hybrid, and full hybrid.

The electric vehicle corresponds to approach B) in Fig. 2.5. An energy storage system stores electrical energy, which then supplies an electric motor. As shown in Fig. 2.6, the different power classes of electric cars cover a relatively large energy and power demand. The use of an electric powertrain with an electric storage system allows recuperation—that is, the return of kinetic energy into electrical energy for braking the vehicle. This mechanism has a positive effect on the range that can be achieved by these vehicles.

The mild hybrid realizes the concept C) in Fig. 2.5. The powertrain can be driven by both the combustion engine and the electric storage (Fig. 2.6: 'Mild hybrid (el. storage)'). The storage capacity is 300 Wh–1500 Wh, which is low compared to an electric vehicle with a storage capacity of 10 Wh–100 kWh. However, the available power of 10 kW to 20 kW is in the lower power range of an electric vehicle. It is therefore a power application. The small storage unit of the mild hybrid serves for short-term charging and discharging. Here, the application is to support the combustion engine when starting and to recuperate kinetic energy when braking.

Like all vehicles with an internal combustion engine, mild hybrids also carry another energy storage system: the tank for their fossil fuel. The power and energy content of the combustion engine are also shown in Fig. 2.6 ('Mild hybrid (fossil storage)'. Since in a mild hybrid the majority of the propulsion power and energy consumption is provided by the combustion engine, the storage capacity and the power that can be called up are in the upper range of this figure.

Plug-in and full hybrids also realize concept C), but both have a higher electrical storage capacity and electrical power. They are in a similar power range to that of the electric vehicle, at 100 kW to 250 kW. However, the storage capacity of a full hybrid is significantly lower, with 1 kWh to 2 kWh. The storage of both vehicles is large enough to allow purely electric driving. Due to the larger storage, this operation is longer for a plug-in hybrid than for a full hybrid. In both types of vehicle, the electric storage is also used to reduce the load on the combustion engine in driving conditions in which it is not working efficiently.

Fig. 2.7 shows the E-Rates in the mobility application area. eScooters and eBikes are in the boundary area between energy application and power application. This cannot be explained by the type of application alone, because greater power is only needed when starting up. Here, it is more the cost pressure that comes into play. The storage is designed to be so small that it is more of a power application.

In the case of cars, electric vehicles, mild hybrids, and plug-in hybrids all fall into the category of power application, if one considers the electrical storage system alone. The fossil fuel powertrain, on the other hand, is considered an energy application. Fossil fuels have a higher energy density—that is, more energy can be transported in a vehicle tank. The effect is that the storage tank is larger in relation to the required power. This effect is very clear in the mild hybrid and the full hybrid vehicle. Both systems are characterized by a relatively small battery—that is, the tank can be designed to be correspondingly larger. With the plug-in hybrid, which has a larger battery, the tank is somewhat smaller, and the E-Rate shifts in the direction of power application.

So far, we have only looked at vehicles that are essentially transporting people. None of the vehicles described in Fig. 2.7 are considered working machines. Only the

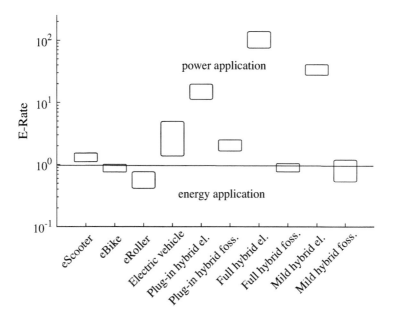

Fig. 2.7 E-Rates for different mobility applications.

transport of weekly shopping or holiday luggage is a working application for these vehicles. Furthermore, they are not designed for continuous operation. A passenger car is designed for a service life of 10 to 20 years. It is assumed that these vehicles are simply stationary most of the time, and neither their storage nor the drive train are used during parking. This assumption can be easily understood. Working commuters drive to work in the morning, then work on site without using the car, and then drive home again. In this example, the car is not used for more than about two to four hours.

The situation is completely different when vehicles are used to perform work. Vehicles and their components are used more intensively and continuously. Mobile machines also differ from the mobility applications mentioned so far, in that the vehicles are not only used for transportation, but also for performing work. In Fig. 2.8, two examples are shown. One is a reachstacker and the other is a straddle carrier. Both vehicles are served in ports to transport containers from one place to another. Both vehicles have engines that are used to move them and they have components that serve to lift the loads. During the lifting process, kinetic energy is converted into potential energy.

A reachstacker can basically use the same drives for traction—that is, movement from A to B—as are used in mobility. As with cars, however, internal combustion engines dominate here. For lifting, lowering, and picking up loads, a hydraulic drive train is often used here. A hydraulic system consists of a pump that builds up very high pressure in a fluid. Oils that can absorb this pressure well are used for this purpose. This pressurized fluid is stored in a tank. So here we have a conversion of kinetic energy (the pump) into potential energy (the compressed liquid that wants to expand again).

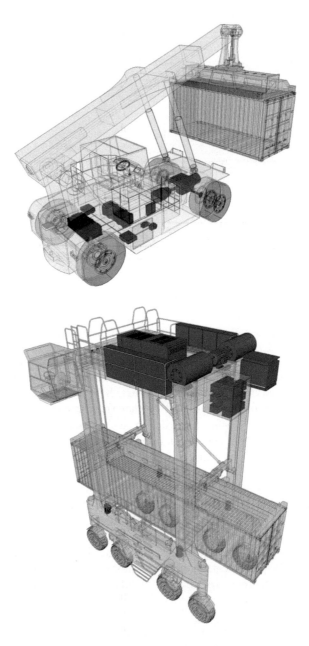

Fig. 2.8 Two examples of electrified mobile working machines: a reachstacker (top) and a straddle carrier (bottom). Both vehicles are is used to transport containers. They have an electric drive train (marked in red) that is used to move the vehicle from A to B. The reach stagger uses an hydraulic system to lift the loads, while the straddle carrier uses an electrical motor to lift the container. (Source: REFUdrive)

To raise and lower, you open a valve, and you use the pressure of the liquid to move a cylinder. The stored potential energy is converted back into kinetic energy.

Just as with traction, the pump can be operated purely electrically or by an internal combustion engine (Immonen *et al.*, 2013; Immonen *et al.*, 2016; Zhang *et al.*, 2019). For the efficiency assessment of this drive train, different efficiencies have to be considered: on the one hand, we must look at the efficiency of the pump drive. This is comparable to the efficiency of a drive train. In addition, there is the efficiency of the pump, which turns kinetic energy into potential energy. The efficiency is about $\eta_{\mathrm{pump}} = 80\%$. The valve and hose system also generate losses. The efficiency here is $\eta_{\mathrm{valve,hose}} = 55\%$ (Immonen *et al.*, 2016).

Exercise 2.7 A commercial vehicle is to be provided with either a hydraulic or an all-electric powertrain. The hydraulic driveline is powered by an internal combustion engine with a 100 l tank. The electric powertrain is equipped with a storage capacity of 250 kWh. The efficiencies of the drive inverter and the electric motor are, respectively, $\eta_{\mathrm{inverter}} = 96\%$ and $\eta_{\mathrm{motor}} = 86\%$.

What is the usable kinetic energy that can be extracted from each of the two drive trains?

Solution: The electric powertrain has an overall efficiency of:

$$\eta_{\mathrm{el.powertrain}} = \eta_{\mathrm{inverter}} \cdot \eta_{\mathrm{motor}} = 0.96 \cdot 0.86 = 0.825.$$

This gives a usable kinetic energy of:

$$E_{\mathrm{kin.}} = \eta_{\mathrm{el.powertrain}} \cdot 250 \text{ kWh} = 206.25 \text{ kWh}.$$

For the hydraulic powertrain, the total efficiency equals:

$$\eta_{\mathrm{hydraulicpowertrain}} = \eta_{\mathrm{ICE}} \cdot \eta_{\mathrm{pump}} \cdot \eta_{\mathrm{Valve,Hose}} = 0.3 \cdot 0.80 \cdot 0.55 = 0.132.$$

The usable kinetic energy is therefore:

$$E_{\mathrm{kin.}} = \eta_{\mathrm{hydraulicpowertrain}} \cdot 100 \text{ l} \cdot 8.8 \, \frac{\text{kWh}}{\text{l}} = 116.16 \text{ kWh}.$$

Exercise 2.7 shows that hydraulic powertrains have a lower overall efficiency compared to electric powertrains. Therefore, there are efforts to partially or fully electrify these powertrains (Immonen *et al.*, 2013; Ponomarev *et al.*, 2015; Zhang *et al.*, 2019).

Although hydraulic powertrains have low efficiency, fully electrified mobile machines are not common. Pure diesel-powered and diesel–electric vehicles dominate. In Fig. 2.9 below, the energies and power of mobile machines are shown. In the case of hybrid machines, the usable energy has been shown: we have not shown the energy content of the diesel here, but have already included the efficiencies of the drive motor, pump, and valve system:

$$\eta = \eta_{\mathrm{ICE}} \cdot \eta_{\mathrm{pump}} \cdot \eta_{\mathrm{Valve,Hose}} = 30\% \cdot 80\% \cdot 55\% = 13.2\%. \tag{2.3}$$

Machines with pure combustion engines are not shown in Fig. 2.9. Since the machines of these vehicles do not differ in design from those of the diesel–electric machines, the power values remain identical, but the energy values are 10–20% below the values of the diesel–electric solution (Wang *et al.*, 2017).

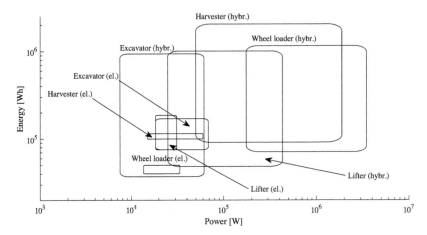

Fig. 2.9 Overview of the energy and power used for mobile machinery. Unlike hybrid cars, hybrid mobile machinery tends to consist of vehicles driven by diesel–electric. The energy values shown here are the usable energy. The energy content of the stored diesel is already corrected by the efficiencies $\eta = \eta_{ICE} \cdot \eta_{pump} \cdot \eta_{Valve,Hose} = 30\% \cdot 80\% \cdot 55\% = 13.2\%$. In this presentation, no vehicle is listed that is operated as a pure internal combustion engine. However, their performance and energy values correspond to those of diesel hybrid vehicles, with the usable energy content being 10–20% lower.

The stored energy of mobile machines ranges from 10 kWh to 250 kWh. This is partly due to the range of applications. Small excavators or wheel loaders, some of which are only used for a few hours a day, naturally need to carry less energy than large, powerful harvesters that harvest a full day's work in the fields.

According to the application, the power ranges are also widely spread. Here, the power range is from 50 kW to 500 kW. So far, only large harvesters and wheel loaders equipped with diesel electric drives reach these upper power ranges. All-electric machines have a power limit that is 100 kW. This is not a technical power limit. Electric motors for mobile applications can reach up to 350 kW. However, even higher outputs would be possible by combining several prime movers. This is due to the fact that the electrical storage capacity is also limited to about 200 kWh. Larger batteries are not used here.

The challenge with the electrification and hybridization of mobile machines is that these are usually power applications rather than energy applications. This can be seen from the E-Rates in Fig. 2.10. One exception can be observed here. Hybrid wheel loaders are energy applications. Wheel loaders use a large amount of their energy to transport goods. They require high power only for lifting the load and in the acceleration phases. This has a positive effect on the ratio between power and the energy being carried.

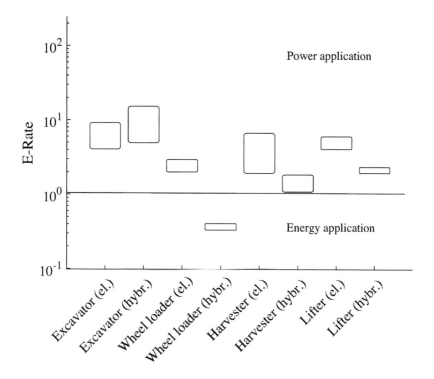

Fig. 2.10 Overview of E-Rates observed in mobile machinery. The energy content of the stored diesel has already been corrected for the efficiency of a diesel–electric hybrid powertrain. Diesel-powered vehicles are not shown here. Their E-Rates are 10–20% lower than those of diesel–electric powered machines.

Mobile machines are used for work. They move large amounts of earth, lift heavy containers, or do harvesting work. They usually do this in places where there is no gas station or charging station. Machines with hybrid drives have a lower E-Rate than electric machines. This is due to the high energy density of fossil fuels. If the efficiency of these drives could be improved, the E-Rate would decrease further, so that a transition to energy applications could be observed here (Immonen *et al.*, 2009; Immonen *et al.*, 2016; Salomaa, 2017; Wang, Wang, and Hu, 2017).

2.3.3 Stationary applications

So far we have dealt with mobile applications. We have stored energy to use it at another location, or on the move: like a cup of hot coffee that we pour into our Thermos in the morning before we go to work. In this section we will now look at stationary applications. In these, energy is stored in order to use it later.

When we are not on the move, we are either at work, engaged in leisure activities, or sitting, living, or working at home. We use energy in different ways: we cook, watch TV, listen to music, use a reading lamp, but we also use it to run machines, computers, or

Table 2.3 Consumers in a private household.

Heating and cooling	W	Work	W	Lighting and information	W
Stove	3,000	Vaccum cleaner	1,200	Lighting	15
Microwave	1,230	Dishwasher	2,500	Television	150
Kettle	1,900	Washing machine	3,000	DVD player	35
Toaster	1,010	Tumble dryer	3,300	Satellite decoder	30
Steam iron	1,235	Washing machine	750	Antenna module	25
Electric water heater	2,600	Water pump	750	Desktop PC	150
Ceiling fan	100	Blender	300	Mobile phone	15
Table fan	35			Laptop	65
Refrigerator	400				

leisure equipment. The energy is provided by the electricity grid. Since the introduction of electric light by Edison and Westinghouse at the end of the nineteenth century, the number of devices and machines powered by electricity has been increasing. Today, there is hardly any household equipment that is powered by muscle or steam power, and very few industrial machines.

Let's have a look at the devices and their power consumption that can be found in a household (Tab. 2.3). The devices can be roughly grouped into three areas: equipment for generating heat or cold, such as the microwave and the refrigerator, but also the iron; equipment that performs work, such as the washing machine or dishwasher; and equipment for generating light and entertainment electronics.

Of course, there are country-specific differences. For example, there are countries where cooking with gas is preferable to electricity, and in warm countries households have fans or air conditioners. Therefore, Tab. 2.3 does not claim to be complete.

Exercise 2.8 On a normal working day, the devices listed in Tab. 2.3 are used for different periods of time or not at all. In Tab. 2.4 the times are listed. What is the total power consumption for the whole day?

Table 2.4 Usage of devices on a working day.

Heating and cooling	h	Work	h	Lighting and information	h
Stove	1	Vaccum cleaner	0.25	Lightings	8
Microwave	0.25	Dishwasher		Television	4
Kettle	0	Washing machine	1	DVD player	2
Toaster	0.05	Tumble dryer		Satellite decoder	4
Steam iron		Washing machine		Antenna module	4
Electric water heater		Water pump		Desktop PC	1
Ceiling fan		Blender		Mobile phone	
Table fan				Laptop	
Refrigerator	24				

Solution: The power requirements of the various equipment must be multiplied by the amount of time the device is switched on for. These are then summed up. For the generation of heat or cold, the energy demand is as follows:

$$E_{\text{heat}} = 3{,}000 \text{ W} \cdot 1 \text{ h} + 1{,}230 \text{ W} \cdot 0.25 \text{ h} + 1{,}010 \text{ W} \cdot 0.05 \text{ h} + 400 \text{ W} \cdot 24 \text{ h} = 12{,}958 \text{ Wh}.$$

The energy required to do work is:

$$E_{\text{work}} = 1{,}200 \text{ W} \cdot 0.25 \text{ h} + 3{,}000 \text{ W} \cdot 1 \text{ h} = 3{,}300 \text{ Wh}.$$

For lighting and consumer electronics,

$$E_{\text{light}} = 15 \text{ W} \cdot 8 \text{ h} + 150 \text{ W} \cdot 4 \text{ h} + 35 \text{ W} \cdot 2 \text{ h} + 30 \text{ W} \cdot 4 \text{ h} + 25 \text{ W} \cdot 4 \text{ h} + 150 \text{ W} \cdot 1 \text{ h} = 1{,}160 \text{ Wh}.$$

The daily consumption is therefore $E_{\text{day}} = 12.9 \text{ kWh} + 3.3 \text{ kWh} + 1.2 \text{ kWh} = 17.4 \text{ kWh}$, of which 75% is for producing heat or cold and 19% is for doing work.

Looking at the results of Exercise 2.8, the total annual consumption equals 17.4 kWh· 365 d/a = 6,351 kWh/a. However, this calculation does not take into account that the consumption of a household varies between the different days of the week. On a working day, many people are not at home or they work at home. They do different things than they do at the weekend. At the weekend they might sleep for longer or read a good book about storage technology; they might cook convenience food during the week and have a three-course meal at the weekend. These variations are noticeable in energy consumption. There are also seasonal fluctuations. In the dark and cold seasons, more light and heat is needed. In the hot seasons, on the other hand, the air conditioning has to be switched on more often.

In private households, energy storage units are used to store electrical energy and heat. The heat storage units serve as buffer storage in the hot water circuit or store water heated by solar thermal energy. This type of energy storage is very common and has been around for many centuries. When there was no electricity and heating systems in homes consisted of a cooker or open fire, there was often a pot of hot water on the heat source so that hot water could be used quickly when needed.

Electric storage is a relatively new technology. It was first used for households that were not connected to the electrical grid. In order for these households to have access to electricity, they were powered by photovoltaic systems or small wind turbines (Blum et al., 2013; Chowdhury et al., 2015; Hong, Abe, and Baclay, 2011; Breyer et al., 2009). In these applications, the storage system served to provide energy to the household at night when there was no sun, or a wind calm when there was no wind.

Until the introduction of these systems, diesel generators were used. These use an internal combustion engine to generate electrical power. Diesel generators are still used today when very large outputs or amounts of energy are needed, including for power supplies on construction sites, in mines, and in large residential complexes. Since the prices of photovoltaic systems have fallen further and further in recent years, more and more hybrid systems are being realized (Williamson et al., 2003; Sopian, Othman, and Rahman, 2005). In these hybrid plants, diesel generators and photovoltaic or wind plants are operated together in order to optimize energy costs. Storage units are used in these plants to realize the energy supply even in windless or sunless phases, or to shave power peaks.

The shaving of power peaks is an important application in hybrid plants. Diesel generators do not work optimally at all power levels and need a certain amount of time before they can provide the full power. In order to avoid power fluctuations at construction sites, mines, and large residential and factory facilities, the output of diesel generators is designed to be larger so that a peak can be supplied by one generator for a short time while other generators start up in parallel. This is not only inefficient, as the large engine does not always work at the optimal operating point, but also costs more to purchase. For this reason, storage units are being installed for such systems. The storage unit does not necessarily need a high storage capacity, but is able to provide the power that is needed at peak times.

Fig. 2.11 shows the energy and power requirements for stationary applications. Looking at off-grid systems for households, we see that they have a power demand between 5 kW and 20 kW and a maximum energy demand of 25 kWh. The storage capacity is therefore selected so that electricity can be used for several days without sun or wind.

With the spread of solar power systems in private households and the drop in the price of battery systems, the possibility of using storage systems for grid-operated solar power also arose. These home storage systems differ from the classic off-grid systems in that they are connected to the grid. When the storage system is completely charged, the surplus solar power can be fed into the grid. The household can also always access the electricity from the grid. Here, the storage unit serves to store energy for the night, or as a power buffer to relieve the grid connection (Maclay, Brouwer, and Samuelsen, 2007; Yoo *et al.*, 2012; Erdinc, 2014; Schmiegel and Kleine, 2014).

The storage capacities realized for household electrical storage range from 2.4 kWh to 16 kWh (Fig. 2.11), with systems ranging in power from 1.2 kW to 16 kW. Compared to off-grid systems, their power output and storage capacity are lower. Since there is no need to always supply all consumers via the solar power system or the storage, the output and energy content of these systems can be smaller.

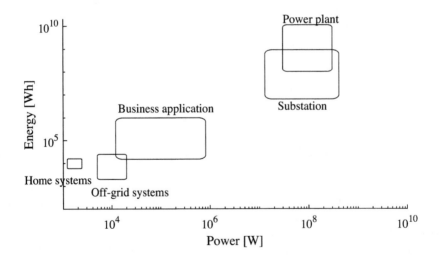

Fig. 2.11 Charging power and storage capacities for stationary energy storage applications.

The same fields of application can be used for companies and factories as for households. If companies have their own solar power system, the storage unit can be used for the intermediate storage of surplus energy. Safeguarding production from a power failure is also an application. Here, the storage unit is part of an emergency power supply concept. Depending on the energy and power requirements of the operation, the storage unit is combined with a diesel generator. As shown in Fig. 2.11, the energy and power requirements are significantly higher than for off-grid systems or home storage systems.

So far we have looked at residential and commercial stationary applications. In both cases, the storage was used to store energy at a certain time to use it later or to store energy with a lower charging power in order to call up a higher charging power later. Auxiliary services are another field of application for stationary storage. These are applications that support the electricity grid in its function and stability.

The basic principle of the electrical grid has hardly changed since Westinghouse. Basically, you can interpret the grid as an electrical circuit in which many small consumers are connected in parallel to a dynamo. This dynamo is located in a power plant and is driven by a steam turbine or a water wheel. The dynamo rotates at a frequency of 50 Hz or 60 Hz and generates a voltage between 400 V and 440 V. If consumers are switched on or off, the frequency changes because the power generated by the turbine or water wheel depends on the moment of inertia T and the rotational speed ω:

$$P_{\text{rot}} = T\omega.$$

If loads are switched on, the required electrical power increases. The generator needs to cover this increase in power demand. Since the moment of inertia cannot change, the speed of rotation must change. The generator rotates slower, which leads to a decrease in the frequency. The same applies when consumers are switched off or the required power is reduced. In this case, the rotation speed increases and the frequency increases.

For some consumers a variation in the frequency is not a problem. A light bulb lights up regardless of the mains frequency. The same applies to the heating coil in an electric cooker. However, there are consumers whose function is frequency-dependent. For example, frequency fluctuations in the first industrial electric motors produced fluctuations in the rotational speed they generated, and this could cause malfunctions in the industrial processes in which the motors were used.

To prevent damage to consumers as well as generators, grid codes have been established. These grid codes determine which frequencies and which voltages are permitted. But how does the grid, or the power plants, react when a frequency fluctuation is observed?

Let's imagine that we produce electricity in our house with the help of an ergometer. A family member sits on the ergometer and uses his muscle power to drive a dynamo that supplies electricity to all the loads in the house. We can't generate much power this way—even well-trained cyclists can only manage a few hundred watts over a longer period of time—but our household is very economical. And so we can easily serve the basic load in the house. We are still alone in the house because the rest of the family is at school or at work. But now the family members come home and switch on one device after the other. Every time a device is switched on, we feel the resistance of the pedals getting stronger and stronger. We have to generate as much movement power as the electrical power that is consumed. The more devices are switched on, the harder it becomes to pedal, and we notice that we are reaching our limits. The lights start to flicker and the driving speed decreases, and so does the frequency of the alternating current. Luckily, however, we have other ergometers in the house, and as soon as the frequency drops below a level we have agreed with family members, someone gets on a free ergometer and generates power as well.

What is described here on a small scale is what takes place in our energy grid. As soon as the frequency in the grid drops, other power plants step in—after prior agreement—and feed in enough power to stabilize the frequency. The coordination takes place via energy and power markets known as the primary, secondary, and tertiary reserve markets. Primary and secondary reserve markets are power reserves provided by power plants. For the tertiary market, power plants provide energy reserves. On these markets, power plant operators auction off the provision of capacity or energy, and whether this is then called up depends on generation and consumption in the grid (Loisel *et al.*, 2010; Bauer, 2019).

Power plant operators sell part of their available power and energy, but energy storage facilities are also used; for example, pumped storage is used for the tertiary markets. In times when too much power is fed into the grid, the pumped storage power plant is charged, and in times when energy is needed, the pumped storage is discharged. Primary and secondary reserve markets require energy storage systems that are able to provide the required power in a very short period of time. Battery storage, flywheel storage, and gas-fired power plants are used in this context (Bouffard and Galiana, 2004). In Fig. 2.11 you can see that the power and energy demands of these power plants are very high.

Energy storage can of course also be used for price arbitrage (Mokrian *et al.*, 2006; Xi, Sioshansi, and Marano, 2014; Hesse *et al.*, 2017). This is when you buy electrical energy when the price of electricity on the power exchanges is low and use it to charge your storage. If the price on the power exchanges rises, you can sell the stored energy at a profit. Of course, this only works if the storage losses are low and the margins are large enough.

Why do energy prices fluctuate on the power exchanges? There are two main reasons for this. Conventional large-scale power plants work with long-term schedules: over a long period of time it is determined how much energy these power plants should produce. These energy quantities are also partly sold in advance. Both large consumers and the power plant operators profit from this. The large consumer gets cheaper energy in the long term, and the power plant operator can better plan for the supply of fuel, including coal, oil, gas, or nuclear fuel elements.

However, it can be the case that too little energy production has been planned. In such cases, energy prices rise because the supply is smaller than the demand. But there is also the other scenario, where too much energy is produced. Especially in energy grids where the share of renewable energy is high, in good weather conditions (a lot of sun and strong wind at the same time) the production of solar and wind power is considerably higher than the planned consumption, and so the energy prices fall.

So there is the possibility of waiting for such conditions and buying energy cheaply. Then, when there is too little energy on the market and prices rise, you can sell it again. If these fluctuations are known, they can be the basis of a business model.

The electrical grid consists of many subgrids (Fig. 2.12), which are connected to each other with lines. It has a hierarchical structure. Large power plants feed their electrical energy into the transmission grid. The lines of the transmission grid distribute the power to the medium-voltage grid, which connects smaller towns and districts. The power in the medium-voltage grid is then distributed in the low-voltage grid to the individual household. The connections between the grids cannot transmit arbitrary amounts of power. Each connection is a piece of cable—that is, a piece of insulated metal whose temperature increases with the amount of electric current until it melts and breaks. Since grids were organized hierarchically in the past, this was not much of a problem. The large power plant was known to be the producer and the grid operators knew the consumption patterns in their grids. If there were consumers who needed more electricity, lines were reinforced and the grid remained stable.

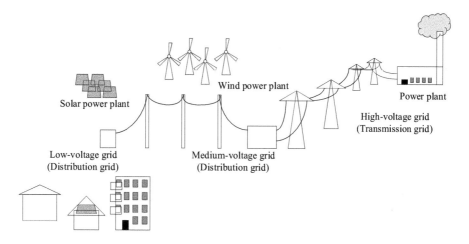

Fig. 2.12 Representation of the electrical grid. It is arranged in hierarchical levels that have different voltages. At the highest level, the high voltage, the production of electrical energy is carried out by large power plants. At this level, the electricity is distributed to various other high-voltage networks. Medium-voltage grids branch off from these high-voltage grids and connect different towns or parts of towns with each other. The low-voltage grid branches off from this and distributes the electricity to the end consumer.

With the increase in decentralized generators feeding into the grid, the organization of the grid has become more complex. Now generators are also feeding in energy at low or medium voltage. Two things can now be observed in the grid: since less electrical power is taken from the power plant but production is not throttled, the surplus kinetic energy is converted into a higher rotational speed. This results in an increase in frequency in the grid. In the low- and medium-voltage grid, on the other hand, the voltage increases because the available power fed in is not immediately consumed where it is produced.

To ensure that decentralized feed-in does not destabilize the grids, specifications for decentralized feed-in are made in the grid codes. In addition, there are technical solutions that allow active voltage control (Stetz, 2013; Stetz *et al.*, 2015). These technical solutions also include energy storage systems. They allow, for example, the temporal shifting of the feed-in or the limitation of the feed-in of power, known as peak shaving (Von Appen *et al.*, 2014; Von Appen, 2018). This even allows for the postponement of necessary grid expansion measures when building larger solar and wind plants (Gonzalez *et al.*, 2004).

In Fig. 2.11 these applications are summarized in the substation section. The grid support applications described are power applications, so the energy storage requirement is less than the power requirement. However, the application area is somewhat larger. This is also shown by the E-Rates in Fig. 2.13. Energy trading needs less power but large amounts of energy that can be traded. Grid support requires only power or a smaller share of energy for many applications. Therefore, a transition between energy applications and power applications can be seen here.

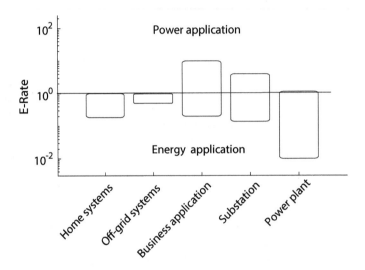

Fig. 2.13 E-Rate for different stationary energy storage applications.

If we look at the E-Rates of the various stationary applications (Fig. 2.13), we see that energy applications dominate here. Some of the E-Rates are significantly lower than those we observed in other applications. For example, power plants have E-Rates as low as 0.01. Since weight and volume do not play such a dominant role in stationary applications, storage systems can be designed more generously. Applications such as energy trading are also based on buying large amounts of energy cheaply and storing it. When this is sold, the performance is then less decisive than the quantity.

2.4 Conclusion

In this chapter we discussed the question of where energy storage is used and how it is used. We have seen that storage is always used when the production and consumption of energy should not take place at the same time. In mobile applications, this means that we store energy in order to transport the required kinetic energy along with it during the journey. In stationary applications, this is to achieve a temporal shift of production and consumption. Another application we looked at was buffering energy consumption—that is, delaying its use in time or space.

The different applications of storage can be well characterized by the quantities of energy and power demand. The energy demand tells us how much storage capacity is needed. The power demand tells us how much energy is needed per amount of time. The ratio of power demand and energy demand is called the E-Rate; this can serve as a simple orientation. In Fig. 2.14 the energy demand and the power demand were shown for different applications. The applications lie in a diagonal band, meaning we do not see applications that have a power demand of a few watts and a storage demand of megawatt-hours. We also do not see applications that have a storage requirement of a few watt-hours but a power requirement in the megawatt range.

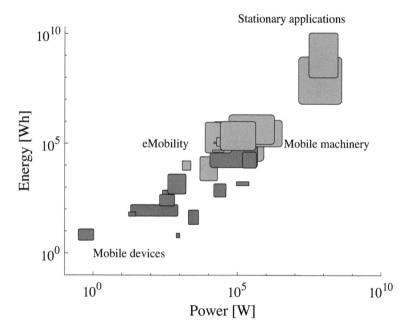

Fig. 2.14 Power and energy requirements of the energy storage applications presented so far.

Energy storage systems convert different forms of energy into each other for storage and use. In vehicles and mobile machines, for example, thermal energy or electrical energy is converted into kinetic energy. In mobile applications such as the electric lawnmower, electrical energy is converted into kinetic energy. Losses occur during these conversions, which can be described by the efficiency. While the task remains the same, the technical realizations differ in the efficiencies. A vehicle discharges its storage in order to move. If we neglect the possibility of recuperation, the only difference between a purely electrically driven vehicle and a vehicle with an internal combustion engine is the efficiency.

In the next chapter, we will elaborate on this idea. We will develop a description that allows us to describe energy storage systems in a technology-independent way, in order to be able to evaluate different technologies and realizations in a second step.

Chapter summary

- Storage is always needed when production and consumption are separated spatially or temporally.
- If the consumption of an energy production is not fast enough, storage can be used to temporarily store the production, with the storage serving as a buffer.
- Storage systems deal with the storage of mechanical, electrical, or thermal energy.
- The different forms of energy can be partially converted into each other. The quality of the conversion is described by the efficiency.
- A storage application can be characterized by the required storage capacity and the charging and discharging power.
- Mobile applications can be classified by the way in which thermal and electrical energy generate the kinetic energy.
- An application can be characterized by the E-Rate, the ratio of power demand to storage capacity. E-Rates above one are power applications, i.e. the storage is used to provide power. E-Rates below one are energy applications. Here, the focus is less on power and more on the provision of energy.
- Storage applications are distributed over the entire energy and power range within a band (Fig. 2.14). Very high or very low E rates are not observed in the application fields mentioned here.

3

The general description of energy storage systems

3.1 Introduction

In Chapter 2, we saw how diverse energy storage systems are. Despite this diversity, the application was limited to two essential tasks: the production and consumption of energy should be separated spatially or temporally. The specific technological realization, the energy content, and the charging and discharging power to be provided may have differed considerably, but the task always remained comparable.

In this chapter we want to follow up this idea. It will help us to develop a generally valid, technology-independent description of energy storage systems. This description will help us to plan and design energy systems and to decide which structure and which technology should be used.

There are a number of design considerations which have to be taken into account if you want to design an energy storage system. Let's consider the design of a DC fast charging station for electric buses (Fig. 3.1). It's called an opportunity charger, because the intention is to charge the bus whenever the user has the opportunity to charge. Therefore, this station shall be located at bus stops. Whenever the bus stops, the battery of the bus is to be charged with up to 400 kW during the short period when people are getting on or off. However, there is a technical challenge: the power connection at the bus stops are not always strong enough to provide 400 kW. For this reason, the charging station shall be equipped with energy storage. This will be charged with the accessible power of the electricity connection, but can be discharged with 400 kW. To further reduce the load on the power connection, a solar power plant is also to be connected to the storage unit.

There are a number of possibilities to realize this station. Two possible realizations are shown in Fig. 3.2. Both solve the task. However, they differ in their technical implementation. The first version consists of three AC/DC converters. These are devices which are able to transform a direct current into an alternating current. In this version, both the storage unit (11) and the photovoltaic (PV) system (12) are connected to the AC grid and the charging process takes place via a charge unit (13).

The second version uses one AC/DC converter and three DC/DC converters. A DC/DC converter transforms the direct current of one voltage level, for example $850V_{DC}$ into the direct current of another voltage level $250V_{DC}$. In Chapter 5, we'll discuss its properties in more detail.

Energy Storage Systems. Armin U. Schmiegel, Oxford University Press. © Armin U. Schmiegel (2023).
DOI: 10.1093/oso/9780192858009.003.0003

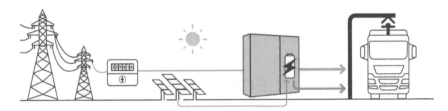

Fig. 3.1 An opportunity charger is used to charge an electric bus during bus stops. An energy storage system shall allow charging rates which are larger than the accessible electrical power of the grid connection. In order to further reduce energy cost, a PV system is also connected to the energy storage. (Source: Heliox)

In version 2, the rectifier (21), storage unit (22), PV system (23), and charging unit(24) are connected to each other via a direct current network (DC bus). A rectifier feeds energy onto this DC bus, as does the PV system. The charge unit takes the energy from the storage unit or the grid via the DC bus and charges the vehicle.

How can we judge between these two solutions? Various aspects have to be considered:

- What are the power capabilities of the PV system and the grid? The description of the grid is straightforward. You just need to know the maximum amount of electrical power and the associated price, which may vary over time. The price might also differ from power level to power level. To describe the PV system, you need some information about the expected production per time. A PV system only produces electricity when the sun is shining and its output also depends on the season, the location, and its orientation.

- How much energy is needed for charging? What are the charging intervals? For this purpose, you need to determine when which bus passes the station and how much energy it needs to at least be able to drive to the next station.

- Which rectifiers and inverters should be used? In both designs, different direct current and alternating current networks are coupled together. Power electronic converters are used to enable the energy transport. Each converter costs money and each conversion also consumes energy. Converters are also not equally suitable for all current and voltage ranges (more on this in Chapter 5).

- Which storage technology should you use and how much energy needs to be stored? There are a number of storage technologies that are suitable for the task. In the upcoming chapters the most important technologies are described. Each technology has its own characteristics and requirements that must be taken into account in the overall design.

- What should the energy management look like? The energy management determines whether the storage system is to be charged from the grid or via the PV system. There are many different approaches and strategies to control the energy flows. Should the energy flow be optimized to generate a maximum of profit? Should it optimize the energy flow to assure that the buses are always getting more than enough energy? Will the station participate in the energy market?

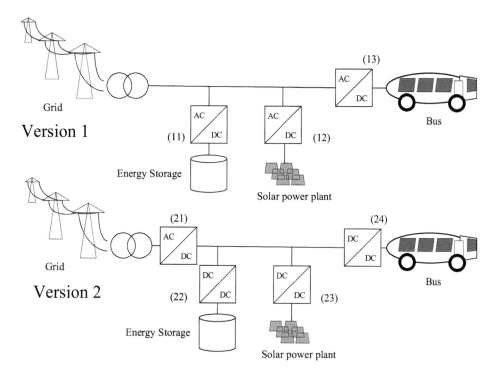

Fig. 3.2 Two possible realizations of the charging station shown in Fig. 3.1. In version 1, the storage unit (11), solar power system (12), and charging station (13) are connected directly to the AC grid. In version 2, the current from the grid is first rectified (21), and then the storage (22), solar power system (23), and charging station (24) are connected to the DC bus.

The analysis of these questions seems to be a very complex task. Even if the input data—that is, the PV production profile, the needed load for the buses, and the energy prices—are all known, there are a lot of tasks involved in designing a system, deriving an optimal energy management strategy, and comparing the different designs. Luckily, we can structure these tasks in a way that reduces complexity. In this chapter, we will show how all these different aspects can be described in a uniform mathematical framework.

Although many different storage system technologies are addressed in this book, their formal description is comparable. Whether electric vehicles, pumped-storage power plants, or residential PV storage systems, all these systems can be described by the same tools. This makes it easier to design individual systems and to combine technologies to create more complex, hybrid storage systems. The formal description is independent from the storage technology and allows us to compare different realizations and calculate the optimal energy management strategy.

3.2 Mathematical description of structure and function of storage systems

The structure and function of a storage system can be described mathematically by introducing the power flow diagram (Fig. 3.3). The beauty of this description is its independence from the applied technology. The architecture of the storage system can be designed without the need to make a decision regarding the type of technology to be used in the design stage of the development process. This enables the architect to compare different storage system solutions later.

3.2.1 Charging and discharging

Let's have a look at the charging station shown in Fig. 3.1. If you were to describe its function, you may use words like: 'power is being transferred from the grid into the storage', or 'the storage is being discharged to charge the bus', or 'the energy is stored during the night to charge the first bus in the morning shift'. Hence, the natural way to describe the system is to describe the power flow between the different components. Therefore, we interpreted an energy storage system as a network of sources and sinks which have interconnection between some of them. The connections define the way in which power can be transferred from one node to another. Fig. 3.3 provides a simple example. This storage system consists of three elements: a power source $Q(t)$, which produces a surplus of power; the storage system $S(t)$, which is able to store this power; and the load $L(t)$, which consumes this power. Since power production, storing, and consumption is a dynamic process, all these quantities are time dependent.

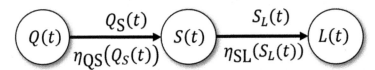

Fig. 3.3 Formal description of a storage system: the storage $S(t)$ is charged by a power source $Q(t)$ with a charging power of $Q_S(t)$. During the discharging process, the power $S_L(t)$ flows towards the load $L(t)$. Since both processes are not ideal, both power flows have some losses, which are described by the efficiency η_{QS} and η_{SL}. Both efficiencies depend on the amount of transferred power.

In case of the charging station, we have two energy sources: the grid and the PV plant. We might label them as $Q^{\mathrm{grid}}(t)$ and $Q^{\mathrm{PV}}(t)$; the energy storage is labelled as the storage $S(t)$ and the bus charging unit can be described as the load $L(t)$.

A power flow is labelled using arrows. In Fig. 3.3 one arrow, labelled as $Q_S(t)$, represents the power transfer from Q to S, and another arrow, labelled as $S_L(t)$, represents the power transfer from S to L. The source of the power flow is always the big letter and the sink is the sub-index of it.

In Exercise 3.1, we show another example, highlighting how the power transfer can be used to describe the charge and discharge process of an energy storage system. The physics is different in comparison to the charging station shown in Fig. 3.1, but the formal description is the same.

Exercise 3.1 Let's consider the simple process of filling up a car. In this case the power source $Q(t)$ is the fuel station. We pump petrol into a fuel tank, which is our storage $S(t)$. The amount of power transported from the source into the storage is $Q_S(t)$. We want to compare two different charging technologies: the petrol station, which pumps petrol directly into the tank, and a person who has only a bucket with some holes.

  Solution: In both cases, the transferred power can be easily estimated. The energy density of petrol is roughly about 10 kWh per litre. Let's assume that the petrol station is able to pump 60 l per five-minute period. The corresponding charging power equals:

$$Q_S(t) = \frac{60 \text{ l} \times 10 \text{ } \frac{\text{kWh}}{\text{l}}}{\frac{1}{12}\text{h}} = 7{,}200 \text{ kW}.$$

The man with the bucket is not so fast. He is only capable of transporting 5 l per five-minute period. The corresponding charging power equals:

$$Q_S(t) = \frac{5 \text{ l} \cdot 10 \text{ } \frac{\text{kWh}}{\text{l}}}{\frac{1}{12}\text{h}} = 600 \text{ kW}.$$

Regardless of the technology used, every storage system has a limited capacity. In order to know how much energy a storage is able to provide at a given time, it is necessary to determine its energy content, $\kappa(t)$ (kWh). For the example shown in Fig. 3.3, the energy content $\kappa(t)$ is determined by:

$$\kappa(t) = \kappa(0) + \int_0^t Q_S(\tau) \cdot \eta_{QS}(Q_S(\tau)) - S_L(\tau) \, d\tau. \tag{3.1}$$

The energy content $\kappa(t)$ is given by the sum of all power in- and outflows.

  Each power transfer is not ideal. There are losses due to the transport or conversion of power from one form of energy to another. Consider the different ways the PV power is transferred in the examples shown in Fig. 3.2. In version 1, the PV power generated in (12), is transformed from DC into AC, and then it is transformed back from alternating current into direct current, either to charge the storage or to charge the battery of the bus.

  In version 2, the solar power is transformed from one direct current power level into another direct current power level. Without a more detailed analysis of the used converter systems, it is difficult to judge which of these two approaches is the better one. In order to describe the power transfer, not only do we need to describe the power transfer from the source to the sink, but we also need to describe the losses of this power transfer.

  In general, these losses are described in terms of efficiency η. The sub-index of η marks the source and the sink. Hence η_{QS} is the efficiency of the power transfer from the source $Q(t)$ to the storage $S(t)$. The efficiency is defined as the ratio between the amount of power reaching the sink Q_S' and the power originally sent from the source

Q_S. The efficiency might depend on the amount of power which is being transferred, hence we need to add the power transfer as an argument for the efficiency $\eta_{QS}(Q_S)$:

$$\eta_{QS}(Q_S) = \frac{Q'_S}{Q_S}. \tag{3.2}$$

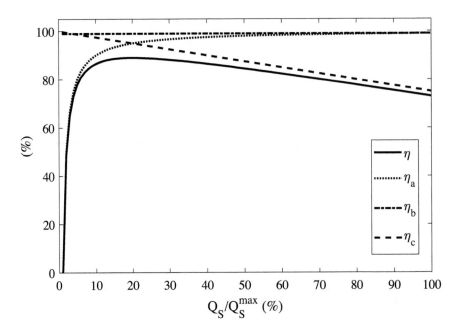

Fig. 3.4 Example of an efficiency curve resulting from the approach shown in eqn 3.3. η represents the overall efficiency, $\eta_{\{a,b,c\}}$ the efficiency resulting from the individual terms.

The solid line in Fig. 3.4 shows the efficiency $\eta_{QS}(Q_S)$, which is a non-linear function of Q_S. If you compare different power conversion technologies, you will see that the curves always look very similar. For low power, there is a very strong reduction in efficiency. At higher power levels, there is also a slight drop. In between, there is a more or less significant maximum. This generic form of the efficiency curve can be easily understood by looking at the power conversion losses and separating them into a constant portion, a linear portion, and a quadratic portion:

$$Q'_S = Q_S - \left(a_{Q_S} + b_{Q_S} \cdot Q_S(t) + c_{Q_S} \cdot Q_S(t)^2\right). \tag{3.3}$$

Here a_{QS}, b_{QS}, and c_{QS} describe the constant, linear, and quadratic conversion losses.

Exercise 3.2 A power transfer technology has the following losses: $a = 1\,\text{kW}$, $b = 0.07$, $c = 10^{-6}\frac{1}{\text{kW}}$. How large is the loss if $P_{QS} = 54\,\text{kW}$ is to be transferred from Q to S? What is the efficiency of this transfer?

Solution: We first need to calculate the losses. So we can calculate Q'_S:

$$Q'_S = Q_S - (a + b \cdot Q_S + c \cdot Q_S^2)$$
$$= 54 \text{ kW} - \left(1 \text{ kW} + 0.07 \cdot 54 \text{ kW} + 10^{-6} \frac{1}{\text{kW}} \cdot 54^2 \text{ kW}^2\right)$$
$$= 54 \text{ kW} - 4.78 \text{ kW} = 49.22 \text{ kW}.$$

Now we can calculate η_{QS} by dividing the output power Q'_S with the input power Q_S:

$$\eta_{QS} = \frac{Q'_S}{Q_S}$$
$$= \frac{49.22 \text{ kW}}{54 \text{ kW}} = 91.15\%.$$

If these losses are included in the definition of efficiency (eqn (3.2)), the efficiency is given by:

$$\eta_{QS}(Q_S(t)) = 1 - \left(b_{Q_S} + \frac{a_{Q_S}}{Q_S(t)} + c_{Q_S} Q_S(t)\right). \tag{3.4}$$

To better understand the influence of the individual terms in eqn (3.4) on the shape of the efficiency curve, the terms are added to Fig. 3.4 individually. η_a corresponds to the shape of the efficiency curve when b_{Q_S} and c_{Q_S} are set equal to zero, and η_b when a_{Q_S} and c_{Q_S} are set equal to zero accordingly. The same applies to η_c. The constant proportion a_{Q_S} provides the characteristic $\frac{1}{x}$ shape at low power. These losses are responsible for low efficiency at low power transfers. The b_{Q_S}, which is linear in terms of the transferred power, remains constant for the entire range and limits the maximum. c_{Q_S}, which is square to the power, dominates the efficiency at high power levels.

Exercise 3.3 Aside from the fact that the person using the bucket in Exercise 3.1 is only able to fill up the car with a charging power of 600 kW, the charging process is not very efficient: the bucket has some holes. If we assume that the time for filling up the bucket, walking to the car, and filling up the car is always the same, these holes result in a linear loss of petrol, for example 5%. During the filling up process, some of the petrol is dropped to the ground. This amount of petrol seems to be roughly 100 ml and constant. What is the total efficiency of this filling process?
  Solution: The efficiency of this filling process equals:

$$\eta = \frac{5 \text{ l} - (5 \text{ l} \cdot 5\% + 0.1 \text{ l})}{5 \text{ l}} = \frac{4.65 \text{ l}}{5 \text{ l}} = 93\%.$$

This curve is typical for all transport and conversion technologies. Each technology has constant losses, whether it is heat losses or simply energy consumption of some auxiliary drives. The linear components have their origin in standard operation. The quadratic components, on the other hand, are caused by heat losses that increase quadratic in power. In the case of an electronic component, a_{Q_S} corresponds to the sum of the constant losses that arise from maintaining the operating voltage (power supply unit), supplying the control electronics, operating the fan, etc. Linear losses, b_{QS}, are caused by the electrical resistance of the lines and by the resistance generated by capacitors or inductors. Square losses, c_{QS}, result from non-linear effects in the inductors and capacitances involved. One can see that the approach shown in eqn (3.3) can describe the different physical processes.

Exercise 3.4 An empty storage is charged with 15 kW for one hour and with 5 kW for half an hour. The efficiency of the charging process is 80% for a 15 kW charge, and 10% for a 5 kW charge.

What is the energy content at the end of the charging process?

Solution: The power of the energy source reads as follows:

$$Q_S(t) = \begin{cases} 15 \text{ kW}; t \in [0 \text{ h}, 1 \text{ h}] \\ 5 \text{ kW}; t \in]1 \text{ h}, 1.5 \text{ h}]. \end{cases}$$

The efficiency depends on the power flow:

$$\eta_{QS}(Q_S(t)) = \begin{cases} 80\% \; ; Q_S(t) \geq 10 \text{ kW} \\ 10\% \; ; Q_S(t) < 10 \text{ kW}. \end{cases}$$

To get the energy content of the storage, we need to integrate the power flow, corrected by the efficiency over the time:

$$\kappa(T) = \int_0^T Q_S(t) \cdot \eta_{QS}(Q_S(t)) \mathrm{d}t$$

$$= \int_0^{1\,\text{h}} 15 \,\text{kW} \cdot 80\,\% \; \mathrm{d}t + \int_{1\,\text{h}}^{1,5\,\text{h}} 5 \,\text{kW} \cdot 10\,\% \; \mathrm{d}t$$

$$= 12 \text{ kWh} + 0.25 \text{ kWh} = 12.25 \text{ kWh}.$$

The definition of efficiency in eqn (3.2) has two limitations: as already shown in Fig. 3.4, its shape is non-linear. This makes its use for further calculations uncomfortable. It is more convenient to directly use the losses in the flow equations. Another alternative is to define an average efficiency, which is independent from the transferred power. The value is given by averaging the efficiency weighted at different set points. In this way, the loss calculation is reduced to multiplying the input power by a number. The later method is used to evaluate the efficiency of solar inverters using the European efficiency standard (Häberlin, 2007; Smets et al., 2015) or for determining the efficiency of residential solar storage systems (Munzke et al., 2017a; Munzke et al., 2017b).

For an experimental determination of the efficiency, it is necessary to measure the Q_S and Q_S'. For storage technologies, this is not always feasible, as not every storage technology allows the storage content to be measured directly. In this case, the efficiency can only be evaluated by charging and discharging the storage in a defined way. Therefore, it must be determined how much energy can be taken from a storage unit to cover the load, and this must be put in relation to the amount of energy stored:

$$\eta_{QS} = \frac{\int_0^{T_1} S_L(t)dt}{\int_0^{T_2} Q_S(t)dt}. \tag{3.5}$$

In eqn (3.5), $Q_S(t)$ is the power transferred into the storage and $S_L(t)$ the amount of power taken from the storage. Both are time series. This measurement procedure has the disadvantage that the result depends on the history of the charging and discharging process. In Exercise 3.4 the measured efficiency equals $\eta_{QS} = \frac{12.5 \text{ kWh}}{17.5 \text{ kWh}} = 71.4\%$. If the charging rate stayed at 15 kW, the efficiency would be 80%. Therefore, if we want to compare the efficiency of different storage systems, it is necessary to define the charging and discharging process.

If we look at the power flow in Fig. 3.2 from the grid to the charging station, we see that in the two realizations in Fig. 3.2 the transmission paths are different. In version 1, the grid current is converted directly into a direct current (13). In this case, it is sufficient to look at the efficiency or losses of the charging station. In version 2, the power is converted twice: first the AC power is transformed into DC power (21), then the vehicle battery is charged (24); here the losses of two conversion stages must be taken into account. We see that the information about which technology is used for the realization of the power flow Q_L is only included in the efficiency η_{QL}. We thus have a description that is generally valid, but becomes specific with simple adjustments.

3.2.2   Self-discharging of storage systems

The Bible itself describes the main task of a storage system: collect grain in the seven good years and spend them in next seven bad years. In terms of the power flow diagram, this process corresponds to a power flow from the past to the present $S_S(t - \Delta t)$ and a power flow into the future $S_S(t)$ to $S(t + \Delta t)$ (Fig. 3.5). In this book we work with discrete time steps. Δt is the time increment between two time steps. This approach is more suitable for the numerical modelling of storage systems.

Of course, losses occur during this virtual transfer of power from the past to the present or future. In the power flow diagram, those nodes that represent a storage $S(t)$ have two additional connections: one that connects the storage node from the past to the present $S(t - \Delta t)$ with the transfer $S_S(t - \Delta t)$ and one that leads from the present to the future $S(t + \Delta t)$. The transfer here is labelled $S_S(t)$. While both are marked with a power flow, only the node from the past also has a value for its efficiency $\eta_{SS}(S(t - \Delta t))$, since we do not want to calculate the losses twice per time increment. An example of this description is shown in Fig. 3.5.

Because we can assign losses to the power transport from the past, we are also able to describe the phenomenon of self-discharge of a storage device (see Exercise 3.5). Self-discharge describes the phenomenon in which a storage loses part of his energy content even when it is not used. Again, it is sufficient to use the approach described in eqn (3.3). Constant losses are more likely to occur with technologies that require additional auxiliary systems; this is the case, for example, with flywheel storage

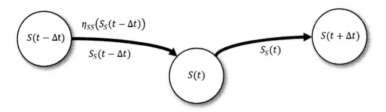

Fig. 3.5 Description of the self-discharge of a storage: the energy content of the storage at time $t - \Delta t$ flows into the storage at time t. If the storage at time t is not completely discharged, its energy content flows to the storage at $t + \Delta t$. η_{SS} is the efficiency of the storage process and is called self-discharge.

(Chapter 6) or redox flow batteries (Section 8.7). A linear loss, on the other hand, can be observed in batteries. This is due to the fact that the probability of an unwanted chemical reaction increases in proportion to the storage content.

Exercise 3.5 Let's assume that we completely charge our mobile phone once. From the nameplate on the battery, we have learned that this battery has about 2,000 mAh storage capacity and an average voltage of 3.65 V. We therefore note $S(0) = 2,000$ mAh $\cdot$ 3.65 V $= 7.3$ Wh. Now the mobile phone is switched off, and we put it in the drawer. Thirty days later, we switch the phone on again and see that the battery has lost $\frac{1}{4}$ of its capacity. What is the daily efficiency η_{SS}, which represents this self-discharge?

   Solution: $S(t = 30\,days) = 7.3$ Wh $\cdot$ 75% $= 5.475$ Wh. We interpret this as follows: a power transfer of $S_S = \frac{7.3\text{Wh}}{30 \cdot 24\text{h}} = 0.01$ W has taken place. The efficiency was $\eta_{SS} = 25\%$.

3.2.3   Description of a storage system via power flow networks

Fig. 3.6 combines the different aspects of an energy storage system into one power flow diagram. An energy source $Q(t)$ provides power to the storage system $S(t)$. A storage system provides power to the Load $L(t)$. The energy content of the storage system is determined by the amount of power transferred from the past to the actual time, and surplus energy not needed now will be transferred to the future.

   The power flow diagram is the main tool to describe and compare different storage technologies. Different power sources, sinks, and storages are described in capital letters. In general, the power transport from A to B is labelled $A_B(t)$. The power transfer losses of this process are labelled $\eta_{AB}(A_B(t))$.

   The power flow diagram allows us to combine different elements to describe complex storage systems, incorporating different energy sources, storage technologies, and loads. It is a graphical representation of the storage system. All technical details are localized in properties of the power transfer paths. To transfer this diagram into a mathematical description, the following steps are necessary:

- Formulate the balance equation for the system components.
- Define properties of the power transfer.
- Define the objective function.

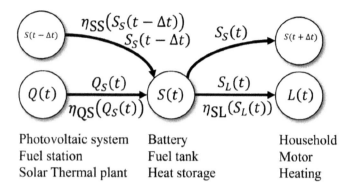

Photovoltaic system   Battery   Household
Fuel station   Fuel tank   Motor
Solar Thermal plant   Heat storage   Heating

Fig. 3.6 The basic power flow diagram consists of a source $Q(t)$, the storage $S(t)$, and the load $L(t)$. Different storage systems can be described by the same power flow diagram.

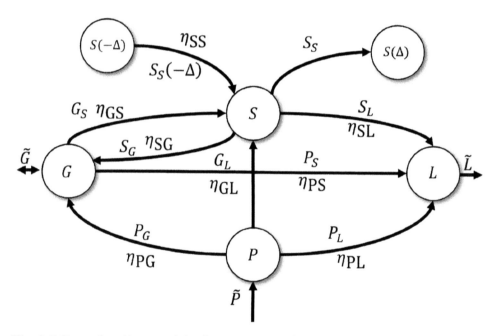

Fig. 3.7 Power flow diagram of the charging station shown in Fig. 3.1. There are four nodes: the grid G, the storage S, the PV system P, and the charge point L. The load profile, i.e. the information on how much power will be taken from the station, is represented as the power flow $\tilde{L}$ from the load node. The information on how much solar power is available is represented by a power flow $\tilde{P}$ into the PV production node. The power flow $\tilde{G}$ describes the amount of power needed from the grid or injected into the grid. (We neglect the arguments and time dependencies of the variables, so as to better read the diagram.)

We will now carry out the various steps for the charging station that we presented at the beginning of this chapter (Fig. 3.1). We have four nodes: the grid G, the storage S, the PV system P, and the charge point L. (We omit the time argument to keep the formula simple.) Fig. 3.7 shows these nodes and the possible power flows.

To formulate the balance equation of the storage, we just need to sum over all power flows. Their sum equals zeros, i.e. we accept only power losses via the transfer:

$$0 = \underbrace{\eta_{SS}\ S_S(-\Delta) + \eta_{GS}\ G_S + \eta_{PS}\ P_S}_{\text{Chargingpower}} - \underbrace{(S_S + S_L + S_G)}_{\text{Dischargingpower}}. \tag{3.6}$$

Note that we drop the arguments of $\eta_{AB}(A_B)$. We have also reduced $S_S(t - \Delta t)$ to $S_S(-\Delta)$ for a better reading.

To formulate the balance equations for the PV system, we proceed analogously. However, we have to describe the possible amount of PV power that is available at a certain point in time. To do this, we use a time series $\tilde{P}(t)$, which describes which amount of power can be taken from the node at which point in time. We can interpret this as an inflow of power. This equation is an inequality, because it may well be that more PV power is available than is used by the three power flows:

$$\tilde{P} \geq P_G + P_S + P_L. \tag{3.7}$$

Again, we have omitted the time argument for better readability of the formula. Similar to the PV power production profile $\tilde{P}$, we can define a load profile $\tilde{L}(t)$ here. This contains information regarding the time at which the various amounts of power are drawn from the bus:

$$\tilde{L} \geq \eta_{SL}\ S_L + \eta_{PL}\ P_L + \eta_{GL}\ G_L. \tag{3.8}$$

Eqn (3.8) is formulated as an inequality. This is reasonable, since we cannot expect the charging station always to be able to provide sufficient power for charging the bus.

For the balance equation of the grid we also introduce an auxiliary variable, $\tilde{G}(t)$. This sums up the inflows and outflows from the grid connection point:

$$\tilde{G} = \eta_{SG}\ S_G + \eta_{PG}\ P_G - (G_L + G_S). \tag{3.9}$$

Next, the boundary conditions for the various power flows are defined. All power transfers are considered positive quantities, so that the orientation of the arrows is maintained. A maximum and minimum value can be specified for each power flow.

For a storage system S we cannot assume that the maximum charging power is equal to the maximum discharging power. We also need to have a look at the power capabilities of the power converters used in the specific realization. For example, to determine what the maximum charging power of the storage is in version 1 (Fig. 3.2), we need to look at the three conversion stages and their power capabilities—that is, what is the minimum maximum power of the AC/DC converter and the storage (11)? In version 2, on the other hand, we would have to look at the minimum maximum power of the rectifier (21), DC/DC converter, and storage (22) – that is, the smallest value that the maximum power can have. Hence, the information about the technology used is therefore contained in the definition of the boundary conditions for the power flow:

$$S_S \in [0, S_{S_{max}}] \tag{3.10}$$
$$S_L \in [0, S_{L_{max}}] \tag{3.11}$$
$$S_S \in [0, \kappa_{max}/\Delta] . \tag{3.12}$$

Eqn (3.10) reflects the limitation of the storage capacity. κ is the storage capacity in Wh. This value is divided by the time increment Δ to have a virtual power transfer (see Exercise 3.5).

For the power transfers from the grid, the boundary conditions read as:

$$G_S \in [0, G_{S_{max}}] \tag{3.13}$$
$$G_L \in [0, G_{L_{max}}] \tag{3.14}$$
$$G_S + G_L - (P_S + S_G) \in [-G_{max}, G_{max}] . \tag{3.15}$$

The first two conditions combine the power capabilities of the grid connection and the power capabilities of the converter involved in the power transfer from the grid to the storage and to the charging point. The third condition reflects the power capability of the grid connection and sums over all power transfers.

The PV generator has three power transfers, which are limited by the power capabilities of the power conversion technologies:

$$P_S \in [0, P_{S_{max}}] \tag{3.16}$$
$$P_L \in [0, P_{L_{max}}] \tag{3.17}$$
$$P_G \in [0, P_{G_{max}}] . \tag{3.18}$$

Using eqns (3.6) to (3.16), we can now describe the behaviour of the system. At each point the load demand of the charging station $\tilde{L}$ and production of the PV generator $\tilde{P}$ are given. The task is to choose the different power flows appropriately. Eqns (3.6) to (3.16) must be satisfied. We have seven possible power flows and four equations: in other words, the system of equations is underdetermined. So there is always a multitude of possible power flows. In order to evaluate these against each other, a suitable criterion is necessary: the objective function. In the case of the charging station (Fig. 3.1) we require, for example, that the grid consumption should be minimized:

$$\min Y = G_S + G_L. \tag{3.19}$$

3.2.4 System design with the aid of power flow considerations

We have developed a mathematical framework with the help of which we are now able to compare the various realizations of a storage system. The basic idea is to consider the storage system as a network in which power is distributed from different nodes. The technological boundary conditions and properties of certain realizations are represented by boundary conditions and efficiencies for the different power flows. The network remains the same; only the properties of the transport paths change. The storage facility is highlighted as a node, in that it can receive power from an earlier node in time and transmit it to a later node in time. The result of such a description is usually an underdetermined inequality system. To evaluate the different possibilities of distributing the power flows, an objective function is used. There are different possibilities for solving this mathematical task. It is possible to solve the inequalities numerically, in rare cases even analytically, or by heuristically determined strategies. The task of implementing a suitable solution strategy is usually assigned to an energy management system (EMS). The methodology used is versatile (Koot, 2006; Nge *et al.*, 2010; Schmiegel *et al.*, 2010; Acone *et al.*, 2015).

With the power flow diagram in Fig. 3.7, we have a tool in hand to determine the system design of our charging station. We can now evaluate which of the two technical solutions shown in Fig. 3.2 is the most sensible. Suitable decision criteria must be defined for the system design. We will start here with the evaluation with regard to power dissipation. In Section 3.3 we will then carry out an evaluation with regard to the operating costs after we have introduced the financial mathematical tools.

In this analysis we restrict ourselves to considering the two realizations shown in Fig. 3.2. (Of course, there are a number of alternative ways to solve the same technical task. For example, the battery and solar power plant could be connected to a common DC bus, so that only one DC/DC controller or one AC/DC inverter would be needed. Cascaded solutions could also be considered, where smaller units are connected together. However, analysing this variety is beyond the scope of this section. Such analyses have been carried out for other applications in Schimpe *et al.*, 2018 and Bauer, 2019). In this analysis we also neglect die storage technology itself. Therefore, we are not investigating the self-discharge efficiency of the storage system. We do this to stay focused on the power transfer and topology discussions and avoid starting to analyse storage technologies, which we have not introduced yet.

In general, the analysis of different topologies is divided into four steps:

- determination of the technical boundary conditions
- determination of the transfer losses
- definition of an optimized operation management
- comparison of technical solutions.

We now want to carry out these steps.

3.2.5 Determination of the technical boundary conditions

The first step is to determine the technical boundary conditions. It is very important to know which boundary conditions and requirements are requested by the user. It makes a significant difference in the design if, for example, our charging station is to be installed at a motorway service station to refuel coaches, or if the charging station is to charge trolleybuses in the inner city area that have no contact with the overhead line for a very short distance. Therefore, the definition of the technical boundary conditions should always be done at the beginning of a consideration. (One should always know the rules of the game before starting to play.)

In Chapter 4, we will learn a methodology that structures this step. In this section we limit ourselves to those constraints or requirements that are needed for the design.

Table 3.1 Summary of technical requirements for the charging station.

Minimum interval between charging operations:	10–15 min
Charging time:	2.5 min
Max. charging power:	500 kW
Max. power of the grid connection point:	90 kW
Max. power of the solar power plant:	90 kW
Max. charging power of the battery:	200 kW
Max. discharge power of the battery:	410 kW
Feed-in tariff:	$0.25\frac{\$}{\text{kWh}}$
Electricity tariff:	$0.31\frac{\$}{\text{kWh}}$

These are the power specifications for grid, battery, solar power plant, and buses, as well as the energy contents of bus battery and battery, and some commercial aspects:

- The charging station is used in rural areas, i.e. the buses come at longer intervals. The interval between two charges is 10–15 minutes from 07:00 to 18:00, 25–30 minutes from 18:00 to 22:00 and 55–60 minutes from 22:00 to 07:00.
- Connect and disconnect to the charging station needs in total 30 seconds; hence the total charging time is about $2\frac{1}{2}$ minutes.
- The bus battery has a maximum charging capacity of 500 kW.
- In order to save costs, no reinforcement of the distribution network shall be necessary. The charging station is connected to the low-voltage grid, the distribution stations provide a maximum power of 90 kW.
- The solar power plant should be dimensioned so that it can also feed its entire output into the grid. This ensures that the solar power is also used when, for example, the battery is already full. This limits the output of the solar power plant to 90 kW.
- Together with the grid and the battery, the charging power of approximately. 500 kW should be able to be provided. Therefore, the maximum discharge power of the battery is about 410 kW.
- The used battery technology cannot be charged with the same power. This is due to the electro chemistry used. The charging power of the battery is limited to 200 kW.
- The feed-in tariff for the solar power plant equals $0.25\frac{\$}{\text{kWh}}$.
- The electricity tariff is $0.31\frac{\$}{\text{kWh}}$.

Tab. 3.1 summarizes these requirements.

3.2.6 Determination of transfer losses

In order to determine the transfer losses, we must first consider the losses of the various power electronic components. Version 1 contains three components: the battery inverter (11), the solar inverter (12), and the charging station (13).

Fig. 3.8 shows the efficiencies of the three components. On the left side is the efficiency of version 1 and on the right side is the efficiency of version 2. In order to be able to compare the efficiencies better, the input power was normalized to the

Table 3.2 Losses of power electronic components of the two technical realizations shown in Fig. 3.2.

Device-ID	$a[\mathrm{W}]$	$b[\cdot]$	$c[\frac{1}{\mathrm{W}}]$
Version 1			
(11)	450	$9 \cdot 10^{-3}$	$8 \cdot 10^{-8}$
(12)	250	$2 \cdot 10^{-3}$	$6 \cdot 10^{-7}$
(13)	1,800	$8 \cdot 10^{-3}$	$1 \cdot 10^{-8}$
Version 2			
(21)	450	$3 \cdot 10^{-3}$	$4 \cdot 10^{-8}$
(22)	300	$5 \cdot 10^{-3}$	$5 \cdot 10^{-8}$
(23)	250	$1 \cdot 10^{-3}$	$3 \cdot 10^{-7}$
(24)	1,200	$8 \cdot 10^{-3}$	$2 \cdot 10^{-8}$

maximum power. Let us first look at the efficiencies of version 1. The battery inverter (11) has a high efficiency over a wide power range. This is very useful for the application. The battery can be charged with a maximum of 50% of the nominal power, but should ideally be discharged with 100%. The solar inverter (12) has a lower efficiency at higher powers than at lower powers. The charging station (13), on the other hand, is optimized for the maximum charging power.

In version 2, we first have a rectifier (21). It is most efficient in the case of a charging power of 90 kW. The battery inverter (22) is efficient over a wider power range, as is the solar inverter (23). The attenuation at (23) is less at higher powers. The charging station itself (24) is optimized for high power.

Next, we determine the transfer efficiency of the power flows in the power flow diagram Fig. 3.7. For this purpose, we consider via which power electronic components the power has to be guided. For a first simplified representation, we just add the configurations, i.e. we describe the powerflow from battery to the bus η_{SL} by (11)+(13). This means that the power has to be transferred from the battery to the AC bus and then via the charging station to the battery of the bus. We start with version 1:

- $\eta_{GS} = (11)$
- $\eta_{GL} = (13)$
- $\eta_{SG} = (11)$
- $\eta_{SL} = (11) + (13)$
- $\eta_{PG} = (12)$
- $\eta_{PS} = (12) + (11)$
- $\eta_{PL} = (12) + (13)$.

To determine the transfer efficiency of the powerflows in version 2, we need to have a closer look at how we calculate the power losses on a two-stage transfer.

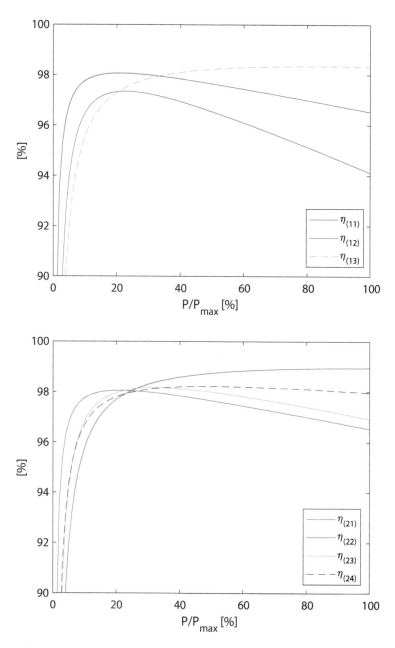

Fig. 3.8 Efficiencies of the different power electronic components from Fig. 3.2. The upper figure shows version 1, which connects all components via an AC bus: battery inverter $\eta_{(11)}$, solar inverter $\eta_{(12)}$, and the charging station $\eta_{(13)}$. The lower figure shows the components of version 2, which uses a DC bus: rectifier $\eta_{(21)}$, battery inverter $\eta_{(22)}$, solar inverter $\eta_{(23)}$, and charging station $\eta_{(24)}$.

Exercise 3.6 The efficiency η_{SL} depends on the efficiency of the battery inverter (11) and the efficiency of the charging station (13). As can be seen from Fig. 3.8, both efficiency curves are non-linear functions of the input power; as such, for the efficiency η_{SL} it must be taken into account which power is transmitted by the grid and by PV. Determine the formula for the power loss S'_L and the combined efficiency η_{SL}. For the power from the grid and PV, set a combined value P_{row} (row = rest of world).

Solution: The power to be transmitted from the storage is S_L. The losses of the inverter are given by:

$$\hat{S}_L^{(11)} = a_{(11)} + b_{(11)} \; S_L + c_{(11)} \; S_L^2.$$

$a_{(11)}$, $b_{(11)}$, and $c_{(11)}$ are the lost coefficient of stage (11). The power transmitted onto the AC bus is the original power S_L reduced by the losses, and thus amounts to:

$$S_L^{(11)} = S_L - \hat{S}_L^{(11)}$$
$$= S_L - \left(a_{(11)} + b_{(11)} \; S_L + c_{(11)} \; S_L^2\right).$$

In order to consider the losses during the power transfer of the charging station (13), the sum of $S_L^{(11)}$ and P_{row} must be used as input variable. The total losses are then given by:

$$\hat{S}_L^{(13)}(P_{row}) = a_{(13)} + b_{(13)} \; \left(S_L^{(11)} + P_{row}\right) + c_{(11)} \; \left(S_L^{(11)} + P_{row}\right)^2$$
$$= a_{(13)} + b_{(13)} \; \left(S_L - \left(a_{(11)} + b_{(11)} \; S_L + c_{(11)} \; S_L^2\right) + P_{row}\right)$$
$$+ c_{(11)} \; \left(S_L - \left(a_{(11)} + b_{(11)} \; S_L + c_{(11)} \; S_L^2\right) + P_{row}\right)^2.$$

The efficiency is given by:

$$\eta_{SL}(S_L, P_{row}) = \frac{\left(S_L - \hat{S}_L^{(13)}(P_{row})\right)}{S_L}.$$

In Fig. 3.9 the total efficiency $\eta_{SL}(S_L, P_{row})$ of the power transfer from the battery (11) through to the charging station (13) dependent on surplus power provided by the grid and the solar plant is shown. One can see that in this configuration the η_{SL} is between 90% and 95%. Above a battery discharge power of $S_L > 50$ kW the efficiency is rather constant. Fig. 3.9 also indicates that the non-linear terms $c_{(xx)}$, which are relevant for the losses at higher power transfers, do not play a major role in the overall efficiency. As we investigate further, we therefore ignore the non-linear terms. (Of course, such a simplification must be considered with care. For an analytical consideration, the terms simplify considerably, as we will see in the following. However, as soon as the systems under consideration have reached a level of complexity that they can only be analysed numerically, the non-linear terms should be taken along, as they only slightly increase the complexity of the numerical analysis (Ried, Schmiegel, and Munzke 2020).)

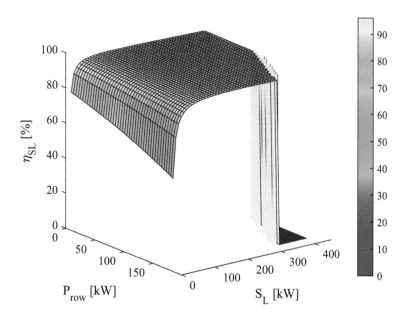

Fig. 3.9 The total efficiency $\eta_{\mathrm{SL}}(S_{\mathrm{L}}, P_{\mathrm{row}})$ of the power transfer from the battery (11) through to charging station (13) dependent on surplus power provided by the grid and the solar plant. Note, that the charging power S_{L} of the charging station is limited to 500 kW. Power flows which violate this boundary condition are marked with $\eta = 0$.

Exercise 3.7 The efficiency η_{SL} depends on the efficiency of the battery inverter (11) and the efficiency of the charging station (13). In this exercise, we want to make the same investigations as in Exercise 3.6, but we want to neglect the non-linear terms, i.e. $c_{\mathrm{xx}} \to 0$. We want to determine the formula for the power loss S_{L}' and the combined efficiency η_{SL}. For the power from the grid and PV, set a combined value P_{row} (row = rest of world).

Solution: The power to be transmitted from the storage is S_{L}. The losses of the inverter are given by:

$$\hat{S}_{\mathrm{L}}^{(11)} = a_{(11)} + b_{(11)}\ S_{\mathrm{L}}.$$

$a_{(11)}$ and $b_{(11)}$ are the lost coefficient of stage (11). The power transmitted onto the AC bus is the original power S_{L} reduced by the losses, and thus amounts to:

$$
\begin{aligned}
S_{\mathrm{L}}^{(11)} &= S_{\mathrm{L}} - \hat{S}_{\mathrm{L}}^{(11)} \\
&= S_{\mathrm{L}} - \left(a_{(11)} + b_{(11)}\ S_{\mathrm{L}}\right) \\
&= S_{\mathrm{L}}\left(1 - b_{(11)}\right) - a_{(11)}.
\end{aligned}
$$

In order to consider the losses during the power transfer of the charging station (13), the sum of $S_{\mathrm{L}}^{(11)}$ and P_{row} must be used as an input variable. The total losses are then given by:

$$
\begin{aligned}
\hat{S}_{\mathrm{L}}^{(13)}(P_{\mathrm{row}}) &= a_{(13)} + b_{(13)}\ \left(S_{\mathrm{L}}^{(11)} + P_{\mathrm{row}}\right) \\
&= a_{(13)} + b_{(13)}\ \left(S_{\mathrm{L}}\left(1 - b_{(11)}\right) - a_{(11)} + P_{\mathrm{row}}\right) \\
&= S_{\mathrm{L}}\ b_{(13)}\left(1 - b_{(11)}\right) + a_{(13)} + b_{(13)}\left(P_{\mathrm{row}} - a_{(11)}\right).
\end{aligned}
$$

The efficiency is given by:

$$\eta_{SL}(S_L, P_{row}) = \frac{\left(S_L - \hat{S}_L^{(13)}(P_{row})\right)}{S_L}$$

$$= \frac{S_L \left(1 - b_{(13)} \left(1 - b_{(11)}\right)\right) - \left(a_{(13)} + b_{(13)} \left(P_{row} - a_{(11)}\right)\right)}{S_L}$$

$$= \left(1 - b_{(13)} \left(1 - b_{(11)}\right)\right) - \frac{\left(a_{(13)} + b_{(13)} \left(P_{row} - a_{(11)}\right)\right)}{S_L}$$

$$= \hat{b} - \frac{\hat{a}}{S_L}.$$

If the non-linear terms are dropped, the mathematical structure of the loss equation remains the same. The losses break down into a constant part, which has an influence on the efficiency that is independent of the power, and a reciprocal part, whose influence disappears at high powers.

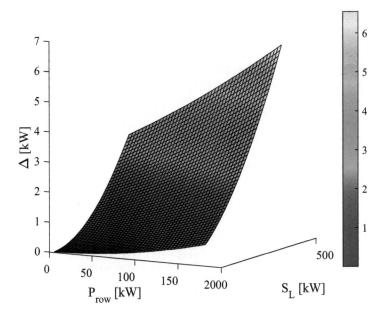

Fig. 3.10 The difference in the calculation of the transfer losses considering the non-linear terms and the linear approximation Δ. Note that the charging power S_L of the charging station is limited to 500 kW.

The effects of this simplification can be seen very well in Fig. 3.10. We do not consider the error in efficiency here—that is, the deviation between the efficiency with non-linear losses and the linear approximation—but consider the absolute power loss here. This is more meaningful, because a high relative deviation at low power values has a small effect on absolute power values. As can be seen in Fig. 3.10, the error increases rapidly with the power to be transmitted. Since the non-linear components contribute more to the losses at high powers, this loss is also in line with our expectations. However, we also realize that in absolute values, the approximation is quite acceptable. For a charging power of 410 kW, the difference is 4 kW. This corresponds to the heat output of approximately four kettles. But in relation to the charging power, this error is negligible in this case.

The linear approximation is very helpful for an initial system analysis, as it can be used without much numerical effort. However, an error analysis should be carried out to assess whether this approximation is acceptable.

We now want to summarize the transfer losses for the power flows in version 1. The power flows that have only one power electronic component can be determined directly from Tab. 3.2. For power flows that have multiple stages, where other power flows also enter, the calculation is more complex. Let us first create a general equation.

Exercise 3.8 We have seen that the linear approximation is sufficient for an initial consideration of power losses. Determine a general equation for a power transfer over two stages, (1) and (2). In both stages, there is the possibility that an additional power flow $P_{(1),(2)}$ affects the operating point.

Solution: We want to transfer the power S. The losses at the first stage $\hat{S}_{(1)}$ are given by:

$$\hat{S}_{(1)} = a_{(1)} + b_{(1)}\left(S + P_{(1)}\right).$$

Thus, the power arriving at the second stage equals:

$$\begin{aligned} S_{(1)} &= S - \left(a_{(1)} + b_{(1)}\left(S + P_{(1)}\right)\right) \\ &= S\left(1 - b_{(1)}\right) - \left(a_{(1)} + b_{(1)}\ P_{(1)}\right). \end{aligned}$$

The losses at the second stage $\hat{S}_{(2)}$ are given by:

$$\begin{aligned} \hat{S}_{(2)} &= a_{(2)} + b_{(2)}\left(S\left(1 - b_{(1)}\right) - \left(a_{(1)} + b_{(1)}\ P_{(1)}\right) + P_{(2)}\right) \\ &= S_{(1)}b_{(2)}\left(1 - b(1)\right) + a_{(2)} + b_{(2)}\left(P_{(2)} - \left(a_{(1)} + b_{(1)}\ P_{(1)}\right)\right). \end{aligned}$$

The power transferred from (1) to (2) is thus:

$$S_{(2)} = S\left(1 - b_{(2)}\left(1 - b_{(1)}\right)\right) - \left(a_{(2)} + b_{(2)}\left(P_{(2)} - \left(a_{(1)} + b_{(1)}\ P_{(1)}\right)\right)\right).$$

We summarize the coefficients:

$$\begin{aligned} \tilde{a} &= a_{(2)} - a_{(1)}\ b_{(2)} \\ \tilde{b} &= \left(b_{(2)}\left(1 - b_{(1)}\right)\right) \\ \tilde{c} &= b_{(1)}\ b_{(2)}. \end{aligned}$$

Thus, the equation simplifies to:

$$S_{(2)} = \left(1 - \tilde{b}\right)\ S - \left(\tilde{a} + \tilde{c}\ P_{(1)} + b_{(2)}\ P_{(1)}\right).$$

Table 3.3 Loss coefficients for power transfers with two power electronic components for the two technical realizations from Fig. 3.2. The nomenclature corresponds to that in Exercise 3.8. Since in all power flows the operating point of the first stage is not influenced by another power flow, $P_{(1)}$ and $\tilde{c}$ are not shown. For power transfers that also have no entry by $P_{(2)}$, $b_{(2)}$ and $P_{(2)}$ have been omitted from the table.

Power flow	$\tilde{a}[\mathrm{W}]$	$\tilde{b}[\cdot]$	$b_{(2)}$	$P_{(2)}$
Version 1				
S_L	1,796	0.0079	0.008	$G_L + P_L$
P_S	447.7	0.009	0.009	G_S
P_L	1,798	0.008	0.008	$G_L + S_L$
Version 2				
G_S	446.85	0.009	0.009	P_S
G_L	1,197	0.008	0.008	$S_L + P_L$
S_L	1,196	0.0079	0.008	$G_L + P_L$
S_G	348.65	0.003	0.003	P_G
P_G	349.25	0.003	0.003	S_G
P_S	447.75	0.009	0.009	G_S
P_L	1,198	0.008	0.008	$G_L + S_L$

In Tab. 3.3 we have shown the loss coefficients for the power transfers of version 1 and version 2 where two stages are involved. In version 2, all power is distributed via a DC bus—that is, two power electronic components are always involved in each power transfer. Accordingly, the efficiencies result from the following combinations:

- $\eta_{GS} = (21) + (22)$
- $\eta_{GL} = (21) + (24)$
- $\eta_{SG} = (21) + (22)$
- $\eta_{SL} = (22) + (24)$
- $\eta_{PG} = (23) + (21)$
- $\eta_{PS} = (23) + (22)$
- $\eta_{PL} = (23) + (24)$.

We have no power flow in either version 1 or version 2 whose losses are affected by a power flow to the source node $P_{(1)}$. Therefore, in Tab. 3.3, $p_{(1)}$ and $\tilde{c}$ were omitted. For power transfers that also have no contribution from $P_{(2)}$, $b_{(2)}$ and $P_{(2)}$ have also been omitted from the table.

With the determination of the loss coefficient, we now have all the elements together to calculate the power flow equations. The question now arises of which criteria should be used to select the power flows. Two criteria are available: minimization of power losses or minimization of costs.

Let us first consider the minimization of losses: a decision is always necessary when two or more power paths can transmit their power into one node. Let us consider this in Fig. 3.11. The node C is supposed to deliver a power of $\tilde{C}$. This power can be obtained by a mixture of the two power flows A_C and B_C. In this case, power is transmitted from A, or B to C.

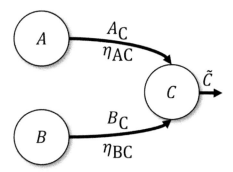

Fig. 3.11 Example of a power flow mixed from two power flows. C receives from each of the two nodes A and B a part of the required power $\tilde{C}$ that the node is to deliver.

The total loss L results in a linear approximation to:

$$L = a_{AC}\gamma_A + b_{AC}A_C + a_{BC}\gamma_B + b_{BC}B_C. \tag{3.20}$$

Here we have introduced the two auxiliary variables γ_A, γ_B. For γ_A the following holds:

$$\gamma_A = \begin{cases} 1 & A_C > 0 \\ 0 & A_C = 0. \end{cases} \tag{3.21}$$

γ_B is defined similarly. The fact that we need this auxiliary variable becomes clear when we look at eqn (3.20). Assuming A_C would be zero—that is, the node A is not used at all—there is still a power loss of a_{AC}. However, if we can ensure that the node is not used, then the constant loss should not be observed either.

Since we must always satisfy the power $\tilde{C}$, it follows that:

$$\tilde{C} = A_C + B_C. \tag{3.22}$$

We can thus calculate the power flow B_C from the eqn (3.20) substitute.

$$B_C = \tilde{C} - A_C. \tag{3.23}$$

We thus obtain a loss equation that depends only on A_C:

$$L = (a_{AC} \, \gamma_A + a_{BC} \, \gamma_B) + b_{AC}A_C + b_{BC} \left(\tilde{C} - A_C\right) \tag{3.24}$$

$$= (a_{AC} \, \gamma_A + a_{BC} \, \gamma_B) + b_{BC}\tilde{C} + (b_{AC} - b_{BC}) \, A_C. \tag{3.25}$$

Eqn (3.25) can be viewed as a straight line equation. If $\tilde{C} \to 0$ but $A_C \neq 0$ and $B_C \neq 0$, the losses approach $(a_{AC}\gamma_A + a_{BC}\gamma_B)$. This, of course, becomes zero when both power sources are switched off.

Whether power should be transferred from A or from B depends on the sign of the factor $(b_{AC} - b_{BC})$. If $(b_{AC} - b_{BC}) < 0$, the losses decreases if A_C has a higher fraction in the mixture. On the other hand, if $(b_{AC} - b_{BC}) > 0$, the power of B_C should be used preferentially. However, a mixture of two power nodes should only be used if one node alone cannot provide the power. This is because a mixture of the two nodes always results in higher constant power losses.

Exercise 3.9 We assume that for power flow A_C the constant loss is $a_{AC} = 10$ W and the linear component is $b_{AC} = 0.002$. For the power flow B_C, the constant loss is $a_{BC} = 12$ W and the linear component is $b_{BC} = 0.0025$. A power of $\tilde{C} = 1,000$ W is to be transmitted. A_C is limited to 1,000 W. What are the losses at:

$$a)\ A_C = 1{,}000\ W,\ B_C = 0\ W$$
$$b)\ A_C = 0\ W,\ B_C = 1{,}000\ W$$
$$c)\ A_C = 500\ W,\ B_C = 500\ W?$$

Solution: In all three cases we can use eqn (3.25):

a) $L = a_{AC} + b_{AC} A_C = 10\ \text{W} + 0.002 \cdot 1{,}000\ \text{W} = 12\ \text{W}$

b) $L = a_{BC} + b_{BC} B_C = 120\ \text{W} + 0.0025 \cdot 1{,}000\ \text{W} = 14.5\ \text{W}$

c) $L = (a_{AC}\ \gamma_A + a_{BC}\ \gamma_B) + b_{BC}\ \tilde{C} + (b_{AC} - b_{BC})\ A_C$
$\qquad = (10\ \text{W} + 12\ \text{W}) + 0.0025 \cdot 1{,}000\ \text{W} + (0.002 - 0.0025)\ 500\ \text{W}$
$\qquad = 24\ \text{W}.$

We see that solution $a)$ has the lowest losses. Solution $b)$ has slightly higher losses. For $c)$ the losses are almost twice as high. This is because both power sources produce a high constant loss.

Next, we consider a cost-optimized choice of power flows. A power transfer A_C has a cost of c_{AC} and a power transfer B_C has a cost of c_{BC}. We should always be satisfied that the constraint that $\tilde{C} = A_C + B_C$ is still valid. Thus, for the total cost c_{sum}, the following applies:

$$c_{\text{sum}} = c_{AC}\ A_C + c_{BC}\ B_C$$
$$= c_{AC}\ A_C + c_{BC} (\tilde{C} - A_C)$$
$$= A_C (c_{AC} - c_{BC}) + c_{BC}\ \tilde{C}.$$

A similar criterion can be derived here as before for the losses: if $(c_{AC} - c_{BC}) < 0)$, the power flow via B_C is to be preferred; otherwise, A_C should be used.

Exercise 3.10 For power flow A_C, the cost is $c_{AC} = 0.1\frac{\$}{W}$. For power flow B_C, the cost is $c_{BC} = 0.05\frac{\$}{W}$. A power of $\tilde{C} = 1{,}000$ W is to be transmitted. A_C is limited to 1,000 W. What is the cost at:

$$a)\ A_C = 1{,}000\ W,\ B_C = 0\ W$$
$$b)\ A_C = 0\ W,\ B_C = 1{,}000\ W$$
$$c)\ A_C = 500\ W,\ B_C = 500\ W?$$

Solution: The costs can be determined by eqn (3.10):

$$a)\ c_{\text{sum}} = A_C c_{AC} = 0.1\frac{\$}{W}\ 1{,}000\ W = \$100$$

$$b)\ c_{\text{sum}} = B_C c_{BC} = 0.05\frac{\$}{W}\ 1{,}000\ W = \$50$$

$$c)\ c_{\text{sum}} = A_C\,(c_{AC} - c_{BC}) + c_{BC}\ \tilde{C}$$

$$= 500\ W\left(0.1\frac{\$}{W} - 0.05\frac{\$}{W}\right) + 0.05\frac{\$}{W}\ 1{,}000\ W,$$

$$= \$75$$

We see that solution *a*) has the highest cost. Solution *b*) has the lowest cost. For *c*) the costs are in between.

The question of selecting the power path for supplying a node C is reduced to two equations. If we want to reduce the power dissipation, we compare the quantities b_{AC} and b_{BC} and choose the power flow that has a smaller b component.

If we want to minimize the costs, we consider the two quantities c_{AC} and c_{BC} and choose the node that generates the lowest costs.

3.2.7 Determination of optimal operational management

In this section, we will address the question of how the power flows for the charging station are chosen. The operation of the charging station is influenced by three variables: the current solar power production $\tilde{P}$, the power demand at the charging station $\tilde{L}$, and the state of charge of the storage κ. We now want to look at different system states and define a reasonable operation for each. In doing so, we will consider both loss-optimized and cost-optimized control.

$\tilde{P} = 0$, $\tilde{L} = 0\text{kW}$, $\kappa <= \kappa_{\text{min}}$: From Tab. 3.1 we know that we have a time of 10 to 15 minutes to charge the battery. In this time the battery can be charged. The maximum charging power is $P_S + G_S <= 180$ kW. We first want to determine how much capacity we need for at least one charging process.

Exercise 3.11 According to the specification, the charging process of the bus takes 2.5 min. We assume that a maximum charging power of 500 kW is required throughout. How much capacity must be contained in the battery if we assume that no solar power is available to charge the bus?

Solution: From Tab. 3.1 we know that the discharge power of the battery is 410 kW. The battery should provide this power for 2.5 min. In this example, we are not interested in the amount of energy that arrives at the vehicle. The battery is the source in this consideration; therefore, no power losses are taken into account. Thus, the required capacity is calculated from:

$$\kappa_{\text{min}} = \int_{t=0}^{t=2.5\text{min}} G_L\ dt$$

$$= [G_L]_{t=0}^{t=2.5\text{min}} = 410\ \text{kW} \cdot 2.5\ \text{min}$$

$$= 1{,}025\ \text{kWmin} = 17.08\ \text{kWh}.$$

17.08 kWh is the minimum storage capacity we need if we want to charge a bus at maximum power within the full 2.6 min. In this system state, only the grid is available as a source. So we have no choice, and only have to ask the question of whether we can get the battery charged.

Exercise 3.12 There is 10 min of charging time available. It can only be charged from the grid, i.e. $P_S = 0$. How much energy can be charged into the battery in this time? How many losses occur in the procedure? Calculate this for version 1 and version 2 from Fig. 3.2.

   Solution: Only the power flow from the grid to the storage needs to be integrated. Since we are interested in the state of charge of the battery, i.e. the power that arrives in the storage, the losses must still be taken into account:

$$\kappa = \int_{t=0}^{t=10\text{min}} \left((1-b)G_S - a\right)\ dt$$

$$= \int_{t=0}^{t=10\text{min}} (1-b)G_S\ dt - \int_{t=0}^{t=10\text{min}} a\ dt$$

$$= (1-b)G_S\ 10\ \text{min} - a\ 10\ \text{min}.$$

For a realization with version 1, this results in a maximum storage capacity of:

$$\kappa = \left((1 - 9\cdot 10^{-3})80\ \text{kW} - 0.45\ \text{kW}\right)\ 10\text{min}$$

$$= (79.28\ \text{kW} - .45\ \text{kW})\ 10\text{min}$$

$$= 788.3\ \text{kWmin} = 13.13\ \text{kWh}.$$

For a realization with version 2, this results in something similar:

$$\kappa = \left((1 - 9\cdot 10^{-3})80\ \text{kW} - 0.446\ \text{kW}\right)\ 10\text{min}$$

$$= (79.28\ \text{kW} - .446\ \text{kW})\ 10\text{min}$$

$$= 788.34\ \text{kWmin} = 13.14\ \text{kWh}.$$

The difference between the two versions is not that big, but we can see that without solar power support, the battery would not reach the maximum required storage capacity of 17.08 kWh.

$\tilde{P} = \tilde{P},\ \tilde{L} = 0\text{kW},\ \kappa <= \kappa_{\min}$: in this case we have two power sources and we need to decide which of the power paths P_S, P_G, G_S should be used.

   Let us first consider the question of whether P_G should be fed into the grid or sold. Since the feed-in tariff $c_{PS} = 0.25\frac{\$}{\text{kWh}}$ is smaller than the electricity price $c_{GS} = c_{GL} = 0.31\frac{\$}{\text{kWh}}$, this only makes sense if no additional electricity will be drawn from the grid in the future. So as long as $\kappa <= \kappa_{\min}$, it makes sense for cost savings to charge the battery.

   If we want to minimize losses, we have to compare the loss coefficients of the two power paths P_S and G_S with each other.

Exercise 3.13 To charge the battery, both solar power P_S and grid power G_S are available. What power flow must we choose so that in version 1 and version 2 (Fig. 3.2) the losses are minimal?

Solution: We first compare the loss coefficients b_{PS} and b_{GS} with each other:

$$\text{Version 1:} b_{PS} = 0.009; b_{GS} = 0.009$$
$$\text{Version 2:} b_{PS} = 0.009; b_{GS} = 0.009.$$

We see that the linear coefficients are equal; hence, they do not provide a decision criteria. We have to look at the constant losses:

$$\text{Version 1:} a_{PS} = 447.75 \text{ W}; a_{GS} = 450 \text{ W}$$
$$\text{Version 2:} b_{PS} = 447.75 \text{ W}; a_{GS} = 446.85 \text{ W}.$$

In version 1, the use of P_S would be preferred, as the losses are lower here. In version 2, on the other hand, G_S would be preferred.

$\tilde{P} = \tilde{P}$, $\tilde{L} = 0$ kW, $\kappa > \kappa_{\min}$: in this case, we have three power sources and need to decide which of the power paths P_S, P_G, G_S, S_G should be used.

We first consider S_G, as it is related to the question of the value of stored energy. We have a storage of surplus energy in this state. Does it make sense to feed the power into the grid with a feed-in tariff of $c_{SG} = c_{PG} = 0.25\frac{\$}{\text{kWh}}$? This really depends on our expectation for the future. Suppose we have too little stored power to charge the bus. In this case, we would have to buy power from the grid at a price of $c_{GL} = 0.31\frac{\$}{\text{kWh}}$. So we would have made a loss of $0.06\frac{\$}{\text{kWh}}$. It Therefore, in order to avoid this loss, it makes sense not to feed the stored power into the grid. However, this only applies if we also ensure that we discharge the storage. Stored power that is never used is worth nothing in monetary terms. On the contrary, the self-discharge of the storage unit causes a monetary loss.

Exercise 3.14 We want to estimate how much storage capacity we need if we do not want to feed solar power into the grid. For simplicity, we assume that the solar power system produces 90,000 kWh annually. We also assume that the daily amount of energy produced is $\frac{90,000 \text{ kWh}}{365 \text{ day}} = 246.58$ kWh. Is it possible to operate the charging station using the solar power system alone? How large would the storage have to be if the solar power plant were to be used only to compensate for the grid connection point?

Solution: The requirements state three operating phases with different intervals. We take the smallest interval in each case. The total charging and discharging cycle is made up of the waiting time of 10, 25, and 55 minutes respectively and the charging process, which takes a total of 3.5 minutes. This results in the number of cycles:

$$07{:}00\text{--}18{:}00 : \frac{11 \text{ h}}{0.225 \text{ h}} = 48.8 \approx 49$$
$$18{:}00\text{--}22{:}00 : \frac{4 \text{ h}}{0.475 \text{ h}} = 8.42 \approx 9$$
$$22{:}00\text{--}07{:}00 : \frac{9 \text{ h}}{0.975 \text{ h}} = 9.23 \approx 10.$$

From Exercise 3.11 we know that we need a total of 17.08 kWh. So the solar power is only enough for $\frac{246.58 \text{ kWh}}{17.08 \text{ } kWh} = 14.84 \approx 15$ charging cycles. Operation from the solar power plant alone is not possible with an output of 90 kW.

From Exercise 3.12 we also know that we can only charge the storage from the grid up to 13.13 kWh or 13.14 kWh, as the power of the grid connection point would not be sufficient to fill the storage in time. Therefore, we would only have to compensate for the missing 3.95 kWh with the solar power plant. This would still work for $\frac{246.58 \text{ kWh}}{3.95 \text{ } kWh} = 62.42 \approx 62$ charging cycles.

Only in the period from 07:00 to 18:00 is support from solar power necessary. In all other operating phases, we have enough time to charge the battery with electricity from the grid. Therefore, the storage capacity really needed is $49 \cdot 3.95$ kWh $= 193.55$ kWh.

Let us now consider loss-optimized control. If we want to minimize losses, the choice $S_G = 0$ is a good choice, because a power flow that is not active cannot generate losses.

Since there is no more reason to charge the battery, $G_S = 0$. This is both the cost-optimal and the loss-optimal solution. What about P_S and P_G? Since the feed-in tariff is smaller than the electricity costs, it does not make sense to feed solar power into the grid when considering costs. Therefore, $P_G = 0$ applies. But what about loss optimization?

Exercise 3.15 Solar power can be transferred to the storage as well as to the grid. Which power flow would have the lower loss in version 1 and version 2 (Fig. 3.2) ?

Solution: We first compare the loss coefficients b_{PS} and b_{PG} with each other:

$$\text{Version 1:} b_{PS} = 0.009; b_{PG} = 0.002$$
$$\text{Version 2:} b_{PS} = 0.009; b_{PG} = 0.003.$$

In both versions, direct injection of solar power into the grid would be the lowest-loss option.

The previous states were all states in which no bus should be charged. This time was used to charge the battery so that enough power was available for a test charge. In the following we look at the charging situation.

$\tilde{P} = 0$, $\tilde{L} = \tilde{L}$, $\kappa > 0$: we have no solar radiation, but the battery is not empty. Two power paths are relevant: G_L and S_L. Since we need to provide the required 500 kW and G_L cannot provide the required power, the assignment of the power flows is clear. Both deliver the maximum power for as long as they can and charge the bus battery.

$\tilde{P} = \tilde{P}$, $\tilde{L} = \tilde{L}$, $\kappa > 0$: in this case, solar power is also available for the charging process, so the load can be met by the three power paths P_L, G_L, and S_L. The paths with the lowest costs are P_L and S_L. As long as solar power and battery power are sufficient to supply the required 500 kW, the use of grid power should be avoided. But what about a loss-optimized approach?

Exercise 3.16 S_L feeds the battery of the bus. However, 90 kW is still missing, which can be drawn from the solar power plant or the grid. Which power flow do we have to choose so that in Version 1 and in Version 2 (Fig. 3.2) the losses are minimal?

Table 3.4 Summary of the cost- or loss-optimized choice of power flows for different system states.

System state	Cost-optimized control	Loss-optimized control version 1	Loss-optimized control version 2
$\tilde{P}=0,\ \tilde{L}=0\text{kW},\ \kappa <= \kappa_{min}$ $\tilde{P}=\tilde{P},\ \tilde{L}=0,\ \kappa <= \kappa_{min}$	$G_S>0$ $P_S>0,$ $G_S = 90\ kW - P_S$	$G_S>0$ $P_S>0,\ P_G=0,$ $G_S = 90\ kW - P_S$	$G_S>0$ $P_G>0,\ G_S>0$
$\tilde{P}=\tilde{P},\ \tilde{L}=0,\ \kappa > \kappa_{min}$	$P_S>0$	$P_S=0,\ P_G>0,$ $G_S,\ S_G=0$	$P_S=0,\ P_G>0,\ G_S,\ S_G=0$
$\tilde{P}=0,\ \tilde{L}=\tilde{L},\ \kappa>0$ $\tilde{P}=\tilde{P},\ \tilde{L}=\tilde{L},\ \kappa>0$	$G_L>0,\ S_L>0$ $P_L>0,\ S_L>0$	$G_L>0,\ S_L>0$ $P_L>0,\ G_L=0,$ $S_L>0$	$G_L>0,\ S_L>0$ $P_L=0,\ G_L>0,\ S_L>0$

Solution: We first compare the loss coefficients b_{PL} and b_{GL} with each other:

$$\text{Version 1: } b_{PL} = 0.008;\ b_{GL} = 0.008$$
$$\text{Version 2: } b_{PL} = 0.008;\ b_{GL} = 0.008.$$

We see that the linear coefficients do not provide a decision characteristic. Therefore, we need to look at the constant losses:

$$\text{Version 1: } a_{PL} = 1798\ \text{W};\ a_{GL} = 1{,}800\ \text{W}$$
$$\text{Version 2: } b_{PL} = 1198\ \text{W};\ a_{GL} = 1{,}197\ \text{W}$$

Accordingly, in version 1, the use of P_L is preferred, since the losses are lower. In version 2, on the other hand, G_L is be preferred.

We have worked out decision recommendations for the most important states, which can be used for operational management. In Tab. 3.4, these are summarized.

3.2.8 Comparison of the technical solutions

After the power losses have been determined and the operation management has been defined, we want to compare the different solutions and operation management with each other. The comparison should consider the costs and losses over a full year, as we also want to take seasonal changes into account. First, we need time series for $\tilde{P}$ and $\tilde{L}$. For the production of solar power $\tilde{P}$, we can use time series of solar power plants. These time series cover a full year's run. When determining the load profile $\tilde{L}$, we face the challenge that the system does not yet exist. Hence, we cannot fall back on historical data at this point in time. For our considerations, therefore, we want to generate a synthetic load profile.

To do this, we first define the period Δt. The time interval allows us to write down power flows over time in a table. The choice of Δt is oriented to the time scale on which the dynamics we want to observe take place. This is comparable to the question of whether we want to paint a portrait on a centimetre or millimetre or micrometre resolution. If we only want to recognize the face, a centimetre resolution is usually sufficient, because the human brain can recognize faces very well. If, on the other hand, we want to use the portrait to analyse nuances of the skin, we should use a micrometre resolution. When choosing the resolution, we have to be guided by the structure of the things we want to observe. For solar, a resolution between one minute and a quarter of an hour is sufficiently accurate when analysing yields over the course of a year (Dragoon and Schumaker, 2010). For the load profile, a resolution of a quarter of an hour is too low, because we know that the charging process lasts about two minutes. A resolution of one minute should be sufficient and is quite common in analyses of energy storage systems (Bauer, 2019).

The design rule for generating a synthetic load profile is relatively simple. Once the time scale is defined, we need to look at each time to see which loads are applied at time t and sum over their consumption (Armstrong *et al.*, 2009). In our case, using the timetable and the requirements from Tab. 3.1, we have all the information. We only need to start looking when a bus arrives and then enter the loading process as a time series in the table $\tilde{L}$.

We only got interval information from the requirements. From this we can conclude that there is a certain statistical variation. Sometimes a bus arrives a little faster, sometimes a little slower. It is reasonable to assume that this is also the case with the required charging power. Therefore, we randomly vary the arrival time of the bus within an interval t_1, t_2. The times t_1 and t_2 represent the given limits from the timetable. We use a random generator $\mathtt{random}(t_1, t_2)$, which determines a random value in the interval t_1 to t_2. The distribution of the random values is equal—that is, the occurrence of a certain value within the interval is equally probable. Thus, in the load profile $\tilde{L}$, the intervals between different charging processes are slightly different.

We proceed analogously with the charging power. We know that it should not be greater than 500 kW, but we have no information on how much lower it can be. We set the interval of possible powers to 300 kW to 500kW and use a random generator to determine a power value random $(300, 500)$ for each charging process.

The load profile can now be generated with these elements. The procedure is simple. Start on day one at minute one of a year, determine the arrival of the next bus, and assign the values of the load profile with the charging power.

Exercise 3.17 We want to simulate the function of the charging station for a full year. The required charging power at a stop is between 300 kW and 500 kW. The charging duration is 2.5 min. The interval between two charges is 10–15 minutes from 07:00 to 18:00, 25–30 minutes from 18:00 to 22:00 and 55–60 minutes from 22:00 to 07:00. The time series should have a resolution of $\Delta t = 1$ min. What is an algorithm that generates such a time series?

Solution: We start the time series at day one at 00:00. We determine the arrival time using a random variable that lies between 55 and 60. As long as the arrival time has not yet been reached, we set $\tilde{L}(t) = 0$. Once the arrival time is reached, the charging power is determined. Here we use a random variable that lies between 300 kW and 500 kW—that is, $\hat{L} = \mathtt{random}(300, 500)$. Since half a minute elapses with the connection of the bus, the load profile is first set to $\tilde{L}(t) = \frac{\hat{L}}{2}$. The two subsequent entries in the load profile are given the value $\hat{L}$, and then the disconnection from the charging station takes place, which also takes half a minute; hence, again, $\tilde{L}(t) = \frac{\hat{L}}{2}$. Next, the new waiting time $\hat{t}$ is determined. $\hat{t} = \mathrm{random}(t_1, t_2)$; here, the two times t_1 and t_2 result from the timetable, which depends on the time.

Algorithm 1 Creating a time series for $\tilde{L}$.

$\hat{t} \leftarrow \mathrm{random}(t_1, t_2)$
$t \leftarrow 1$
for day < 365 **do**
    for hour < 24 **do**
        $t \leftarrow t + 1$
        if $t >= \hat{t}$ **then**
            $\hat{L} \leftarrow \mathrm{random}(300, 500)$
            $\tilde{L}(t) \leftarrow \frac{\hat{L}}{2}$
            $L(\tilde{t} + 1) \leftarrow \hat{L}$
            $L(\tilde{t} + 2) \leftarrow \hat{L}$
            $L(\tilde{t} + 3) \leftarrow \frac{\hat{L}}{2}$
            $t \leftarrow t + 4$
            $\hat{t} \leftarrow \mathrm{random}(t_1, t_2)$
        else
            $\tilde{L}(t \leftarrow 0$
        end if
    end for
end for

In Fig. 3.12 (below) the load profile $\tilde{L}$ and the solar power profile $\tilde{P}$ are shown for day 100. For better visual comparison, the load was divided by 10. $\tilde{P}$ starts to increase with sunrise at around 05:00. At around 12:00, the solar production reaches its maximum, with a power of 78 kW; around 19:00 the sun has set again or the solar modules are no longer illuminated strongly enough. The curve of $\tilde{P}$ is not smooth, but has small peaks. This is because clouds reduce the irradiation for a short time.

The load profile $\tilde{L}$ generated with Algorithm 1 shows, on the one hand, the different charging powers. Since the charging process only lasts 2.5 minutes, we only recognize a series of small peaks whose height varies. We recognize the different timetables from the different intervals. In the period from 07:00 to 18:00, the intervals are very small. In the time from 22:00 to 07:00, the intervals are very large.

With the load profile $\tilde{P}$ and the production profile $\tilde{P}$, we now have all the elements to simulate an annual cycle of the charging station. The procedure is very simple and formally represented in Algorithm 2. For every minute of the year, the current production and consumption is determined from the profiles. Then it is determined how the power flows are to be allocated.

Fig. 3.13 shows an example of the power flows calculated in this way for one day. A cost-optimized method was selected for operation, so the power flow P_G is not used. The solar power is either used to charge the storage $P_S > 0$ or to serve the load $P_L > 0$.

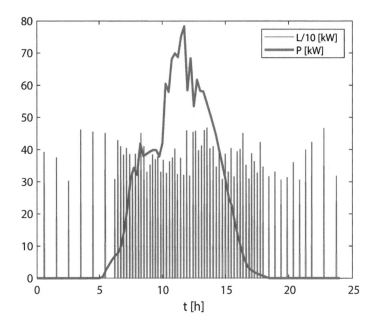

Fig. 3.12 Example of a daily profile of the solar power P and the load L. For better compa-rability, the load was divided by 10.

In the times when the sun is not shining, we see that the storage is first charged from the grid $G_S > 0$. If a load arises, the storage is discharged: $S_L > 0$. If the power of the storage unit is not sufficient, either the solar power system P_L or the grid G_L supports the charging process.

Let us consider the power flow S_S, which correlates with the current state of charge of the storage tank. As can be seen in Fig. 3.13, at night the storage is charged from the grid to a value of $\kappa_{min} = 10$ kWh. However, the storage is not discharged in the evening. Therefore, it would make sense to dimension the storage smaller here. We also see that during the day, when the charging intervals become much smaller, there are situations when the storage is empty. However, this means that the bus can only be charged from the solar power system and the grid—that is, with a maximum power of $P_L + S_L = 180$ kW.

We can measure how this happens. To do this, we need to see the amount of energy used for charging from the grid, solar plant, and storage in relation to the total demand:

$$R = \frac{\int_{t=0}^{t=T} G_L + S_L + P_L dt}{\int_{t=0}^{t=T} \tilde{L} dt}.$$

The question now arises as to what effect an increase in κ_{min} has on R. In Fig. 3.14, R is shown as a function of κ_{min}. Both technical realizations and operating modes were tested. The differences with regard to R are minimal and only occur with small amounts of storage.

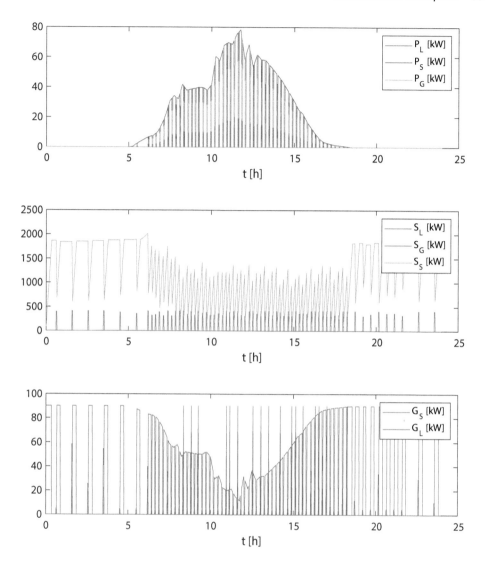

Fig. 3.13 Power flows for a full day. A cost-optimized operation management with version 1 as technical realization was calculated. In this example, κ_{min} was set to 10 kWh.

R varies for storage sizes up to 25 kWh. Already at 30 kWh we have R of 95%. This means that in the course of the year, the load cannot be fully covered on 5% of the charging events. R increases approximately linearly from 30 kWh and reaches approximately 98% at 100 kWh.

Algorithm 2 Simulation of the operation of the charging station.

Ensure: $0 = \eta_{\mathrm{SS}}\, S_{\mathrm{S}}(-\Delta) + \eta_{\mathrm{GS}}\, G_{\mathrm{S}} + \eta_{\mathrm{PS}}\, P_{\mathrm{S}} - (S_{\mathrm{S}} + S_{\mathrm{L}} + S_{\mathrm{G}})$

Ensure: $\tilde{P} \le P_{\mathrm{G}} + P_{\mathrm{S}} + P_{\mathrm{L}}$

Ensure: $\tilde{L} \le \eta_{\mathrm{SL}}\, S_{\mathrm{L}} + \eta_{\mathrm{PL}}\, P_{\mathrm{L}} + \eta_{\mathrm{GL}}\, G_{\mathrm{L}}$

Ensure: Power transfer limitation of all power flows.

Ensure: Capacity limits $\kappa \in [0, \kappa_{\max}]$

   for day < 365 **do**

      for hour < 24 **do**

         for minute < 60 **do**

            $t \leftarrow t + 1$

            $P \leftarrow \tilde{P}(t)$

            $L \leftarrow \tilde{L}(t)$

            Determine Powerflow for $P_L(t), P_S(t), P_G(t), S_G(t), S_L(t), G_L(t), G_S(t), S_S(t)$

         end for

      end for

   end for

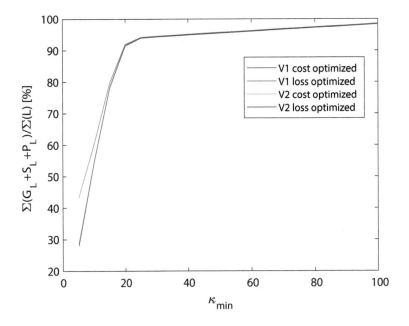

Fig. 3.14 Ratio between the amount of energy that is used from the grid, solar power system, and storage for charging and the amount of energy that the buses would expect in total, dependent on $\kappa_{\min}$, the minimum requested capacity. The ratio R was determined for both technical realizations and both operation management modes.

Table 3.5 Relative losses and relative costs for the two technical versions and the two operating modes. (Mode 1 = cost-optimized, Mode 2 = loss-optimized)

Realization	Rel. losses Mode 1	Rel. costs Mode 1	Rel. losses Mode 2	Rel. costs Mode 2
Version 1	1.48%	83.09%	1.47%	83.08%
Version 2	1.35%	82.93%	1.21%	85.74%

Let's look at the time series of one day at a value of $\kappa_{\min} = 100$ kWh (Fig. 3.15). Our operation control works in such a way that it first tries to charge the storage to a value of $\kappa_{\min}$. Unfortunately, this is only possible at the beginning of the day or in the evening, because there is enough time between the charging intervals. From 07:00 onwards, the storage is further discharged every time the bus is loaded, until the supply is exhausted around 12:00 and the storage can only be charged for a short time in the short intervals. From 18:00 the intervals are longer again, and the storage manages to reach a value of $\kappa_{\min}$ again by 20:00.

Exercise 3.18 We are faced with a design contradiction. Operating and investment costs depend on $\kappa_{\min}$. A small storage would be cheaper than a large one. However, if we want to ensure that every bus gets the required amount of energy, we have to design a storage with very large capacity. We thus pay a lot of money to prepare for a few charging processes. What options are there for resolving this contradiction?

Solution: There are many ways to resolve the contradiction. A technical problem does not always have to be solved using a technical approach. Here are some ideas:

- It is likely that the lack of charging power occurs mainly in the winter months. So we could configure $\kappa_{\min}$ depending on the season, so that more charging is done from the grid in the winter months.

- The solar power plant is designed for 90 kW peak power. We could increase the size of the solar power plant so that we have more solar power available during the day. The advantage of this would be that surpluses could be fed into the grid and sold.

- Without battery storage, the charging power is still $G_L + P_L = 180$ kW. To transfer the same amount of energy as with the storage, we need eight minutes. We could talk to the bus operator and ask whether a slightly longer waiting time or only partial charging of the bus would be acceptable.

Regardless of the choice of $\kappa_{\min}$, the next step is to determine the running costs or the losses that occur. As we can see in Fig. 3.14, R is independent of the technical realization or the operational management—that is, the choice of technology can be made without having made a final decision on $\kappa_{\min}$.

To determine the costs, we need a feed-in tariff $c_{\mathrm{fi}} = \$0.25/\mathrm{kWh}$ and an electricity price $c_{\mathrm{gc}} = \$0.31/\mathrm{kWh}$. For the calculation of the costs, the following applies:

$$C = \int_{t=0}^{t=T} [c_{\mathrm{gc}} (G_S + G_L)) - c_{\mathrm{fi}} [P_G - (a_{PG} + b_{PG} \ P_G) + (S_G - (a_{SG} + b_{SG} \ S_G)]] \, dt.$$

We therefore sum over the power flows over time and weight them in accordance with the tariffs. For better comparability, we normalize the costs and the losses.

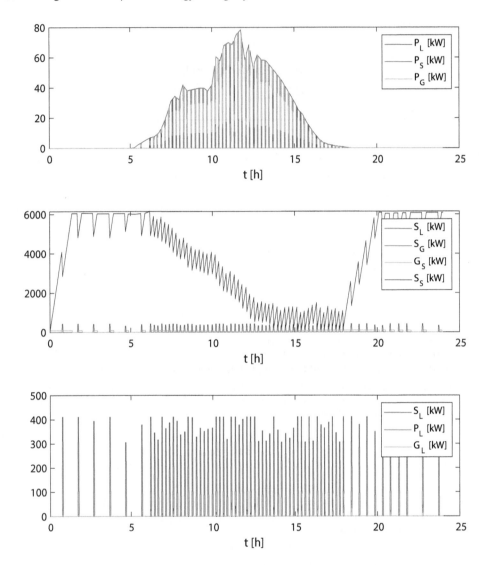

Fig. 3.15 Power flows for a full day. A cost-optimized operation management with version 1 as technical realization was calculated. In this example, $\kappa_{\min}$ was set to 100 kWh.

The losses are related to the total charging power. The costs are based on the reference case in which the entire charging power would have to be taken from the grid and paid for.

Tab. 3.5 shows the different losses and costs. If we compare the differences between cost-optimized and loss-optimized control, we see that the differences between cost-optimized and loss-optimized control are small for version 1. The differences are significantly higher in version 2. While in version 1 the cost difference is 0.01% points, in version 2 we have a difference of 2.81% points. Overall, version 2 with cost-optimized operation management seems to be the preferred solution here. The costs and losses are lower than in version 1, and the loss-optimized solution has significantly higher costs than the cost-optimized solution.

In this section, we were able to show how we could compare two different realizations and operation management with the help of the power flow diagram and the consideration of the losses. For this, it was necessary to use a load profile and a solar power profile and integrate them into a numerical solution. The results showed us that version 2 with a cost-optimized operation management is the preferred realization. The question of how large the storage is, or how large the chosen κ_{min} must be is still open. Unfortunately, we do not have a decision criterion for this. The consideration so far has been a technical one. In the next section we want to deal with the question of how to choose the system design from a financial point of view. For this, we need the data that we have determined in this section.

3.3 Evaluation of system designs from a financial point of view

Economic considerations always accompany the design and operation of storage systems. These considerations are often decisive for the investment. One reason is that storage technologies are expensive. Another reason is that storage technologies are used in economic fields where investments are decided on the basis of profitability analyses. In the energy industry, for example, it is common to use the discounted energy costs, or *Levelized Cost of Energy* (LCOE), and to make the investment decision on the basis of this value.

Roughly speaking, a distinction can be made between two types of economic efficiency considerations: in the simplest case, the investment costs or acquisition costs are determined during system design. Examples of this are the design of a storage system for an electric car or a portable video camera. Here, the total system costs must not exceed a certain value, otherwise sales are no longer possible or only possible in limited quantities. The level at which the system costs may lie results from the consideration of the Bill of Materials(BOM)—that is, all components of the product are evaluated in terms of their costs.

The second type of profitability analysis corresponds to the classic investment calculation. All cash flows that occur during the life cycle of a system are evaluated, starting with the acquisition and ending with maintenance and possible returns. In the energy industry, such considerations are usually mandatory, whether it is an investment decision for a pumped-storage power plant or the acquisition of a battery storage system for the primary and secondary control energy market (Figgener *et al.*, 2020; Akhil *et al.*, 2013; Kempener and Borden, 2015).

3.3.1 Bill of Materials, production, and product cost

We first look at cost-effectiveness in relation to investment costs. A rough classification is made between four price tags: BOM, Manufacturing Cost, Product Cost, and Selling Price (Ulrich, 2003). We will describe the four categories in detail using our charging station. In doing so, we will also see what possibilities exist to influence these variables.

Table 3.6 Bill of Materials for version 1 of the two realizations of the charging station. The prices are rough estimates.

System component	Part	Price
power electronics	Battery inverter (11)	$24,600
	Solar inverter (12)	$ 5,400
	Charging station (13)	$30,000
Cables, plugs, and connectors	Power connections	$ 2,000
	Signal connections	$ 1,000
Safety system	Battery safety switch	$ 1,200
	Fuses, interlock, etc.	$ 800
System control	Battery management system	$ 200
	Energy management system	$ 100
Mechanical components	Cooling system	$ 4,000
	Housing	$ 2,500
Energy storage	Lithium ion battery	$ 9,500
Total Bill of Materials		$81,300

The BOM describes the costs of the components of a product. In Tab. 3.6 and Tab. 3.7, we have summarized the different system components and their prices. (Warning! The prices are rough estimates and are intended to illustrate the procedure. Please do not create a start-up on the basis of these prices!) Let us now look at the BOM for the two realizations of the charging station. The charging station consists of a number of DC/DC and AC/DC converter units. The number and their dimensioning depends on the realization. There are a variety of technical realizations and modifications of how version 1 and version 2 can be implemented (Hesse *et al.*, 2017). In a first approximation, we assume that the costs essentially scale with the power. Thus, an inverter that delivers twice as much power as another would have twice the material cost. This is correct in the sense that a doubling of the power output is accompanied by a doubling of the current and in the sense that this higher current must be carried by more copper and more material.

For the inverter, we assume a BOM cost of 6$ct/W. For the DC/DC converters, the BOM cost is 4$ct/W. The BOM price difference is because inverters contain more power semiconductors and filter elements (coils and capacitors). These are necessary to meet the grid codes.

The power electronics are used to convert DC into AC and to regulate the power flow. This is only possible with the second group of system components: cables, plugs, and connectors. We distinguish between power connections, which are all power cables that have to transport a high power, and signal connections. These connections are used to transport analogue or digital signals. In version 1, the cost of these system components is $3,000. In version 2, the cost is $4,540. The difference in cost is due to the fact that in version 2, a total of four components must be connected and their interaction must be coordinated. Therefore, more signal and power connections are necessary.

Table 3.7 Bill of Materials for version 2 of the the two realization of the charging station. The prices are rough estimates.

System component	Part	Price
power electronics	Grid inverter (21)	$ 5,400
	Battery charger (22)	$16,400
	Solar converter (23)	$ 3,600
	Charging station (24)	$20,000
Cables, plugs, and connectors	Power connections	$ 3,000
	Signal connections	$ 1,540
Safety system	Battery safety switch	$ 1,200
	Fuses, interlock, etc.	$ 1,000
System control	Battery management system	$ 200
	Energy management system	$ 250
Mechanical components	Cooling system	$ 5,000
	Housing	$ 2,500
Energy storage	Lithium ion battery	$ 9,500
Total Bill of Materials		$69,590

Since the system is to transport and store large amounts of power, a safety system will be necessary that includes various monitoring and safety functions. In version 1, this will be less complex, as only three components are used, which are connected via an AC bus. For this, the normative requirements are usually lower and the components cheaper, as these safety components are very common and used in many applications. For version 2, slightly higher costs are to be expected, as high-power DC buses are not as common and there are more components in the system. We assume here that the price of these components is $2,000 and $2,200 respectively.

The system control is divided into two units: the battery management system and the EMS. The task of the battery management system is to ensure that the storage is operated safely. The EMS, on the other hand, has the task of regulating the power flows and communicating with the outside world. In terms of price, the system control will be comparable. We assume a cost of $300 for version 1 and $450 for version 2. Version 2 will be slightly more expensive because more components will need to be controlled, and therefore more interconnection and signalling will be required.

The mechanics include two elements: the cooling system and the housing. The cooling system ensures that the components do not overheat. The housing is responsible for the entire mechanical structure: the mounting of the inverters, the fixing of cables, the installation of contact protection, and mechanical protection measures. For version 1, the cost is $6,500. For version 2, the cost is slightly higher, at $7,500.

The last item is the storage. Our charging station is to be equipped with a lithium ion battery (Section 8.4). The cost of lithium ion batteries is measured in $/kWh. We assume here a price of $95/kWh. The storage capacity should be 100 kWh. This results in a price of $9,500.

The total material cost is thus $81,300 for version 1 and $69,590 for version 2.

Let us look at the distribution of BOM costs among the different components (Fig. 3.16). The largest share of costs is in the power electronics. In version 1, these costs take up almost three quarters of the BOM costs. The main cause is that each power electronics component is equipped with filters that ensure that the grid code of each component is met. In version 2, there is only one interface to the grid. In comparison to version 1, which has a power rate of 500 kW, the power rate of version 2 is relatively small, at only 90 kW. Since line filters scale with the power, this reduction of power also reduces the price.

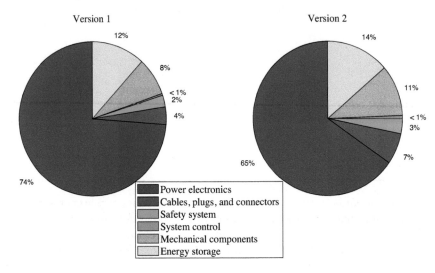

Fig. 3.16 BOM cost distribution for realization version 1 and version 2.

The next system component is the storage. But in comparison to the power electronics, the cost of this component is small. This breakdown is not surprising. We have built a system which is a power application. The storage is charged slowly, but has to deliver a lot of power in a relatively short time. The total amount of stored energy is small in comparison to the power.

While the BOM costs summarize the costs of the material, the manufacturing costs summarize the costs of the actual manufacturing process. In addition, the material overheads and general overheads are added. Tab. 3.8 shows the costs for both versions. When assembling the units, the working time is multiplied by an hourly rate. This gives the costs for each individual work step. The overheads are calculated proportionally to the BOM costs. It is assumed that equipment with high BOM costs also generates more expenses in logistics, purchasing, and other departments. Similarly, the general overheads are also calculated proportionally to the BOM cost.

The pricing of the product depends on various aspects:

- What is the customer willing to pay? Of course, the product price needs to be designed in such a way that customers are willing to pay it.
- How should the product be placed in the market? If we position our product in

Table 3.8 Calculation of production cost

Category		Version 1		Version 2	
Unit assembly					
	Assembly of cabinet	2 h	$160	2 h	$160
	Cooling system	0.5 h	$40	0.75 h	$60
	Power electronics	0.5 h	$40	0.75 h	$60
	Safety system	0.25 h	$20	0.5 h	$40
	Cable harness	1 h	$80	1.5 h	$120
	System control	0.1 h	$8	0.2 h	$16
	Integration of components	4 h	$320	5 h	$400
Device testing					
	Communication tests	0.1 h	$8	0.2 h	$16
	Power flow tests	0.25 h	$20	0.45 h	$36
	Performance tests	0.25 h	$20	0.4 h	$32
Packaging					
		0.05 h	$4	0.05 h	$4
Total assembly cost 9 h		720 $	11.8 h	944 $	
Material overheads	15% of BOM cost		$12,195		$10,438
General overheads	10% of BOM cost incl. assembly cost		$10,234.50		$8,793.15

the low-budget segment, it is automatically associated with a price range. The same applies to luxury goods.

- What are the company's revenue expectations? A company naturally has expectations about the profit it wants to make with a product. For example, the company may want to see a part of its development costs paid back through the sale of the product, or a certain annual profit is to be achieved. These considerations are reflected to the sales force in the selling price policies.
- Do strategic constraints have to be considered? Does the company want to penetrate a new market segment with the new product? Does it want to win some key customers, or take a market share from a competitor?
- Shall the sales department have room to negotiate? Often the company defines a target margin that is below the list price. This price difference should enable the sales staff to meet their negotiating partner in price negotiations within a defined framework.
- What does the sales organization look like? If different distribution channels are involved in the sales process, for example a wholesaler and an intermediary, their margins also have an influence on the sales price.
- How secure are the supplier prices and thus the BOM costs? It must, of course, be ensured that fluctuations in procurement prices do not jeopardize the profitability of the business.

- Are there external conditions, such as government regulations or a special economic situation?

For a product such as an electric lawnmower or a laptop, pricing is based on the considerations outlined above. For capital goods—that is, goods that are considered as an investment, the investment costs play a more significant role. Investment costs combine both the costs of the goods and their installation, but also the operating costs. In the following section, we will present an introduction to investment costing and the notion of LCOE. As the customers for our charging station are municipalities and bus companies, it is important to consider the economic efficiency from these points of view.

3.3.2 Introduction to investment cost analysis

The economic viability calculation analyses the cash flows of the investment over the entire life cycle of the storage system, both the acquisition and the income and running costs. Our charging station is supposed to have a life cycle of 10 years. After 10 years, it must be dismantled and scrapped. The cash flows in this case consist of three groups. First, there are investment costs at the beginning of the procurement. These include the product costs and the installation costs. Next, there are the operating costs, because during the 10 years of active operation, maintenance and repairs are incurred. These costs are offset against the income. The last group is the scrapping costs, which include the dismantling and scrapping of the station.

Expenses and revenues are added up each year, and these annual totals are either discounted at the time of investment or accumulated at the end of the investment's life cycle. When discounting, the value of money is transformed to the value of money at the time of investment. Thus, one considers the value of the investment at the beginning of the investment. This value is the initial value. When accumulating interest, the value of the money is transformed to the value at the end of the investment. This value is the end value.

Exercise 3.19 We consider the following investment: $K = (\$1,000, -\$100, \$10)$. The interest rate is $r = 2\%$. What is the initial value and what is the end value?

Solution: The initial value N_A results in:

$$N_A = \$1,000 - \$100 \cdot 1.02^{-1} + \$10 \cdot 1.02^{-2} = \$911.57.$$

The end value N_E is given by:

$$N_E = \$1,000 \cdot 1.02^2 - \$100 \cdot 1.02^1 + \$10 = \$948.4.$$

The consideration of the initial value or the end value is used to compare two different investments with each other. But which two investments are to be compared? For this, we need to be clear what the respective Null-hypothesis is. This means the alternatives between which the investor can choose. In most considerations, the Null-hypothesis is the investment in the financial market. We compare the value of our investment with the amount of money we would have if we did not invest in a charging station but left the money at the bank.

Even with this comparison, we notice that the reference value for the charging station is not the investment in the financial market. As a municipality or fleet operator who wants to electrify bus transport, we cannot simply invest the money in the financial market. We would not be fulfilling our mandate.

Consider another example of how the Null-hypothesis can be chosen differently. Suppose a shop has two TVs for the price of one ('buy one, get one free'). If we do not currently need a TV, the Null-hypothesis is not to buy a TV. In any case, we will not buy two TVs to save the price of one. The Null-hypothesis here is actually to leave the money in the bank account.

But if our TV has just broken, the Null-hypothesis changes. Now we are faced with the alternative of buying one TV or two for the price of one and selling on the second TV. This would be more advantageous for us.

And if we really urgently need two TVs, the Null-hypothesis would be to purchase two sets. In this case, the offer price must be compared with the price of two individual sets.

We see that the Null-hypothesis depends very much on the investor and cannot be answered in a general way.

The value of an investment over N periods is determined by the Net Present Value (NPV):

$$\text{NPV} = \sum_{i=0}^{N} (1+r)^{-i} K_i, \tag{3.26}$$

r where is the interest rate and K_i is the cash flow in period i. A constant interest rate is assumed. The NPV corresponds to the sum of the cash flows, discounted to the period 0. It is an initial value observation. We can also interpret the NPV as the current value of an investment.

Exercise 3.20 Two investments are available to choose from. $K_1 = (-\$1,000, \$100, \$400, \$800)$ with $r_1 = 8\%$ and $K_2 = (-\$500, \$400, \$200, \$100)$ with $r_2 = 10\%$. Which investment has a higher NPV?

Solution: We calculate the NPV of the two investments according to eqn (3.26):

$$\text{NPV}_1 = -\$1,000 + 1.08^{-1} \cdot \$100 + 1.08^{-2} \cdot \$400 + 1.08^{-3} \cdot \$800 = \$70.59$$
$$\text{NPV}_2 = -\$500 + 1.1^{-1} \cdot \$400 + 1.1^{-2} \cdot \$200 + 1.1^{-3} \cdot \$100 = \$104.06.$$

The comparison of NPV_1 and NPV_2 shows that the second investment has a higher value. Hence, we should choose the second investment.

The NPV represents the financial value of an investment at the time of the investment's beginning. In the process, the future cash flows are discounted to their present value. To compare two investments, it is therefore sufficient to compare the values of the NPVs. The investment with the greater value is considered the more advantageous investment.

If the cash flow K_i of the different periods is also constant, eqn (3.26) simplifies to:

$$\text{NPV} = K_0 \sum_{i=1}^{N} (1+r)^{-i} K_i = K_0 + K_i \frac{(1+r)^N - 1}{(1+r)^N r} = K_0 + \text{AV}(r,N) K_i. \qquad (3.27)$$

$\text{AV}(r,N)$ is called the annuity value (AV). Eqn (3.27) allows an analytical calculation of the NPV, but has the disadvantage that both the interest rate and the capital
flow must be kept constant.

Exercise 3.21 $10,000 is invested for the purchase of a solar power storage system. The
returns amount to $540/a over 20 years. The interest rate is constant at 3%. What is the AV
according to eqn (3.27)? What is the NPV?
  Solution: We first calculate the AV for this investment:

$$\text{AV}(r,N) = \frac{(1+r)^N - 1}{(1+r)^N r} = \frac{1.03^{20} - 1}{0.03 \cdot 1.03^{20}} = 14.877.$$

With this value, we can now calculate the NPV:

$$\text{NPV} = K_0 + \text{AV}(r,N) \cdot K_i = -\$10.000 + 14.877 \cdot \$540 = -\$10{,}000 + \$8{,}033.86$$
$$= -\$1{,}966.16.$$

We see that this investment is not beneficial. How can we adjust the investment to correct
this? For the investment to be profitable, either:

• the investment must become at least $1,966 cheaper;

• the interest rate must fall to about 1%; or

• the return must increase to $672.

If the interest rate changes in the different periods, the NPV is calculated by:

$$\text{NPV} = \sum_{t=1}^{N} \prod_{i=0}^{t} (1+r_i)^{-i} K_t. \qquad (3.28)$$

This more general form is always used when the interest rates are variable. This is
the case, for example, in 'crystal-ball' simulations, where probability distributions are
used instead of defined values (Dittmer *et al.*, 2009).
  We now want to determine the economic efficiency of our charging station. Let the
Null-hypothesis be the extension of the grid capacity so that the charging station can
be equipped without storage and without a solar power plant.

Exercise 3.22 We want to calculate the investment costs for the Null-hypothesis. Tab. 3.9
shows the costs of the system components. The assembly costs are $500. Material overheads
are 15% of the BOM cost. General overheads are 10% of BOM cost plus assembly cost. To
determine the selling price, a margin of 32% must be added to the production cost. This is
the price to be used for the investment calculation.

In addition, the costs for grid expansion must be taken into account. This is based on the connected load $15/kW and the distance to the grid connection point $75/m.

What are the investment costs for our Null-hypothesis? We assume an average distance to the grid connection point of 250 m.

Table 3.9 BOM cost of the simple charging station.

System component	Cost
Power electronics	$30,000
Cables, plugs, and connectors	$1,500
Safety system	$500
System control	$150
Mechanical components	$3,000

Solution: We sum up the costs of the system components from Tab. 3.9 and obtain a value of $35,150. Thus, for the three derived cost types and the margin:

$$\text{Production cost } \$5{,}273$$
$$\text{General overhead } \$3{,}565$$
$$\text{Margin } \$14{,}076$$

The costs for the grid expansion amount to $75/m $\cdot 250$ m $= \$18{,}750$. The total investment cost is therefore $76,813.

The yields Y of a storage system generally result from two parts: a part that depends on the power flows, k_i, and one that depends on the states of the system components, k:

$$Y = \sum_{k=0}^{N} \underbrace{c_k k}_{\substack{\text{costs/income} \\ \text{of a state}}} + \sum_{i=0}^{N} \underbrace{c_{k_i} \eta_{k_i} k_i}_{\substack{\text{costs/income} \\ \text{of a powerflow}}} . \tag{3.29}$$

In eqn (3.29), N power nodes are assumed. c_k are the cost/revenue factors for component k. c_{k_i} are the cost/revenue factors for the power flow from k to component i. We have dropped the time index, since we know that all power flows are time-dependent properties. Note, that also the tariffs can be time-dependent.

In our example of the charging station, we have already learned about the costs of the type c_{k_i}. These are the electricity costs c_{gc} and the feed-in tariff c_{fi}. Both depend on the power flow. The costs of the type c_k did not occur in our example. However, we want to introduce them to also solve the problem that sometimes the storage capacity is not sufficient to fully charge the bus (Fig. 3.14).

After long negotiations with the distribution network operator and the municipality, we found the following solution for the operation of the charging station: it is possible to increase the power very strongly for a short time. However, this increased power consumption costs extra.

$$
c_G = \begin{cases} 0 & G_S + G_L - (S_G + P_G) <= 90 \text{ kW} \\ \$0.95/\text{kWh} & G_S + G_L - (S_G + P_G) > 90 \text{ kW}. \end{cases} \tag{3.30}
$$

In order to compare the economic efficiency of the Null-hypothesis with the realization of the charging station, we have to determine the operating costs in addition to the investment costs. Tab. 3.10 shows the cumulative energy quantities for the Null-hypothesis and the various realizations and operating modes, which are created over the course of a year. Operating costs can be calculated from these data.

Exercise 3.23 Based on the data from Tab. 3.10, we want to calculate the operating costs. In doing so, we need to use the electricity price given in eqn (3.30) and negotiated price for short-term peak load; the feed-in tariff is $c_{fi} = \$0.25/\text{kWh}$ and the grid purchase is $c_{gc} = \$0.31/\text{kWh}$. It should be noted that in the case of a peak load, the electricity is billed twice: once via the electricity price and once via the additional charges for the purchase of a peak load.

Solution: The income and expenses result in:

$$
Y_{fi} = c_{fi} \cdot \int_{t=0}^{T} (P_G + S_G)\, dt
$$

$$
Y_{gc} = c_{fi} \cdot \int_{t=0}^{T} (G_L + G_S)\, dt
$$

$$
Y_{Peak} = c_{fi} \cdot \int_{t=0}^{T} F_{Peak}\, dt.
$$

The following applies:

$$
F_{Peak} = \begin{cases} 0 & G_S + G_L - (S_G + P_G) <= 90 \text{ kW} \\ \$0.9/\text{kWh} & G_S + G_L - (S_G + P_G) > 90 \text{ kW}. \end{cases}
$$

Tab. 3.11 shows the income and costs. The total income results in

$$
Y = Y_{fi} - (Y_{gc} + Y_{Peak}).
$$

Table 3.10 Annual energy quantities transported via the various power transfers. In addition, in cumulative $G > 90$ kW, the annual amount of energy has been determined where the power transfer has loaded the grid node with more than 90 kW.

Realization	cum. P_L[kWh]	cum. P_G[kWh]	cum. P_S[kWh]	cum. S_G[kWh]	cum. S_L[kWh]	cum. G_S[kWh]	cum. G_L[kWh]	cum $G >$ 90 kW[kWh]
Null-hypothesis	0	0	0	0	0	0	469,604.93	469,604.93
Version 1 Mode 1	3,785.9	13,953.06	65,669.48	0	457,451.88	397,985.11	14,331	397,985.11
Version 1 Mode 2	3,785.9	79,622.54	0	0	457,451.88	462,220.33	14,331.89	445,879.64
Version 2 Mode 1	3,785.9	13,953	65,669.48	0	457,451.88	397,968.15	13,430.99	397,968.15
Version 2 Mode 2	3,785.9	79,622.54	0	0	457,451.88	462,206.66	13,430.99	445,868.89

Table 3.11 Operating costs of the various realizations.

Realization	$Y_{\text{fi}}[\$]$	$Y_{\text{gc}}[\$]$	$Y_{\text{Peak}}[\$]$	$Y[\$]$
Null-hypothese	$0	$117,401.00	$466,123.80	$-$563,524.80$
Version 1 Mode 1	$3,488.27	$103,079.03	$378,085.85	$-$477,676.62$
Version 1 Mode 2	$19,905.64	$199,138.05	$433,085.66	$-$532,318.08$
Version 2 Mode 1	$3,488.25	$102,849.79	$378,069.74	$-$477,431.28$
Version 2 Mode 2	$19,905.64	$118,909.41	$423,575.45	$-$522,579.22$

If we look at the operating costs that we have determined in Tab. 3.11, we see that the cost-optimized operation has a clear advantage over the loss-optimized operation. Compared to the Null-hypothesis, both operating modes are more favourable, but the cost advantage is considerably greater for the cost-optimized operating mode.

We now want to determine the LCOE for the different realizations. The LCOE results from the NPV and the required amount of energy:

$$\text{LCOE} = \frac{\text{Total amount of Energy}}{\text{NPV}} \qquad (3.31)$$

$$= \frac{\sum_{i=1}^{N} \int_{t=0}^{T} \tilde{L}_i \, dt}{\sum_{i=0}^{T} (1+r)^{-i} Y_i}, \qquad (3.32)$$

where $\tilde{L}_i$ is the load profile in year i and Y_i is the revenue or cost in period i. r is the market interest rate. To calculate the LCOE for our charging station, we still have to consider two aspects. First, we have not yet considered the cost of the solar plant. This is $90,000 and still has to be considered for the investment. Furthermore, we have considered the electricity costs as a fixed amount so far. However, it can be assumed that the electricity costs will increase in the course of the life cycle. Therefore, the electricity costs increase by r_{gc} in each period.

The NPV of the charging station is thus given by:

$$NPV = \sum_{i=0}^{T} (1+r)^{-i} \left(Y_{\text{fi}} - (1+r_{\text{gc}})(Y_{\text{gc}} + Y_{\text{Peak}}) \right). \qquad (3.33)$$

In Tab. 3.12 the respective NPV has been determined for the different realizations and the Null-hypothesis. In addition, the discounted costs are shown for the different periods. The discount factors $(1+r)^i$ and $(1+r_{\text{gc}})^i$ are also shown. In order to be able to compare two investments with each other, only the NPV needs to be compared. The NPV represents the value of an investment at the time of the investment—that is, in year 0. We evaluate the different realizations and operating modes by looking at ΔNPV. The values are listed in Tab. 3.13. The higher the value of ΔNPV, the more beneficial the investment. We can see that operation mode 1, the cost-optimized operation mode, indeed has significant advantages over loss-optimized operation mode. We can also see that version 2, the realization with the help of a DC bus system, is more advantageous than the AC bus version.

Table 3.12 NPV for the different realizations and the Null-hypothesis. The table also shows the discount factors applicable for the different years, as well as the discounted costs of the respective year.

year	NPV	0	1	2	3	4	5	6	7	8	9
$(1+r)^i$; $r = 1\%$		1	0.990	0.980	0.971	0.961	0.951	0.942	0.933	0.923	0.914
$(1+r_{gc})^i$; $r_{gc} = 2\%$		1	1.020	1.040	1.061	1.082	1.104	1.126	1.149	1.172	1.195
Null-hypothesis	$-5,893,069	$-563,525	$-569,104	$-574,739	$-580,429	$-586,176	$-591,980	$-597,841	$-603,760	$-609,738	$-615,775
Version 1 Mode 1	$-4,998,420	$-477,677	$-482,475	$-487,321	$-492,213	$-497,154	$-502,142	$-507,180	$-512,266	$-517,403	$-522,589
Mode 2	$-5,584,470	$-532,318	$-537,983	$-543,700	$-549,469	$-555,292	$-561,169	$-567,100	$-573,086	$-579,128	$-585,226
Version 2 Mode 1	$-4,995,854	$-477,431	$-482,227	$-487,070	$-491,961	$-496,898	$-501,885	$-506,919	$-512,004	$-517,137	$-522,321
Mode 2	$-5,482,626	$-522,579	$-528,147	$-533,767	$-539,438	$-545,162	$-550,938	$-556,768	$-562,652	$-568,590	$-574,584

Table 3.13 LOCE and comparison of the NPV of the realizations with the Null-hypothesis. A market interest rate of $r = 1\%$ and an annual price increase for electricity costs of $r_{gc} = 2\%$ over a period of 10 years were assumed.

Realization	ΔNPV[\$]	LCOE[\$/kWh]
Null-hypothesis		1.27
Version 1 Mode 1	672,536.33	1.11
Version 1 Mode 2	86,486.44	1.23
Version 2 Mode 1	694,130.74	1.10
Version 2 Mode 2	207,359.40	1.21

In Tab. 3.13 values for the LOCE are also shown. We can interpret the LCOE as the average electricity price. As a bus operator, if we do not want to invest in a battery-based charging station, we would have to pay an average of \$1.27/kWh for each charge. With version 2 and its cost-optimized operation, we would pay \$1.10/kWh.

With the calculation of the economic efficiency, or more precisely with the comparison of the different realizations with the Null-hypothesis, we have carried out and finalized all the steps for the design of an energy storage system. In the end, we were able to make a selection that takes into account both the technical and the economic aspects.

In the analysis of the charging station, we neglected two aspects. They come under the headline of parameter variation or parameter sweep. This technique is always used when a model contains various parameters whose values are not clearly defined. These are, for example, the tariffs and the interest rates. It has become common practice here to look at the results of the analysis for different pairs of values to see if there are tipping points where the decision needs to be revised. The same applies to technical parameters such as power classes and storage capacity. Since both parameters have an influence on the yields as well as on the investment costs, a lot of time is invested here in the system design. The focus is on balancing technical performance and costs against yields (Lin, Pan, and Wang, 2008; Badescu, 2007; Weniger *et al.*, 2016; Schmiegel and Kleine, 2013).

3.4   Conclusion

In this chapter, we have learned the basics for the description and design of storage systems. The core element is the description of a storage system in the power flow diagram. This diagram describes the power flows occurring in the application of the storage system. It is independent of the later realization and also allows us to describe different operating modes. Using a battery-supported charging station for buses, we have simulated the application of this tool. We have learned about the different steps and the corresponding recipe:

Determining the technical requirements: The first step is to determine the technical requirements of the overall system. Then we can determine the requirements from the user's point of view. Additional requirements resulting from the choice of technology will only be considered at a later stage.

Creation of the power flow diagram: We now create the power flow diagram and its equations.

Identification of technical realizations and calculation of transfer losses: We determine which technical realizations are suitable for the system and determine the transfer losses from the component losses. As the technical solutions may have further requirements for power flows or operation control, these should be collected and added to the list of requirements.

Determine the operation management strategies: Using the power flow diagram, we define the operation management strategies that we want to consider in the system design.

Determining the Null-hypothesis: For the evaluation of the investment, the Null-hypothesis has to be determined. This represents the investment costs and returns that occur if the investor does not act. After determining the Null-hypothesis, it may also be necessary to describe it as a special realization and operation and to include it in the pool of realizations to be analysed.

Getting the load and production profiles: We acquire load and production profiles that are hopefully representative of the use of the system. If measured profiles are not available, a generated profile may be an option.

Yield simulations of the different realizations and operations: Load and production profiles are now used to calculate the yields for the different realizations and operations. The storage requirement should also be varied, as this is usually one of the largest cost items for the investment costs.

Calculating the investment costs: The investment costs must be calculated for the various implementations. For this purpose, the product costs as well as installation costs and other costs must be compiled.

Economic efficiency calculation: The investment costs and the annual returns can now be used to determine the profitability. For the profitability calculation, an assumption must be made about the development of interest rates. The differences in NPV or LCOE can be used as a decision criterion.

With this, our toolbox for the design of storage systems is well filled. However, we still need another tool, which is introduced in Chapter 4: requirement management and systems engineering. Already in this chapter we have seen how important an analysis of the requirements is and how these requirements influence the design of the system. The next chapter will provide us with some tools that will help us to solve this task more easily.

Chapter summary

- In order to describe energy storage system independently from its technical realization, we use the power flow diagram.
- A power flow diagram consists of nodes, which are either power sources, sinks, or storage components.
- Nodes are connected by links, which are directed and describe the power flow from one node to another. Power from node A to node B is labelled A_B.

- Each power transfer is connected to losses. Hence, we link an efficiency to each power transfer. The efficiency for the power transfer from A to B is labelled η_{AB}. The efficiency depends on the amount of transferred power. Therefore, $\eta_{AB} = \eta_{AB}(A_B)$.

- We always calculate the losses on the receiver side, not on the sender side.

- We interpret the capability of a storage to shift energy from the past to the future by power flows from the storage node from time $t - \Delta t$ to t and from t to $t + \Delta t$. Self-discharge is a transfer lost during this transport.

- For each node, the power flow equation is generated by summing up the inflow and the outflow. In case of an power source an auxiliary variable is added, which reflects the production profile. The same is true for power sinks, where the auxiliary variable equals the load profile.

- The description of the power flow is technology-independent. If we choose a specific technology and system architecture, we have to adjust the efficiency of the power transfer and the boundary conditions of the power transfer, for example the maximum transferable power.

- To determine the efficiency of a power transfer for a given technical realization, we have to identify the involved power electronic components and their efficiency. This efficiency depends on their set point.

- The power flow diagram, the efficiency, and the boundary conditions define the relations between the nodes, but for a given time step, the system of equation is underdetermined. Different power distributions will satisfy the equation. An operating strategy is needed, which will choose the correct one. This strategy is determined by the energy management system (EMS).

- In general, two different strategies are common: we can either maximize the economical yield of the system or we can minimize the losses.

- To calculate the economical yield of the system, the investment cost, the operational cost and the yields from the business model are needed.

- Investment costs consist of the material costs—that is, the Bill of Materials (BOM), the production cost, and the product cost. We also need to include the installation cost of the system.

- The operational costs are determined by replacement of damaged components and preventive maintenance.

- To compare different investments, the Net Present Value (NPV) is a reasonable quantity. Alternatively, we can choose the Levelized Cost of Energy (LCOE) to compare two realizations.

- It is important to have a clear view about the base hypothesis: we need to know our investment alternatives.

4

Introduction to requirement engineering and system design

4.1 Introduction

In Chapter 3, we introduced the power flow diagram. It reduced the complexity in the description of the function and allowed a simpler description that abstracted from the technical details. In this way, storage systems could be described in a technology-independent way. This made it possible to focus on the functional and structural core. In fact, a storage system is more complex in its structure. This chapter now introduces another tool that is used to describe the complexity of a technical realization.

This chapter follows the idea of *Systems Engineering* (Delligatti, 2013; Friedenthal, Moore, and Steiner, 2014). Based on requirements, components that are responsible for the realization of these requirements are identified. To then describe these in detail is the task of the specification, which in this chapter is only done on the system level, and not on the component level.

In practice, the terms 'requirement' and 'specification' are very often mixed up. A common mistake is that specifications are formulated instead of requirements. This is because requirements are perceived as too 'unspecific', or because one already knows what one actually wants to do. Yet the distinction is very simple: in the requirements phase, an attempt is made to understand what the customer wants: 'In this decade, we'll send a man to the moon and bring him back home safely.' In the specification phase, an attempt is then made to fulfil the customer's request. With the above example, it may also become clear why listening in the requirements phase is so important. People tend to think of the realization after the very first sentence—in the above example, they focus on sending a man to the moon. In the process, the second part of the sentence, '...and bring him back home safely,' is no longer heard.

We could easily see the value of a structured and technology-independent analysis in Chapter 3. By following our recipe, we were able to select a system design that met not only the technical but also the economic requirements. Similarly, the tools we will discover in this chapter are of great use.

4.2 Requirements and system components

When we open up a device and look inside, we see a variety of parts: circuit boards, coils and capacitors, cables and plugs, screws and nuts, pipes and tubes, and much more. Nothing that is included in such a device is there for no reason. Each part has its task, its use, its purpose; it also has responsibilities. Some parts only fulfil their task

Energy Storage Systems. Armin U. Schmiegel, Oxford University Press. © Armin U. Schmiegel (2023).
DOI: 10.1093/oso/9780192858009.003.0004

by working together with other parts. Therefore, we speak of system components or simply blocks instead of parts. Please be aware that a system component can consist of hardware as well as software, so it is a more general term.

In this chapter, we talk about systems. A system can be a device, but it can also be a sum of devices. For us, a system is first of all just something that has a set of tasks that it is supposed to solve. In order to accomplish these tasks, the system consists of system components that have the responsibility to solve these tasks partially or completely. In systems engineering, the task is to identify the appropriate system components and to distribute the tasks in such a way that no tasks are left undone and no system component is useless.

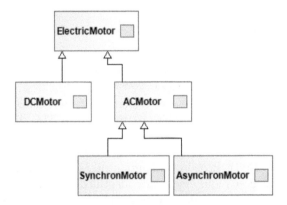

Fig. 4.1 Graphical description of specializations of system components. Electric motors are divided into two groups: DC motors and AC motors. Additionally, the AC motors are subdivided into synchronous and asynchronous motors.

There are different relationships between system components. Naming these helps to structure the system. A component can be a generalization of other components. For example, an electric motor is a generalization from a direct current (DC) motor or an alternating current (AC) motor. Both can be derived from the more general component the electric motor. Synchronous and asynchronous motors are also derived as specializations from the AC motor. In Fig. 4.1, this situation is shown graphically. We call such a diagram a block diagram. It allows statements to be made about the relationships between the system components. Specialization is a useful tool when we want to derive components from existing components. We can thus extend the function of a component. Suppose we want to develop an electric screwdriver that is equipped with a new type of battery. We do not want to change anything about the screwdriver; the batteries also fit into the existing mechanics. However, a monitoring circuit needs to be extended by a function. In that case, this would be represented by such a relationship. The new monitoring circuit is a specialization of the old one.

System components can consist of other system components. Let's take the example of an asynchronous motor. This consists of a stator and a rotor. Since the stator and rotor are system components that only occur in the asynchronous motor and are not

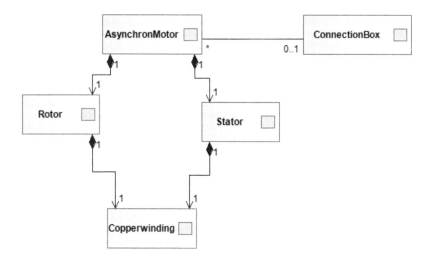

Fig. 4.2 Graphical description of an aggregation and an association between different system components. An asynchronous motor consists of a rotor and a stator, both consisting of copper windings. This is called an aggregation. The relationship between the connection box and the motor is an association. Both components can also exist without the other.

independent, we speak of an aggregation. In Fig. 4.2, this is represented by an arrow with a diamond. Numbers are attached to this arrow. These represent the cardinality. It tells how many components appear. Since an asynchronous machine has only one stator and one rotor, a 1 is shown at the end of the arrow. Conversely, one stator and one rotor can only be installed in one motor at a time, so a 1 is also entered on the motor side.

Both rotor and stator contain a copper winding. We have shown this component in Fig. 4.2. The relationship between the copper winding and the rotor, or stator, is also an aggregation with a cardinality of 1 in both directions. The copper winding appears only once in the diagram. The arrows of the rotor and stator point to the same block. The diagram should be understood as showing that they both need a copper winding, not that they both use the same copper winding.

Each asynchronous motor also has a connection box. In this connection box the power connections for the installation of the motor are collected and easily accessible. In Fig. 4.2, we have not used an aggregation arrow, but a simple line. This stands for an 'association'. While the aggregation represents a 'consists of' relationship, the association represents a 'has' relationship. Both system components can get along well without each other. There are electric motors with no connection box; conversely, there are connection boxes that have no connection to an electric motor at all. This different type of relationship is also reflected in the cardinality. An asynchronous motor has either no or one connection box. However, it is never connected to more than one connection box. Therefore, the connection box is marked '0..1'. A connection box can combine the connections of many asynchronous motors. The number is arbitrary, which is represented by '*'.

Associations are weak relationships. Therefore, the direction is not always clearly defined. A 'has' relationship can be interpreted in one direction or the other.

We have learned a few elements to describe the relationships between system components. System components have tasks within a system. Where do these tasks come from? Tasks come from what someone wants to do with a system. But a system component could also be responsible for a system having certain properties. Since we do not want to distinguish between 'tasks' and 'properties', in the following we will only speak of requirements: tasks can often be distinguished as functional requirements and properties as non-functional requirements.

In order to develop a system, it is definitely recommended to talk to all people who have requirements for the system. In Fig. 4.3, some of these people or organizations are shown. We are talking about actors. The actors have very different interests and, accordingly, have different requirements and expectations of the system. Here is a brief description of the roles that these actors have:

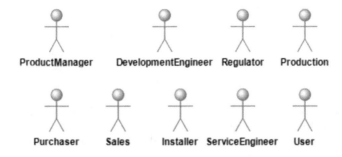

Fig. 4.3 A collection of actors that may have requirements for an asynchronous motor.

- **Product Manager:** The product manager is responsible for defining the characteristics of the product. He collects the requirements from different people, from the market, and from different departments. Sometimes he is also responsible for a larger set of products, which is called the product portfolio.
- **Development Engineer:** The development engineer is the person who is responsible for the realization of the system within the development department.
- **Regulator:** Usually there are not only technical requirements, but also normative and regulatory requirements. The regulator stands here for those actors who define norms and standards that a system must comply with. In the case of a solar storage system, this includes the grid operator.
- **Production:** This actor represents all those persons who formulate requirements in production. This concerns, for example, the manufacturability of a product or the materials used for production.
- **Purchaser:** Purchasing is responsible for the subsequent procurement of the material from which the system is to be built. Purchasing therefore has requirements regarding the procurability of the material or their prices.
- **Sales:** The sales department wants to sell the system. It therefore has requirements regarding the sales price, availability, and delivery times.

- **Installer:** The installer installs the system at the customer's site. This person will therefore have requirements relating to the tools to be used, the installation process, and the handling of the system.
- **ServiceEngineer:** Once the system is installed, the service engineer will be called whenever the system stops working. This person will therefore have requirements regarding the accessibility of defective parts or fault diagnosis.
- **User:** The user groups together all those actors who want to work with the system during its life cycle. This person is the source of most of the functional but also the non-functional requirements.

For the formulation of the requirements, the following format has proven itself:

AS A <actor>, I WANT <description> SO THAT <goal>

It first describes who wants something from the system. Then it describes the goal it wants to achieve with this request; after this comes the description. Why this formulation is useful can be seen well in the following two formulations:

The rear window of the car should be able to withstand a driving speed of 100 km/h.

This requirement seems nonsense at first sight. Why should the rear window of a car be designed to withstand the airflow of a driving speed of 100 km/h? Especially since most cars are not capable of moving that fast in a backwards direction!

Before we remove this requirement from the specifications, let's use a different wording here:

AS A head of the shipping department, I WANT the rear window of the car to withstand a driving speed of 100 km/h SO THAT we can also ship cars by train, where they are transported with the rear in the driving direction of the train.

When we read this formulation, it becomes clear why we had better not delete this requirement. Otherwise, the head of the shipping department would have had a surprise when the product was completed during the first rail transport.

Exercise 4.1 In Fig. 4.3, we got to know the different actors. What could be the requirements of these actors regarding the asynchronous machine?

Solution: There are many possible requirements that can be placed on a motor. Here is a small selection of possibilities:

- AS A ProductManager, I WANT a maximum weight of 5 kg SO THAT one person can carry the motor.

- AS A ProductManager, I WANT an engine power of 250 kW SO THAT the engine can be served as a drive for a truck.

- AS A DevelopmentEngineer, I WANT to use the rotor of the previous model SO THAT the development time is shorter.

- AS A DevelopmentEngineer, I WANT to use the new winding machine SO THAT the windings are positioned more accurately.

- AS A Regulator, I WANT the efficiency of the motor to be greater than 95% SO THAT the carbon footprint of the propulsion system is lower.

- AS A Regulator, I WANT warning signs to be clearly displayed SO THAT everyone who works with the engine recognizes the dangers.

- AS A Prodcution, I WANT only two types of screws to be used SO THAT fewer different materials are stored in the production line.

- AS A Production, I WANT the engine to be assembled within 10 minutes SO THAT we can produce 46 engines in one shift.

- AS A Sales, I WANT a green housing SO THAT at trade fairs every customer immediately recognizes our motors.

- AS A Sales, I WANT a sales price of $450 SO THAT we have a price advantage over the competitors.

- AS An Installer, I WANT hand grips on the chassis SO THAT I can easily lift and mount the motor.

- AS An Installer, I WANT an easily accessible connection box SO THAT I can make all the connections quickly.

- AS A ServiceEngineer, I WANT a diagnostic interface SO THAT I can immediately read out the engine data on site with my service PC.

- AS A ServiceEngineer, I WANT a modular design SO THAT I can easily exchange parts.

- AS A User, I WANT a low noise level SO THAT I don't have to take any noise-reducing protective measures.

- AS A User, I WANT high torque, even with low battery voltage, SO THAT I can also use the motor with smaller batteries.

Like system components, requirements can also be related to each other. Fig. 4.4 shows one such example. The regulator requires a certain efficiency from the engine. This efficiency is specified by law. The product manager generalizes this requirement by demanding that the motor should fulfil more than the legal requirement—namely, the motor should ensure the higher efficiency for all standard applications. This relationship is represented by the generalization arrow. This generalization also includes the requirement that the efficiency at steady state has the higher efficiency. This requirement has not been part of the legal standard, but has always been included in the specifications in the organization, and the product manager has written this requirement down again, but makes it clear that this is part of the more general requirement. Therefore, this requirement is marked with an <<include>> arrow.

In addition, there is now a requirement from Sales. They want efficiency to be present, especially in the applications of Meier Ltd, one of their biggest customers. Since the application of Meier Ltd is not one of the standard applications, this requirement extends the original one. This is represented by an <<extend>> arrow.

We will apply the methodology to a simple storage system, an electric screwdriver. Fig. 4.5 shows an exploded view. We will identify the most important system components and show for which requirements they have been used.

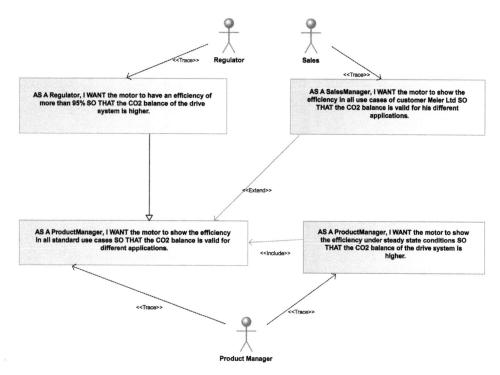

Fig. 4.4 Representation of the relationship of requirements to each other and to the actor. Requirements, like system components, can be derived from more general requirements and more specialized ones. In addition, there is also the possibility that a requirement extends or contains another requirement. For traceability, requirements and actors are linked by a trace relationship.

Exercise 4.2 If we look at the illustration in Fig. 4.5, different components, with and without software parts, can be summarized. We have already learned about some system components from the previous chapter. These and others are now to be found again in the illustration:

- mechanics
- storage
- mechanical drive
- operational control
- user interface.

    Solution: In this exercise, we just need to collect the different parts from Fig. 4.5 and allocate them to the system component.

- Mechanics: 1–6
- Storage: 91
- Mechanical drive: 802, 30, 45, 25
- Operation control: 4
- User interface: 4, 5, 60

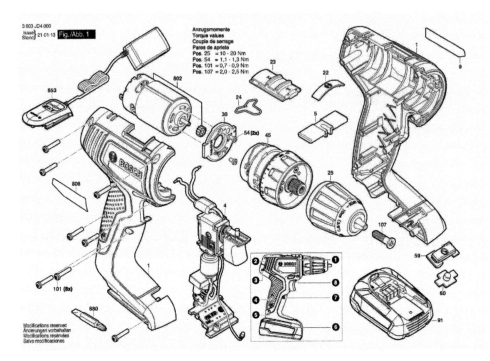

Fig. 4.5 Exploded drawing of a cordless screwdriver. (Source: Bosch Copyright: Robert Bosch Power Tools GmbH, Leinfelden-Echterdingen)

The creation of the system components is not always clear. Otherwise, all electric screwdrivers would consist of the same parts. However, there are also many similarities that are common to all screwdrivers. In Fig. 4.6, a possible division of the system components is shown. A screwdriver has exactly one housing and one drive train. The drive train consists of a rotating axle, a drive motor, and perhaps a gearbox. We assume that we want to install a DC motor, so we can name this system component directly as such. Since a screwdriver does not always have to be equipped with a gearbox, we have entered the cardinality 0..1.

For the energy supply, we need a storage. This consists of a set of battery cells. The cardinality is marked 1..*, because we are basically free to choose the number of battery cells....

We have named the Operation control as an independent component. This component consists of sensors, microcontrollers, and software. The number of microcontrollers and sensors can be arbitrary. For cost reasons, of course, we want to limit the number. Even if it is distributed on different microcontrollers, the software is modelled here as a system component.

Looking at Fig. 4.6, we can identify three system levels. The first system level is the device itself: the electric screwdriver. This is the level that someone sees when he has the screwdriver on his table. The second system level is the system components roughly grouped together: `Housing`, `Driveline`, `Storage`, `Operation control`, and

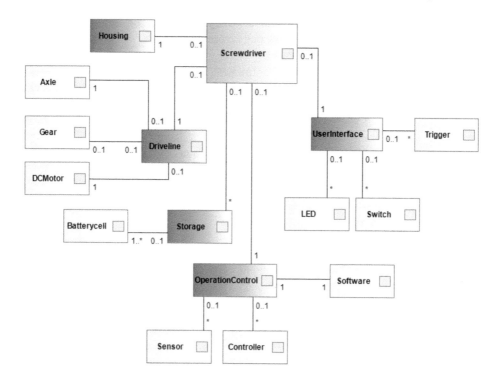

Fig. 4.6 System components of an electric screwdriver. The screwdriver system is marked in grey, which is the first system level. The system components of the second level are marked in green.

User interface. This second system level is the level that someone sees when they have unscrewed the device. The last level is the parts that make up the elements of the second system level. This hierarchical structure helps us to analyse the system. At the beginning, one can focus on the first and second level and only include deeper levels in the more detailed realization.

Let us now compile some requirements for an electric screwdriver:

- R 1 AS A User, I WANT the screwdriver to be able to work for at least four hours in continuous operation SO THAT you only have to recharge the screwdriver during your lunch break or at the end of work.
- R 2 AS A User, I WANT the screwdriver to have a drill attachment SO THAT I can also drill holes with the screwdriver and do not have to carry another tool.
- R 3 AS A ProductManager, I WANT the battery module to fit other tools in our tool series SO THAT we don't have to open a new production line and the customer can use the same battery modules for all tools.
- R 4 AS A User, I WANT to be able to use the battery modules of my cordless drill SO THAT I only have to take one charger and one type of battery pack with me.

- R 5 AS A Regulator, I WANT the battery cells to be monitored SO THAT there is no danger to life and body for the user if the battery cell is faulty.
- R 6 AS A DevelopmentEngineer, I WANT our battery modules to use our standard monitoring components SO THAT we do not have a new development with new certification.
- R 7 AS A User, I WANT to be able to choose between screwing and unscrewing via a switch that I can reach with my thumb SO THAT I can switch between these modes without setting it down.
- R 8 AS A User, I WANT to know the charge level of my battery module SO THAT I always know how much longer I can use the screwdriver before recharging.

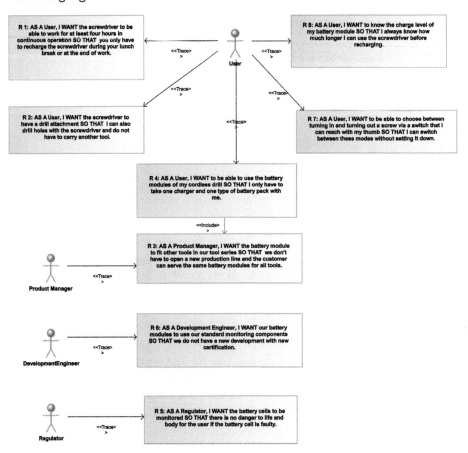

Fig. 4.7 Requirements for the electric screwdriver and their relations to actors.

In Fig. 4.7, the requirements are shown again graphically. The creators of the requirements are also shown and linked to the requirements with a «trace» relationship. R 4 is included in R 3; therefore, there is another <<include>> relationship here.

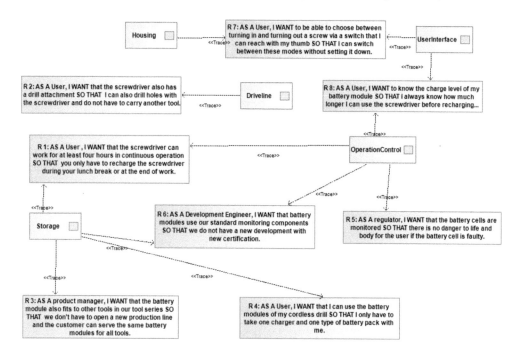

Fig. 4.8 System components and requirements for the electric screwdriver. The system components are linked to some requirements via a «trace» relationship. This means that they are responsible for meeting the requirement.

Listing the requirements and listing the system components alone does not add much value to the system design. At least having a list of requirements is very important, because it allows us to check later that we have met all these requirements. Using the described format makes it easier to read and consult. The structural representation of the system components is also valuable, as it allows us to capture an inventory of our system. But how do we make sure that all the requirements are actually implemented? And how do we ensure that there are no unnecessary system components in our system? To ensure this, we first look at which system components can implement which requirements, or, in other words, which system components are responzible for the realisation of a requirement. These responsibilities are shown in Fig. 4.8.

Mechanics, and more precisely the driveline, is responsible for the possibility of turning a screwdriver into a drill (R 2). The user interface is responsible for everything that concerns human–machine interaction—that is, the requirement to switch between screwing in and screwing out with the thumb (R 7) or to display the state of charge (R 8). In the case of the state of charge, the user interface can certainly not do this alone; here, the operation control is also partly responsible, and is therefore also connected to request R 8 via a «trace» relationship.

Operation control is also connected to a number of other requirements. Firstly, it has responsibility for the safety monitoring mentioned in R 5 and R 6. In addition, there is the requirement that the utilization should be as long as possible (R 1). This requirement can be implemented by the storage in cooperation with the operation control.

For the storage component, there are additional requirements. R 3 and R 4 are constructional requirements that concern the mechanical, but also the electrical interfaces of the storage. R 6, the requirement for the reuse of an already-existing module, is, of course, also a requirement that the storage must fulfil.

By dividing responsibilities in this way, we have achieved several things: on the one hand, we can identify system components that are unnecessary. A system component that has no task does not need to be part of a system. Furthermore, we have now also ensured testability, because in order to check a requirement, we only need to check whether the system components that have this responsibility also fulfil the requirement. This makes it easier to diagnose errors and ensure quality.

The system components shown in Fig. 4.8 reach their limits when systems become larger and more complex. One then uses diagrams that cover partial aspects of the overall system and the requirements. An evaluated procedure is the thematic grouping of requirements. The requirements are divided into topics and then the system components and their responsibilities are assigned to the requirements.

Table 4.1 Requirement traceability matrix for the electrical screwdriver.

	R 1	R 2	R 3	R 4	R 5	R 6	R 7	R 8
Housing							X	X
Driveline		X						
Storage	X		X	X		X		
Operation control	X				X	X		X
User interface							X	X

After this assignment has taken place, it must be checked if all requirements are assigned to a system component and that no system components exist in the system without a requirement. The requirement traceability matrix (RTM) has been established for this overview. In Tab. 4.1, this is shown for the screwdriver.

An RTM is structured in such a way that the requirements are listed in the columns and the system components are listed in the rows. Where a system component is responsible for meeting a requirement, a cross is entered to mark it. For example, Housing is responsible for complying with R 7 and R 8. Therefore, a cross can be found in those columns.

Not only is this representation clear, but it also makes it easier to recognize in large systems whether individual requirements are not fulfilled because there are no components that are responsible for their compliance.

The RTM also makes it possible to identify weaknesses in the architecture or in the formulation of requirements. For example, if several system components are responsible for meeting one requirement, this indicates unnecessary redundancy. This may be intentional; for example, some safety standards require redundant safety systems to be in place for important safety functions. It can also be that the two components are looking at different aspects of the requirements. For example, if the requirement reads 'The components may only be operated in the approved temperature range', there could be both a cooling system and a heating system, with each responsible for different system components. In this case, it makes sense to split the requirements appropriately so that it is clear which component is responsible for which requirement.

Exercise 4.3 A multiple assignment in a column of the RTM may indicate a redundancy or that a requirement contains multiple aspects that should be split. As shown in Fig. 4.8 and Tab. 4.1, this is the case for requirements R 1, R 6, R 7 and R 8. How should these requirements be subdivided so that multiple allocation no longer occurs?

R 1: This requirement has been assigned to storage and operation control because the number of hours of operation depends on the storage capacity and the method used in the operation control. We can represent these two aspects using the following two requirements:

R11: AS A DevelopmentEngineer, I WANT to choose a storage capacity that is sufficiently large SO THAT R1 can be met without having unnecessary capacity built into the device.

R12: AS A DevelopmentEngineer, I WANT to define a discharge strategy that works particularly efficiently in terms of energy SO THAT I can meet R 1 with a small storage.

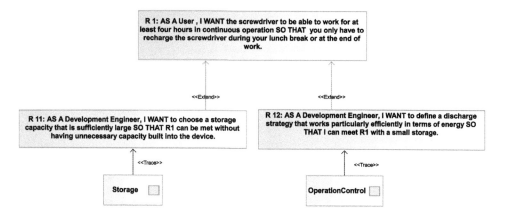

Fig. 4.9 Use case diagram showing the extension of the requirement R 1. By extending R 1 with R 11 and R 12, the responsibilities of the system components are split.

Fig. 4.9 shows the extension in the use case diagram. R 11 and R 12 extend the original requirement R 1. They divide this requirement into two aspects that can now be clearly assigned to the individual system components.

R 6: This is a requirement for the technical implementation of the system components Housing and Operation control. It is a design requirement that must be fulfilled by each system component. Therefore, this multiple assignment is kept.

R 7, R 8: Both requirements are first of all requirements for the user interface. The system components Housing and Operation control are assigned to these requirements because they are involved in the implementation. R 71: AS A DevelopmentEngineer, I WANT to realize the housing in such a way that the switch is fitted ergonomically and cost-effectively SO THAT R7 is fulfilled and no additional costs are created.

We do not split the requirement, but we extend it so that the system component Housing fulfils a very specific aspect of the requirement R 7. In R 8 we again divide the requirement into three aspects: data acquisition, presentation and implementation.

R81: AS A DevelopmentEngineer, I WANT the Operation control to determine the data about the state of charge by measurements on the battery module and processes them for the User Interface SO THAT the User Interface only has to display the statements of the Operation control.

R82: AS A DevelopmentEngineer, I WANT to realize the housing in such a way that the interface between operation control and user interface is integrated SO THAT the display of the state of charge can be done through the user interface.

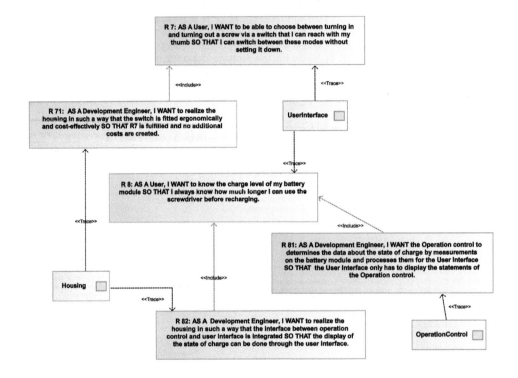

Fig. 4.10 Use case diagram showing the extension of R7 and R8 to have a better separation of the responsibilities of the system components.

Fig. 4.10 shows the extension and division of the requirements R7 and R8. The user interface remains responsible for the higher-level requirements, **Housing** and **Operation control** receive the requirements for derived requirements. This improved structure is also found in the RTM.

Table 4.2 Revised RTM integrating the modifications shown in Figs 4.9 and 4.10.

	R1							R7		R8		
	R11	R12	R2	R3	R4	R5	R6	R7	R71	R8	R81	R 82
Housing									X			X
Driveline			X									
Storage	X			X	X		X					
Operation control	X	X				X	X				X	
User interface								X		X		

Let us look at the RTM. In Tab. 4.1 and Tab. 4.2, we see that some system components are responsible for several requirements at the same time. In the case of storage, this is due to the fact that different aspects of the application have requirements for this component. It is similar with operation control and the user interface: as we can see in Fig. 4.6, these components contain subcomponents. They are therefore higher-level groups, and one will have to decide in the realization at the lower system levels whether a further subdivision makes sense.

In principle, aggregation of requirements to individual system components is a good sign, because it leads to system components performing several tasks simultaneously, which leads to cost savings. The aggregation of functions on one system component to fulfil several requirements represents system optimization.

However, it is important to note here that the complexity of the system and the structure of individual system components can increase, which can cancel out the hoped-for savings.

We have now become familiar with the two important elements of systems engineering: requirements and system components and their relationship to each other. In the next section, we will apply this methodology to storage systems. We will see that there are requirements that apply to any storage system, regardless of what technology it is. We will also see that the same basic components occur for every storage system. The technologies that we will learn about later in the book only extend the requirements and the system components.

4.3   General requirements for energy storage systems

We have already shown in Chapter 3 that energy storage systems can be described by the power flow diagram in a technology-independent way. The individual system configurations and the technology used are reflected in the boundary conditions for the power flows and losses. In this section, we will see that there are eight basic requirements that apply to all storage systems. These will be refined and extended in later chapters on a technology-specific basis.

The first requirement describes the basic function of any storage. We already learned about this in Chapter 2: energy should be able to be stored and served again later or at another location.

`G 1: AS A User, I WANT to store energy SO THAT I can use this energy at a later time or at a different place.`

The power flow diagrams of a storage system consist of three types of power nodes: sources, sinks, and storage. The power is transferred between these nodes. The technology used defines the characteristics of this power transfer. Without power transfer, the storage process cannot take place. This obvious fact is described by requirement G 2:

`G 2: AS A System component of an Energy Storage System,I WANT to be able to transfer power to other system components SO THAT, for example, an energy source can transfer power to the load or a storage.`

To build a storage system, it is not enough that energy can be stored somewhere and retrieved later. Assuming we had found a new material that simply absorbs energy like a magnet but can release the stored energy at a later time but only in a huge explosion, we would not have found a material that can also be used in an energy storage system. Only the technical controllability—that is, the possibility to store energy in this material in a controlled way and to release it at a later time—makes the material a suitable component for a storage system. This results in requirement G 3:

G 3: AS A User, I WANT the system to be protected against uncontrolled charging and discharging of the storage component SO THAT the process is controllable and can be used for technical applications.

G 1 to G 3 describe the basic technical requirements for a storage system. G 4 is now a requirement that initially sounds like a usability requirement:

G 4: AS A User, I WANT to have an estimate about the stored energy SO THAT I can adjust my charging and discharging behaviour.

If we interpret the actor 'user' as the end user of a storage system, G 4 is indeed a usability requirement. A car whose fuel gauge is broken would still work, but the driver would use range estimation to determine whether he should fill up or not. However, if we interpret 'user' as other actors in an energy storage system, G 4 takes on an additional meaning. The operation of the charging station in Chapter 3 would not work well without a value for the current state of charge. On what basis should it be decided whether surplus solar power should be transferred to the storage or to the grid? Operation management usually needs this information; therefore, G 4 is one of the basic requirements of an energy storage system.

Since storage systems store a large amount of energy in a limited volume, any storage technology poses a safety risk. Some risks have become familiar; for example, the subjectively perceived danger of a canister of 10 litres of petrol is considerably lower than that of a 10-litre pack of charged lithium ion batteries, although petrol has 10 times more energy stored. However, there is a lot of experience in handling petrol, and technical systems are equipped with a whole range of constructive and technical safety measures. But even if we have become accustomed to the risks associated with some storage technologies, storage systems must still be safe. This leads into the fifth requirement:

G 5: AS A User, I WANT the storage system to be safe SO THAT there is no danger for people and machine.

We have chosen the formulation 'there is no danger for people and machine'. In technical implementations, G 5 is a part of norms and safety standards, which must then also be complied with in the implementation. Depending on the application of the storage system, compliance with these safety requirements is monitored by internal or external auditors.

The sixth basic requirement relates to the mechanical structure of a storage system. The system components of a storage system do not hang in the air or stand freely in space. They are integrated into a housing. There is a clear requirement for this mechanical structure:

G 6: AS A system component, I WANT to be integrated into a stable mechanical construction SO THAT I'm protected and have a dedicated place and
position in the system.

The meaning of 'protection' and 'placement' is dependent on the application. A
vehicle battery must be able to withstand mechanical shocks and vibrations that occur
in different driving situations; it must withstand a crash test and it must be placed in
the vehicle in such a way that connections are easily accessible by the service technician
but unskilled persons do not accidentally touch the contacts. In the case of our charging
station from Chapter 3, the requirement for shock and vibration is limited to transport
to the site and installation. The enclosure does not need to protect the battery from
moisture, sunlight, and contact, as the battery is installed inside the charging station
and the charging station enclosure takes over this protective function.

The next requirement results from the economic efficiency consideration. We have
seen in Chapter 3 that the investment in a storage system must and often can be
economically justified. The tool used to determine this profitability is the Net Present
Value (NPV) or the Levelized Cost of Energy (LCOE). Both parameters consider the
entire life cycle of the storage system and collect revenues and costs over each year
of the storage system's life. Since there are high investment costs at the beginning of
the investment, it is easier to amortize this investment if the storage system lives for
a long time and income can be collected over each year of its life.

G 7: AS An Investor, I WANT the system to have a long life cycle SO THAT
I receive a higher income over the years.

In order to have high yields, the business model must allow them in the first place.
The requirements necessary for this will have to be defined individually as user requirements for each storage system, as these are very specific. For our charging station, for
example, we worked with a simple business model. However, it could be extended by
introducing time-variable electricity tariffs or by selling the charging power to the bus
company, whereby a time-variable electricity price is set that is based on the time,
charging interval, power, or amount of energy. But what all business models have
in common is that they are more profitable if the losses in the storage system are
minimized:

G 8: AS An Investor, I WANT a high system efficiency SO THAT the reduction of my income caused by power losses is minimized.

These eight requirements apply to all types of storage systems. We will see later
how these are extended by technology-specific requirements and user requirements. In
the next section, we will look at the system components that must be present in any
storage system to meet these basic requirements.

4.4 Basic components of a storage system

In this section, we want to get to know the basic components that are responsible
for realizing the basic requirements listed in Section 4.3. In the remainder of the
book, we will see how these basic components are extended, modified, or realized. We
can identify five system components found in all storage systems that fulfil the basic
requirements.

The first and perhaps most obvious component is the `Storage`. This is the component that makes an energy storage system what it is. However, this component includes more than just those components that perform the energy storage process itself. For example, as we can see in Fig. 4.6, the storage component can include a number of individual storage devices, such as a number of battery cells.

We have seen that a storage system can be described as a network of power nodes. We have already mentioned the storage node as a system component. Similarly, we can name the connection between the nodes that realize the power transfer as a system component: The `PowerDistribution`. In a storage system, the power distribution consists of all those parts that transport power. In an electrical storage system, there are cables or busbars. In a thermal storage system, we are dealing with pipes and a transport medium.

However, storage and power distribution alone are not enough, as we have already seen in the mathematical description of storage systems. We need to control the transport of the power. We need `PowerFlowControl`. Every storage system must therefore have components that enable it to control the power flow. The `power flow control` therefore consists of a number of hardware components that change the physical power flow and a software component that determines how the power flow should be changed. In our example from Chapter 3, the `power flow control` consists of the control boards of the inverters and DC/DC converters, as well as switches and relays. The `switch` and `relay` are necessary to control the power flow out of the charging station. In the case of the electric screwdriver, the power flow control is realized in the battery charge controller and in the switch for the screw command. The battery charge controller ensures that the storage is charged correctly. The switch for the screw command controls the discharging of the storage by activating the motor.

The next component is the `MechanicalConstruction`. The obvious task of this component is to separate the inside of the storage system from the outside, the environment, by realizing some kind of housing. A second task is to fix the internal components of the system and to enable connections between different parts. In Exercise 4.3, we saw that mechanical construction has an impact on the user interface. We see that `mechanical construction` is a component that is involved in the realization of very many requirements for a storage system.

We have talked intensively about the fact that a transfer of power from one power node to another is subject to losses. This power loss causes system components to heat up. With the electric screwdriver, we feel this heat at the charger. If we intensively use the screwdriver over a longer time, stressing the battery heavily, it will become noticeably warm. The same applies to the motor of the screwdriver. The heat loss can lead to components being subjected to thermal stress. Batteries and double-layer capacitors in particular have a shortened lifetime if they become too warm. Therefore, to ensure that the system components are not permanently damaged, storage systems require a `TemperatureControl`. The system component `temperature control` includes all components that are responsible for regulating the temperature of the various parts of a storage system. These can be heat sinks for passive cooling of electronic components, fans for active ventilation, or water pipes for water cooling with a heat exchanger. The temperature management, which is realized by the temperature control, is especially challenging at high power ratings. Take the battery charger from Chapter 3. It has a charging power of 500 kW and an efficiency of 98%. This means that at full load it produces 10 kW of heat loss, which has to move outside the system. For comparison, an electric kettle has a heating power of about 700 W. If the battery charger is charging at full load, we could therefore replace about 15 water boilers with the power loss.

Exercise 4.4 The general requirements and the general components have not yet been related to each other. What do these relationships look like in a block diagram or RTM?

Solution: To define the responsibilities of the system components, we proceed like this: we create an empty RTM. The rows list all system components, the columns the requirements. We start with a system component and put a cross wherever we think this system component has a task to do:

`Storage`: The storage will certainly be responsible for G 1. However, it could also play a role in G 3, since the choice of storage technology has an influence on the controllability of the charging and discharging process.

`SafetySystem`: The safety system will also be responsible for G 3 and, of course, G 5.

`MechanicalConstruction`: The separation of the interior from the exterior of a storage system is already a safety function. Therefore, the mechanical construction is certainly responsible for G 5. Furthermore, the mechanical construction must also be responsible for G 6.

`PowerDistribution`: The power distribution is jointly responsible for the storage process and the power transfer, so G 1 and G 2 must be taken into account. In addition, the power path has a safety component, G 5.

`TemperatureControl`: The temperature control has an influence on G 7 and G 8, as it has a strong influence on the lifetime of the storage and the other system component.

`PowerDistributionControl`: The power distribution control is linked to G 2. However, it also has an influence on the lifetime of the storage G 8, as the power distribution control defines, for example, how deeply a storage may be discharged or whether it may also be overcharged. Furthermore, the power distribution control is responsible for communicating the state of charge to the user G 4.

The links described here verbally are described in the Tab. 4.3 or in Fig. 4.11.

Looking at the columns of Tab. 4.3, it is noticeable that several components are responsible for the same requirement. This is desirable for safety-relevant components in order to guarantee single-fault security. The system must remain safe even if one component has a fault. For other requirements, this double allocation is not desired. Here, how these requirements can be distributed must be analysed on a technology-specific basis.

These double assignments indicate that it should be checked whether the requirements can be split up. For example, for requirement G 5, three components—the power distribution, the mechanical construction, and the safety system—are responsible for ensuring that the storage system is safe. The power distribution and mechanical construction focus on different aspects of safety. The mechanical construction, for example, provides protection against contact and mechanical protection. The power distribution, on the other hand, ensures the safety of the power transfer. Here it may make sense to break down the G 5 requirement further on a technology- and system-specific basis in order to create clearer relationships.

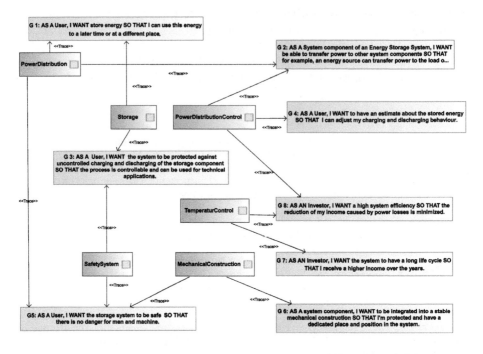

Fig. 4.11 Basic requirements and system components and their responsibilities.

Table 4.3 RTM of the general requirements and system components.

	G 1	G 2	G 3	G 4	G 5	G 6	G 7	G 8
Storage	X		X					
Safety system			X		X			
Mechanical construction					X	X		
Power distribution	X	X			X			
Climate control							X	X
Power distribution control		X		X				X

4.5　Conclusion

In this chapter, we have added another collection of tools to our toolbox. We have learned about describing customer and user needs in terms of requirements. With this technique, which uses the simple form

　　AS A <actor>, I WANT <description> SO THAT <goal>,

we are able to systematically capture what properties and capabilities a storage system should have. We have learned how different requirements can be related to each other so that we can give structure to a collection of requirements.

　　Second, we have learned about system components. System components are parts of a system that have a common responsibility or task. We have learned what relationships system components can have to each other and how these relationships can be represented.

The third tool manifests itself in the RTM. We have learned that the system components must have a relationship to the requirements. If this is not the case, either they are redundant or an important requirement has been forgotten. The RTM not only gives us the possibility to see if all requirements are covered and if there are no superfluous components in the system. It also shows us whether the distribution of responsibilities between the system components makes sense. If several components are responsible for the same requirement, it may make sense to split the requirement. If components are responsible for several requirements, it may make sense to reconsider the tasks of the components.

We then applied these three tools to storage systems in general. We have recognized that every storage system has to fulfil eight basic requirements, and that these eight basic requirements are confronted with six basic components that are responsible for the realization of these requirements.

In the following chapters, we will focus on storage technologies and systems. We will begin with a brief introduction to power electronics. We will focus on those aspects that are important in the development of storage systems.

Chapter summary

- Besides the power flow diagram requirement and systems engineering needed in our toolbox to design energy storage systems in a reasonable way, requirement engineering considers each requirement and the relationships between requirements.

- Requirement engineering considers the requirement and the relationship between the requirements.

- In this book we formulate requirements in the form of AS A <actor>, I WANT <description> SO THAT <goal>.

- Systems engineering considers the system components and the relationship between the system components and their responsibilities to meet the requirement.

- The Requirement Traceability Matrix shows whether all requirements are connected to a system component.

- If several system components are connected to the same requirement, we can assume that the requirement combines different aspects, which presumably can be better described in additional requirements.

- If a system component is responsible for meeting several requirements, we should clarify that this doesn't increase the complexity of the design.

- There are eight general requirements, G 1 to G 8, which are relevant for every storage system and need to be met in all of them.

- There are six basic system components, which are responsible for meeting these requirements. These system components occur in every storage system in its pure form or as a derived form.

5
Power conversion

5.1 Introduction

The previous description of storage systems was still somewhat abstract. In the power flow diagram, we talked about sources, sinks, and storage. The technology was included in the description in the form of boundary conditions. In our example in Chapter 3, we talked about inverters and DC/DC converters and described them mathematically, but we did not go into their structure and function. We want to catch up on this in this chapter. Here, we will introduce technologies that are used for the conversion of different energies. We will show what their properties are and how these properties can be represented in the form of boundary conditions in the power flow diagram.

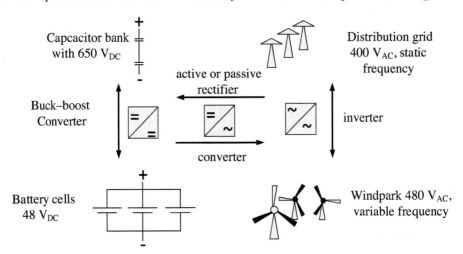

Fig. 5.1 Overview of the various uses of power conversions in electrical applications. On the left part of the illustration, DC current is transferred from one voltage level to another. This is done by a Buck–boost converter. On the right part of the illustration, AC current is transferred in terms of voltage level and frequency. This is done by an inverter. Power can also be transfered between AC current and DC current by using a rectifier or a converter.

In Fig. 5.1, we have shown the different applications of power conversion of electric power. Basically, we distinguish between alternating and direct current (AC and DC). With AC, the sign of the voltage changes periodically. The cause is the origin of this current. This is generated by large, generator-driven motors. The kinetic energy of a

Energy Storage Systems. Armin U. Schmiegel, Oxford University Press. © Armin U. Schmiegel (2023).
DOI: 10.1093/oso/9780192858009.003.0005

turbine is converted into electrical energy. With this type of electricity generation, the rotating machine automatically produces a sinusoidal amplitude curve.

Electrical and electrochemical storages are voltage sources that provide DC. This is due to the type of storage process, in which charge carriers are separated during charging and recombine during discharging. A solar power system is also a DC voltage source. Here, the charge carriers are separated from each other by light.

Power electronic converters are needed to be able to transfer electrical energy between AC and DC systems. The converters that change AC into DC are referred to as rectifiers, and those that do the opposite are known as inverters. In Chapter 3, we used these converters several times to realize the charging station.

Converters are also needed within the DC and AC grid. They are needed to connect networks with different voltage levels or different frequencies.

In the case of AC networks, it happens in various applications that DC must be converted from one voltage level to another. For example, if power at a battery voltage of $48\,V_{DC}$ is to be transferred to a DC grid of $400\,V_{DC}$, the voltage level of the battery is increased to $400\,V_{DC}$ by a DC/DC converter.

A similar case can be found in the control of motors. The AC mains supplies current oscillating at a frequency of $50\,Hz$. However, the motor should rotate at a much higher frequency. A frequency converter is served for this purpose, which realizes an output frequency that corresponds to the desired number of revolutions.

The different types of transformation (Fig. 5.1) can be realized in different ways. The respective circuit topology depends strongly on the concrete application. However, the working s.pdf for the development of power electronics are always the same. First, a suitable topology is selected. This is accompanied by a control concept. Once the topology and control concept have been selected, the basic elements are determined. Power electronic circuits produce heat loss, which makes it necessary to design suitable cooling systems. The third factor to be considered is electromagnetic compatibility—that is, the system must not be disturbed by electromagnetic radiation from outside and, conversely, must not influence other devices.

We will start with an introduction to power electronics and power electronic components; these components are found in all storage systems. Afterwards, we look at electrical drives and focus more intensively on the question of how electrical grids are formed.

5.2 Electronic components for power conversion

In order to get to know the electronic components of a power conversion unit and to understand their structure, we start with a task that has to be solved again and again in storage systems—and also in other devices. A basic element of power conversion is the DC/DC converter: a device that transforms power from one voltage level to another. In Chapter 2, we used this element several times. It has been used to charge and discharge the battery and the bus, and we also used it to transfer the power of the solar plant. We want to start with the realization of a buck converter. We use this device for the following task: We have a voltage source that supplies a voltage U_1. But for our application we need a voltage U_2. It is $U_1 > U_2$. We want to create a system component that ensures that we measure the voltage U_2 at the output of this system component.

Regardless of the technical implementation, the following requirements should be met:

DC Buck 1: AS A user, I WANT the buck converter to be able to cover an output voltage window that is as wide as possible. Ideally, $U_2 \in [0, U_1]$ SO THAT I can better control the power flow.

DC Buck 2: AS A user, I WANT the output current to have a wide range SO THAT I can serve both large and small power with the buck converter.

DC Buck 3: AS A user, I WANT the power transfer to be as efficient as possible SO THAT I have a low-loss transfer.

DC Buck 4: AS A user, I WANT the voltage ripple and current ripple at the inputs and outputs to be low SO THAT the components are not subjected to high voltage and current spikes and no unnecessary electromagnetic disturbances are generated.

5.2.1 Realization of a buck converter with the help of ohmic resistors—the voltage divider

A buck converter is a specialization of a DC/DC converter. A voltage divider is a specialization of a buck converter, which consists of an aggregation of two resistors. This abstract structure is shown in Fig. 5.2. Let us now try to assemble this voltage divider from a clever interconnection of two ohmic resistors. Ohm's law applies to an ohmic resistor:

$$U = R \cdot I. \tag{5.1}$$

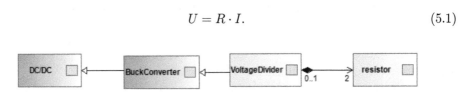

Fig. 5.2 A BuckConverter is a specialization of a DC/DC-converter and a VoltageDivider is a special realization of BuckConverter, which is an aggregation of two resistors.

The voltage U measured across a resistor R is proportional to the current I. Resistors are called passive elements because they change the current and voltage in a circuit without any active intervention or the elements changing their properties or behaviour.

One realization of a DC/DC converter is the voltage divider (Horowitz, Hill, and Robinson, 1989). In Fig. 5.3 we have shown the schematic diagram of a voltage divider in addition to the schematic diagram. A voltage divider consists of two resistors, R_1 and R_2. The output voltage U_2 is tapped at the junction of these two resistors. A current I flows from the input side to the output side. To determine the output voltage U_2, we need Ohm's law on the one hand, and the mesh rule on the other, which says that the voltage in a mesh must always be zero. In case of the voltage divider, we have two meshes. One on the input side and one on the output side. On the input side, the mesh consists of the terminal at the input and the two resistors in series. The following therefore applies to the voltage on the input side:

$$U_1 = (R_1 + R_2) \cdot I. \tag{5.2}$$

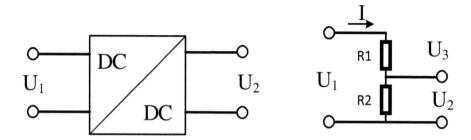

Fig. 5.3 Use of a power conversion unit. In this example, a voltage U_1 is to be converted to a voltage U_2 by a DC/DC converter. In this case, $U_1 > U_2$ applies. This is realized by a voltage divider (right).

On the output side, the mesh consists of the terminals at the output where the voltage U_2 is measured and the resistance R_2. This results in:

$$U_2 = R_2 \cdot I. \tag{5.3}$$

If we combine these two equations, we can find the voltage U_2:

$$U_2 = \frac{R_2}{R_1 + R_2} \cdot U_1. \tag{5.4}$$

The output voltage can therefore be adjusted via the ratio of the two resistors, R_1 and R_2.

Exercise 5.1 We want to build a voltage divider that transfers the input voltage from $U_1 = 12$ V to $U_2 = 4$ V. How must the resistors be chosen?

Solution According to eqn (5.4), to set the correct voltage, only the ratio of the two resistors is relevant. It follows that:

$$\frac{U_2}{U_1} = \frac{R_2}{R_1 + R_2}$$

$$\frac{4 \text{ V}}{12 \text{ V}} = \frac{R_2}{R_1 + R_2}$$

$$\frac{1}{3} R_2 = R_1 + R_2$$

$$R_2 = \frac{1}{2} R_1.$$

One of the basic requirements of energy storage systems is the ability to control the power flow between the power nodes. To meet this requirement, the voltage on the output side must be adjustable. Let us assume that we have connected a battery to the output side of the DC/DC controller. If we use the DC/DC controller to set a voltage that is equal to the battery voltage, no power flow will occur. If we set a voltage with the DC/DC controller that is greater than the battery voltage, a charging current will flow into the battery. The current and thus the power flow can be controlled with the magnitude of the voltage difference. Similarly, the battery is discharged when the voltage is lower than the battery voltage.

The voltage on the output side of the voltage divider can be calculated according to eqn (5.4) by changing the resistors. A potentiometer can be used for this purpose. This is a resistor whose resistance value is adjusted mechanically. If R_1 were replaced by a potentiometer, the power flow would thus be adjustable.

In Fig. 5.3 and eqn (5.4), we have not considered the load. The equations describe an unloaded voltage divider. If we connect a load R_L to the outputs of the DC/DC converter, it will also have an influence on the division of the voltage. Instead of R_2 we would have to consider the total resistance of R_2 and R_L. Since the resistors are connected in parallel, the resulting resistance $R_{2,L}$ is:

$$\frac{1}{R_{2,L}} = \frac{1}{R_L} + \frac{1}{R_2} \tag{5.5}$$

$$R_{2,L} = \frac{R_2\, R_L}{R_2 + R_L}. \tag{5.6}$$

If we express R_L as a multiple of R_2, eqn (5.6) simplifies to

$$R_{2,L} = \frac{R_2\, n \cdot R_2}{R_2 + nR_2} \tag{5.7}$$

$$= \frac{n\, R_2^2}{(n+1)R_2} \tag{5.8}$$

$$= \frac{n}{n+1} R_2 = R_2 \left(1 - \frac{1}{n+1} \right). \tag{5.9}$$

According to eqn (5.9), the larger n becomes—that is, the larger the load resistance becomes compared to R_2—the smaller the difference between R_2 and $R_{2,L}$. For a value of $n = 10$, the difference is 0.91.

For the losses of the voltage divider, the value of the two resistors and the current passing through the voltage divider are relevant. It holds that:

$$P_{\text{loss}} = U_1 \cdot I = (R_1 + R_2)\, I^2. \tag{5.10}$$

Exercise 5.2 We have connected to the voltage divider a load resistor of $R_L = 1{,}500\ \Omega$. $R_2 = 5\ \Omega$, the input voltage is $U_1 = 12$ V, and the output voltage is $U_2 = 4$ V. What is the power consumed across the two resistors R_1 and R_2? What is the power consumed across R_L?

Solution: Using the results from Exercise 5.1, R_1 is 10 Ω. Using Ohm's law, we can find the current I from the voltage source:

$$I = \frac{U}{R_1 + R_2} = \frac{12 \text{ V}}{15 \text{ }\Omega} = 0.8 \text{ A}.$$

The power is

$$P = U \cdot I = 12 \text{ V} \cdot 0.8 \text{ A} = 9.6 \text{ W}.$$

We can determine the load current I_L and the required power in an analogous way:

$$I_L = \frac{U_2}{R_2 + R_L} = \frac{4 \text{ V}}{1{,}505 \text{ }\Omega} = 2.65 \text{ mA}$$

$$P_L = U \cdot I = 4 \text{ V} \cdot 2.65 \text{ mA} = 0.01 \text{ W}.$$

The operation of the voltage divider thus requires 9.6 W of power, while the load requires just 0.01 W. To make this ratio more favourable, the resistors R_1 and R_2 must be given larger dimensions. If we choose $R_1 = 100 \text{ }\Omega$ and $R_2 = 50 \text{ }\Omega$, the required power of the voltage divider would be reduced to $P = 0.96 \text{ W}$. However, we cannot make the resistors R_1 and R_2 arbitrarily larger, because the voltage U_2 no longer remains at 4 V.

As we have seen in Exercise 5.2, the voltage divider fulfils the task of transforming a high voltage into a lower voltage, but the losses are great. Voltage dividers are therefore not used as buck converters, but used to branch off from a primary circuit with a high voltage to a secondary circuit with a lower voltage. A small amount of energy is taken with a small voltage. A typical application is voltage measurement.

5.2.2 Realization of a buck converter using a capacitor and electrical switches: The charge pump

Since the voltage divider is not a suitable realization for a buck converter, we have to think of something else. The voltage divider has the property that it constantly converts a voltage U_1 into a voltage U_2. The conversion depends on the load resistance U_L. However, with constant load resistance, the voltage is also constant. We want to break up this temporal constancy and require that only the mean output voltage equals U_2. But we also require that the fluctuations of U_2 are below a certain value ΔU_2. This way, we can ensure that the load is not damaged.

In order to realize this new buck converter, we need two new components. These new components are the switch and the capacitor. The switch has a simple property: if it is closed, then $U_S = 0$ and $I_S = I$. If it is open, $U_S = U$ and $I_S = 0$. The switching state is defined by an external signal. We discuss the components used to realize such switches in more detail in Section 5.2.5.

The second element we need is a capacitor. In Fig. 5.4 the basic design of a capacitor is shown: two conductive plates with an area A and a distance d are connected to a voltage source with voltage U. The applied voltage causes a quantity of Q charge carriers to redistribute. This redistribution lasts until the voltage at the capacitor corresponds to the voltage of the voltage source. The number of charge carriers is proportional to the voltage. The higher the voltage, the greater the number of charge carriers. However, the number of charge carriers also depends on the area of the two plates.

The larger the surface area, the more charge carriers there are on the plate. If we vary the distance between the plates, we see that this also has an influence on the number of charge carriers. We find that the number of charge carriers decreases with $\frac{1}{d}$. The closer the plates are to each other, the greater the number of charge carriers. If we double the distance between the plates, we halve the number of charge carriers. This is analogous to the electric potential of a charge carrier, which also decreases with $\frac{1}{r}$ as the distance r increases. We also note that the number of charge carriers depends on the material between the two plates. Each material can be assigned a quantity ϵ. This quantity is composed of the electric field constant ϵ_0:

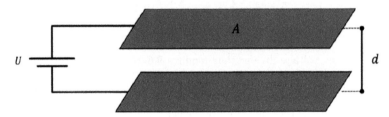

Fig. 5.4 Structure of a capacitor: A voltage U is applied to two conductive plates with an area A and a distance d, this leads to a displacement of Q charge carriers.

$$\epsilon_0 = 8.854 \cdot 10^{-12} \frac{\mathrm{As}}{\mathrm{Vm}}, \tag{5.11}$$

and the material-dependent relative permittivity ϵ_r Tab. 5.1 shows the relative permittivity for different substances. The permittivity ϵ is the product of these two quantities:

$$\epsilon = \epsilon_0 \epsilon_r. \tag{5.12}$$

In summary, the number of charge carriers Q transferred between the two plates of the capacitor are given by:

$$Q = \frac{U}{d} \epsilon A. \tag{5.13}$$

When we build a capacitor, we determine its properties with d, ϵ, and A. These quantities determine how many charge carriers are transported at a certain voltage. Since different values of d, ϵ, and A can produce the same number of charge carriers, we combine these numbers into a constant:

$$C = \frac{\epsilon A}{d}. \tag{5.14}$$

Thus $Q = C\,U$ applies: the higher C, the more charge carriers are stored on the capacitor, which is why C is called the capacitance of a capacitor. The capacitance C is not to be confused here with the previous capacity term, κ. κ represents the energy content of a storage system with the unit watt-second. C, on the other hand, represents a relationship between the applied voltage and the number of charge carriers, and has the unit coulomb.

Table 5.1 Relative permittivity of different substances.

Material	Relative permittivity ϵ_r	Material	Relative permittivity ϵ_r
Vacuum	1	Oil	2.3
Air (dry)	1.00054	Glass	4–10
Hydrogen	1.00025	Quartz glass	1.5
Diamond	5.7	Rubber	2–3.5
Salt	5.9	Polyethylene	2.3
Distilled water	80	Plexiglass	3.4

To determine the energy content of a capacitor, the charge carriers that are displaced by the applied voltage must be counted. The voltage U multiplied by the charge corresponds to the stored energy:

$$E = \int_0^Q U \, dq = \int_0^Q \frac{q}{C} \, dq = \frac{1}{2} \frac{Q^2}{C} = \frac{1}{2} C U^2. \tag{5.15}$$

Exercise 5.3 1 Ws of electrical energy is to be stored. A voltage source with $U = 100\,\text{V}$ is available for this purpose. There is no dielectric between the plates, and the permittivity corresponds to the electric field constant $\epsilon = 8.854 \cdot 10^{-12} \frac{\text{As}}{\text{Vm}}$. What must be the ratio of the area and the distance of the capacitor so that this energy is stored?

Solution: To find the ratio of area and distance, we need to combine eqns (5.14) and (5.15):

$$E = \frac{1}{2} C U^2 \Rightarrow C = \frac{2E}{U^2} = \frac{\epsilon A}{d}$$

$$\frac{A}{d} = \frac{2E}{\epsilon U^2} = 2 \cdot \frac{1\,\text{Ws}}{8.854 \cdot 10^{-12} \frac{\text{As}}{\text{Vm}} \, 100^2 \, \text{V}^2} = 22{,}588{,}660 \; \text{m}^2.$$

With a plate spacing of 10 μm, the required area is $225\,\text{m}^2$.

Different types of capacitors use different dielectrics. These essentially determine the properties of the capacitor. Capacitors cover a very wide voltage and capacitance range, from a few picocoulombs to the millicoulomb range. The dielectric strength goes into the four-digit volt range.

How can we lower the voltage with switches and capacitors? A voltage source provides a certain voltage U_1 throughout. If we charge a capacitor with this voltage source and quickly discharge it again, it should be possible to produce a lower voltage. To illustrate this behaviour, we look at the circuit in Fig. 5.5. We have connected to a voltage source U two switches, S_1 and S_2. A capacitor C and a resistor R are connected between the two switches. If we want to charge the capacitor, S_1 is closed and S_2 is open. If we want to discharge the capacitor, we have to close S_2 and open S_1.

We first consider the charging process. At $t = 0$ we start the charging process; S_1 is closed and S_2 remains open. As soon as S_1 is closed, current I_C flows across the capacitor as charge carriers are redistributed to the capacitor. The magnitude of the

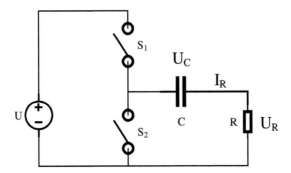

Fig. 5.5 Circuit for charging and discharging a capacitor.

current corresponds to the number of transported charge carriers per time, with eqn (5.14) resulting for the current:

$$I_C = \frac{dQ}{dt} = C\frac{dU_C}{dt}. \tag{5.16}$$

The capacitance C here establishes a relationship between the voltage change per time and the current flowing in the process. Since the same current must also flow across the resistor, the voltage drop across the resistor U_R can be determined using Ohm's law:

$$U_R = RI_C = RC\frac{dU}{dt}. \tag{5.17}$$

The voltage drop across the resistor U_R and the capacitor U_C must be compensated by the voltage source U_0:

$$RC\frac{dU_C}{dt} + U_C = U_0. \tag{5.18}$$

By solving the differential eqn (5.18), the current and voltage of the capacitor can be calculated:

$$U_C = U_0\left(1 - e^{-\frac{t}{\tau}}\right) \; ; \quad I_C = \frac{U_0}{R}e^{-\frac{t}{\tau}} \; ; \quad \tau = RC. \tag{5.19}$$

The current and voltage curve is shown in Fig. 5.6. The voltage approaches that of the voltage source exponentially, while the current decreases exponentially. Here $\tau = RC$ defines the time scale on which the charging process takes place. At $t = \tau$, the charging process is 63% complete. The resistance R limits the initial current during the charging process.

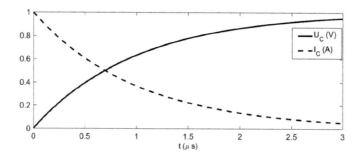

Fig. 5.6 Charging curve of a capacitor according to circuit from Fig 5.5 ($R = 1\,\Omega$, $C = 1\,\mu F$, $U_0 = 1\,V$).

Exercise 5.4 A capacitor is to store 1 Ws. The charging current is to be limited to 10 A. The charging voltage is 12 V. How large must the capacitance of the capacitor and the resistance be? **Solution:**

$$E = \frac{1}{2}CU^2 \Rightarrow C = \frac{2E}{U^2} \Rightarrow C = \frac{2\,\text{Ws}}{(12\,\text{V})^2} = 0.0138\,\text{F}$$

$$I_C = \frac{U_0}{R} \Rightarrow R = \frac{U_0}{I_C} \Rightarrow R = \frac{12\,\text{V}}{10\,\text{A}} = 1.2\,\Omega.$$

After the capacitor has been charged, we open the switch S_1 and close the switch S_2. Now the capacitor discharges via the resistor R. The current and voltage curve can be calculated by setting U_0 to zero in eqn (5.18):

$$U_C - RC\frac{dU_C}{dt} = 0 \Rightarrow U_C = U_0 e^{-\frac{t}{\tau}}; \quad I_C = -\frac{U_0}{R}e^{-\frac{t}{\tau}}. \tag{5.20}$$

Discharging also requires a resistor to limit the current. Both the voltage and the current have an exponential curve, as can be seen in Fig. 5.7. The discharging power, which is the product of these two quantities, thereby decreases at twice the rate:

$$P_c = U_C I_C = -\frac{U_0^2}{R}e^{-\frac{2t}{\tau}}. \tag{5.21}$$

While the voltage or current has reached about 63% of its maximum value at $t = \tau$, the power already reaches 63% of its maximum value at $t = \frac{\tau}{2}$. This must be taken into account in the design.

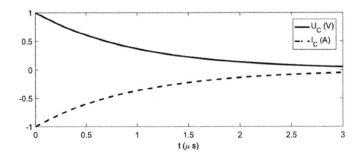

Fig. 5.7 Discharging curve of a capacitor according to the circuit from Fig 5.5 ($R = 1\,\Omega$, $C = 1\,\mu\text{F}, U_0 = 1\,\text{V}$).

Exercise 5.5 An application requires an energy of $E = 30$ Ws. During the discharge process, the power must not fall below $P = 27\,\text{W}$. The voltage applied to the capacitor can be a maximum of $U = 9\,\text{V}$. What must be the capacitance of the capacitor, and what is the energy content of this capacitor?

Solution:

$$E = \tfrac{1}{2}C(U_{\max}^2 - U_{\min}^2) \Rightarrow 30\,\text{Ws} = \tfrac{1}{2}C((9\,\text{V})^2 - (63\,\%\ 9\,\text{V})^2)$$

$$\Rightarrow C = \frac{2 \cdot 30\,\text{Ws}}{(9\,\text{V})^2 - (63\,\%\ 9\,\text{V})^2} = 1.228\,\text{C}.$$

The total energy content of the capacitor is therefore:

$$\kappa = \frac{1}{2}CU_{\max}^2 = \frac{1}{2}1.228\,\text{C}\,(9\,\text{V})^2 = 49.734\,\text{Ws}.$$

The idea for our buck converter looks like this: we charge the capacitor up to a voltage $U_2 + \Delta U$ and then we discharge the capacitor up to a voltage $U_2 - \Delta U$. Thus, on average, the voltage U_1 will appear at the capacitor. However, we will notice that the voltage changes its sign at the load resistor R. The reason is quickly recognized. In eqn (5.20) we see that the direction of the current in the capacitor has a negative sign. The current flows out of the capacitor and loads the resistor with a different sign. This would therefore generate an AC.

To generate a DC voltage, we need to use the circuit from Fig. 5.5 and add some switches, which changes the polarity of the current at the load resistor. We have added this in Fig. 5.8. The switches S_3 and S_4 ensure that the polarity is set correctly.

In Fig. 5.9, we have shown the current and voltage curve of a charge pump. The capacitor had a value of $C = 0.01$ F, and the load resistance was $R = 10\,\Omega$. The input voltage at $U = 12$ V. In this example we have chosen the charge and discharge time to be symmetrical—that is, $D = 50\%$. The duration is $T = 0.1\,\text{s}$. The current curve of I_C in Fig. 5.9 shows very nicely how the sign changes periodically. However, since we switch the polarity at the load resistor during final charging, we do not see this sign change. Instead, the current I_R remains positive and fluctuates between 0.55 A and 0.74 A. Hence, we have an average current of 0.65 A and a current ripple of $\Delta A = 0.1$ A.

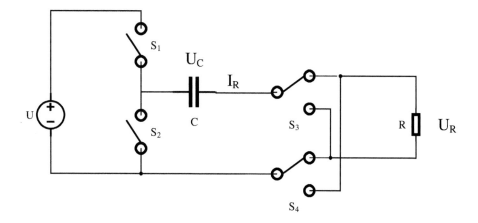

Fig. 5.8 Circuit for using a capacitor as a charge pump. The switches S_1 and S_2 never have the same state and control the charge of the capacitor. The switches S_3 and S_4 control the polarity of the voltage applied to the load resistor R.

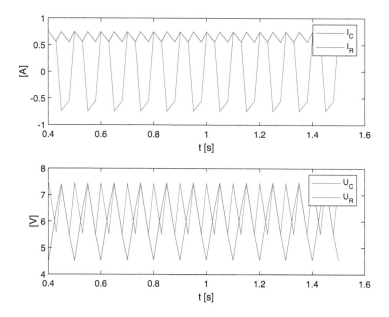

Fig. 5.9 Current and voltage values of a charge pump, as shown in Fig. 5.8. U_C and U_R are the voltage values at the capacitor and at the resistor. I_C and I_R are the current values. It can be seen that the voltage at the resistor varies between 5.5 V and 7.5 V and takes an average value of 6.5 V. The voltage source had a voltage value of $U_1 = 12$ V (Parameters: $C = 1e - 2$F, $R = 10$ Ω, $D = 50\%$, $T = 0.1$ s).

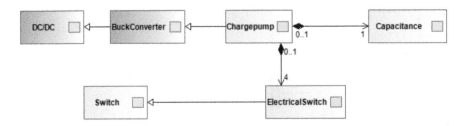

Fig. 5.10 The `Chargepump` is an alternative specialization of the `BuckConverter`. It consists of two system components: a series of `switches` and a `Capacitance`.

For the voltages U_R and U_C the influence of the polarity change can also be seen. The change of polarity reduces the voltage ripple. It can be seen that the voltage at the resistor varies between 5.5 V and 7.5 V and takes an average value of 6.5 V. This corresponds to a voltage ripple of $\Delta U = 1$ V. At the capacitor, the minimum voltage is 4.5 V. The average voltage there is therefore 6 V with a voltage ripple of $\Delta U_C = 1.5$ V.

The charge pump is also a specialization of a buck converter (Fig. 5.10). Let's take a closer look at the properties of the charge pump. For our application, the DC/DC controller must have three properties: 1. On the one hand, we must be able to set the desired voltage levels with the help of the controller, which is important for power flow control, and on the other hand we cannot transfer power between different DC bus voltages; 2. The DC/DC controller must be able to transfer the desired power—even if we can set the voltage so that mathematically the right amount of current can flow from one side of the DC bus to the other, the controller must also be able to transfer this current; 3. The last requirement relates to efficiency. This should be as high as possible.

Let us first consider the question of the voltage range that we can set at the load resistor. The switches S_3 and S_4 do not play a role for the question of the adjustable voltage. They only reverse the polarity so that we do not generate an alternating voltage. We switch within an interval T. Therefore, S_1 is switched on until the time $D \cdot T$. Then we switch to the switch S_2, which is on for a time $(1-D)$. T must be smaller than $\tau = R \cdot C$. Otherwise, we run the risk of completely charging or discharging the capacitor within the switching interval.

Let us first consider the two cases in which $D \to 0$ and $D \to 1$. In both cases, one of the two switches is on almost permanently. With $D \to 0$, the discharge phase is considerably longer than the charging phase. The result is that the voltage at the resistor approaches zero. At $D \to 1$, we have the opposite case: the charging phase is considerably longer than the discharge phase. In this case, the capacitor is almost not discharged. The current flowing through the resistor is therefore very small, which is equivalent to a very small voltage across the resistor. We see that in both $D \to 0$ and $D \to 1$, the voltage is very small. However, the ripple ΔU becomes very large in both cases. In both, the voltage difference between the two switching states is $|U_1|$. So we notice a voltage ripple of $\Delta U_R = |U_1|$.

What happens between these two extremes? Suppose we turn on the charge pump and set $D \approx 0$. We look at our meter, which shows the voltage across the resistor.

As just explained, we will first find that the voltage is close to zero. Now we increase D a little and see what happens to the voltage across the load resistor. We measure that the mean value of the voltage increases because the time in which the capacitor can be charged becomes longer and longer, and thus more energy is available for the final charging time. We also see that the voltage ripple becomes smaller and smaller. This behaviour continues until we reach $D = \frac{1}{2}$, after which the voltage reduces and the voltage ripple increases again. Since at $D = \frac{1}{2}$ the ratios reverse, we see that at this value the maximum voltage and current can be reached. The question is how high these values are.

Exercise 5.6 We set a duty cycle of $D = \frac{1}{2}$ for a charge pump (Fig. 5.8). The switching frequency is $T \leq R \cdot C$. What is the mean current $\bar{I}_R$ and the mean voltage $\bar{U}_C$ at the load resistor? What voltage ripple ΔU_R occurs?

Solution: Let us first consider the average voltage that will appear across the capacitor. For $D = \frac{1}{2}$, the mean voltage across the capacitor is always $\bar{U}_C = \frac{U_1}{2}$. This can be easily illustrated if we consider the case $C \to 0$. So we use a capacitor that has a negligible capacitance. In this case, when $S_1 = 1$, the voltage $U_C = U_1$ would be observed. If the switch is off, the capacitor would be discharged immediately and the voltage $U_C = 0$ would be observed. On average, $\bar{U}_C = \frac{U_1}{2}$ would be observed, because in one half of the switching interval, we can measure a voltage U_1, while during the second half we are measuring no voltage.

Since the switching frequency is smaller than $R \cdot C$, we can assume a linear approximation of the charging and discharging curve. Therefore, the following equation must apply to the ripple ΔU:

$$\bar{U}_C - \Delta U + \frac{1}{2} U_1 \frac{D \cdot T}{R \cdot C} = \bar{U}_C + \delta U.$$

This gives ΔU for the ripple:

$$\Delta U = \frac{1}{4} U_1 \frac{D \cdot T}{R \cdot C}. \tag{5.22}$$

The average load current is obtained directly from Ohm's law. Since we change polarity during the switching process, we have no jump at the load resistor. However, there is a jump at the capacitor. For the average load current we get:

$$\bar{I}_R = \frac{\bar{U}_C}{R}. \tag{5.23}$$

The average voltage across the load resistor is given by the average voltage across the capacitor.

$$\bar{U}_R = \bar{U}_C. \tag{5.24}$$

As we saw in Exercise 5.6, the charge pump has the following properties:

- The voltage can be set between $0\,\mathrm{V}$ and $\frac{U_1}{2}$, although this does not correspond to the ideal requirement **DC Buck 1**, since under certain circumstances there may be the wish that voltages greater than $\frac{U_1}{2}$ should also be realized on the output side.

- The lower the output voltage, the higher the voltage ripple. This increases up to U_1. Its minimum is at $\Delta U = \frac{1}{4} U_1 \frac{D \cdot T}{R \cdot C}$ at a voltage of $\frac{U_1}{2}$. This can lead to a violation of **DC Buck 4**. In particular, if the output voltage is to be much lower than $\frac{U_1}{2}$, the current and voltage ripples will reach very high values.

- The load current results from the load resistance and the capacitor voltage. This property is in contradiction with **DC Buck 2**. We cannot adjust the load current in a charge pump. It is limited by the properties of the capacitor and the voltage. Thus we do not have a flexible solution.
- The losses are very low with this converter. On the input side, charge is transferred to the capacitor and distributed on the output side with a time delay. With an ideal capacitor, losses occur solely through the switching loss and through the load resistance. **DC Buck 4** is thus fulfilled.

The charge pump already fulfils a whole range of requirements that we have for a buck converter. The charge pump is particularly suitable for applications where the output voltage window is stable at $\frac{U_1}{2}$. This is the operating point where current and voltage ripples are at their smallest.

5.2.3 The synchronous buck converter: A realization with coil and switch

The charge pump and voltage divider have not fulfilled all the requirements we have for a buck converter. Since its efficiency is very low, the voltage divider is more suitable for picking up a small voltage from a signal. The charge pump lacks flexibility in the choice of the voltage window and the current window. The reason for the small current window is that a capacitor is a voltage storage device. What is meant by this? When we charge a capacitor, the charge carriers are redistributed. We have a voltage $U_C = U_0$ and no current flows (eqn (5.18)). The applied energy is stored in the electric field between the two plates.

An alternative realization is to use a component that is able to store current. With such a storage, we could generate any voltage, because we would only need to 'branch off' a certain amount of charge carriers in a controlled manner and would thus be able to generate a certain voltage. Unfortunately, we have a fundamental problem here. Electric current by definition comprises moving charge carriers, so we would have to store something that remains in motion all the time. But movement always produces losses. With electric current, we are aware of them as ohmic resistance and perceive these losses in the form of heat. Hence, we have to accept (for the time being) that we can only store current for a short time.

The component that allows us to realize this storage process is the coil. Fig. 5.11 shows its construction. In order to understand the basic principle of a coil, let us remember that a wire through which current flows always also generates a magnetic field whose field lines run in a circle around the wire. If we wind the wire into a coil, the magnetic fields overlap and form a common field H. The strength of the magnetic field H is proportional to the current passing through the coil $H \propto I_L$. If we compare the strength of the magnetic field H of a coil where we have wrapped the wire twice with a coil where we have wrapped the wire 100 times, we find that the field H has also become 100 times stronger. It is therefore valid that H is proportional to the number of windings, n. This is not surprising, because the resulting magnetic field is a superposition of the magnetic field of each individual wire.

Now we do another experiment. We take two coils that have the same number of windings and apply the same current flows. The only difference is that one coil has a length of $l_1 = 1$ cm and the other has a length of $l_2 = 2$ cm. If we now measure the

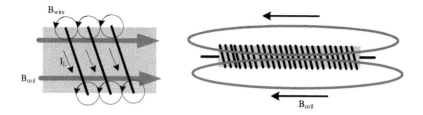

Fig. 5.11 Structure of a coil. A ferrite core is wrapped with a wire. Each individual wire generates a magnetic field when a current flows through it. If the wires are close together, the field lines overlap to form a total magnetic field that runs parallel to the core inside the coil.

Table 5.2 Values for the magnetic permeability of some materials used in electronics.

Material	$\mu \left[\frac{H}{m}\right]$	μ_r
Air	$1.257 \cdot 10^{-6}$	1
Ferrite U60	$1 \cdot 10^{-5}$	8
Iron	$6.28 \cdot 10^{-3}$	5,000
78 Permalloy	0.126	100,000
Supermalloy	1.26	1,000,000

magnetic field, we notice that the magnetic field of the shorter coil is twice as strong as that of the longer coil. The reason can be found in Fig. 5.11. The magnetic field runs through the inside of the coil to the outside and back again. Since the field lines must be closed, with a longer coil the field lines are distributed over a larger area.

In summary, the following applies to the magnetic field H:

$$H = \frac{n\,I}{l}.$$
$$(5.25)$$

In Fig. 5.11, we have not simply wrapped the wire into a coil, but we have wrapped it around an object. We observe that the magnetic field H changes as the material changes. We interpret this to indicate that the magnetic flux B is depending on the material. If we measure B and H, we find that these two quantities are proportional to each other. Analogous to the capacitor, we normalize the relationship between B and H with respect to a reference experiment. This would be the experiment without a bar, in a vacuum. In this case, $B = \mu_0\,H$, where $\mu_0 = 4\pi \cdot 10^{-7}\,\frac{H}{m}$ is described as the magnetic field constant. If we repeat the experiment with other materials, we can determine a relative magnetic permeability μ_r. Similar to the capacitor, we combine these two quantities into the magnetic permeability $\mu = \mu_r \cdot \mu_0$:

$$B = \mu_r \cdot \mu_0 \cdot H = \mu\frac{n\,I}{l}.$$
$$(5.26)$$

We can recall the effect of higher permeability with a simple experiment. If we wrap a wire around a coil and cause current to flow through it, we measure a relatively weak magnetic force. We might be able to move paper clips lying on the desk, but not lift them. Now we wrap the same wire around a piece of iron and let a current flow through the coil. We now observe a much stronger magnetic force that allows us to even lift the paper clips. If we look at the values for the magnetic permeability of iron and the air in Tab. 5.2, we find a quick explanation. The relative permeability of iron is 5,000 times greater than that of air. This means that the force we can generate with this coil is also 5,000 times greater. But there are other materials that increase this factor many times over. Permalloy, a compound of nickel (80%) and iron (20%), is 20 times stronger than iron. If we add molybdenum to the mixture, we get supermalloy, which is made of nickel (75%), iron (20%), and molybdenum (5%), and is 10 times stronger than permalloy.

In a coil that is formed into a rod, the magnetic field lines extend far beyond the coil body (Fig. 5.11); this is necessary because a field line that exits on one side must also re-enter on the other side. Because of the low magnetic permeability of air, the field lines cannot be arbitrarily close together outside the core, so the field lines extend further into space. This effect can be reduced if we work with a ring instead of a rod. The field lines within the ring are close together. There is no reason for the field lines to emerge. That is why coils with closed cores are used in electronics. But what for? So far, we have only seen that a coil generates a magnetic field and, if cleverly chosen, can serve as a powerful magnet that can be switched on and off.

To understand the behaviour or the usefulness of a coil for electronics, let's look again at the basic principle of the coil. If a current flows through a conductor, it can be imagined in a simplified way that charge carriers q flow through it with a speed $\vec{v}$. A location-dependent magnetic field $\vec{B}(\vec{r})$ is created (Jackson, 2009):

$$\vec{B} = \frac{\mu}{4\pi} q\vec{v} \times \frac{\vec{r}}{|\vec{r}|^2}. \tag{5.27}$$

We see in eqn (5.27) that the magnetic field results from a cross-product of the velocity of the charge carriers and a location. Eqn (5.27) also explains why we have circular field lines. Let's look at the field strength outside a circle. Since the distance from the conductor $|\vec{r}|$ remains the same along the circle, the magnitude of $\vec{B}$ does not change. However, for each point on this circle, the location vector $\vec{r}$ and thus the direction of the $\vec{B}$ field changes.

Now, to understand the function of a coil, let's do the following experiment. First, we let a current flow through a conductor. Immediately, a magnetic field is created whose field lines run in a circle around this conductor. Now we simply switch off the current. What will happen? So far, we have understood eqn (5.27) as an equation describing the formation of a magnetic field by moving charges. However, we can also interpret this equation to mean that a magnetic field produces motion in the charge carrier. And this is what happens now. As the current in the conductor decreases, the magnetic field also change. But as it does, it induces a force on the charge carriers. This force, however, goes exactly in the opposite direction to the original direction of movement of the charge carriers. The current is slowed down, so to speak. We measure this by the fact that a voltage builds up.

This experiment can be summarized to the following relationship between voltage and current:

$$U_L = L \cdot \frac{\mathrm{d}}{\mathrm{d}\,t} I_L. \tag{5.28}$$

The voltage across a coil depends on the change in current through the coil. In our experiment, the voltage U_L would initially be zero: the constant current flows through the coil between coil input and output, but the voltage is zero. If the circuit is switched off, the current changes abruptly, and we measure a voltage. The measured voltage depends on the material and the geometry of the coil. Let us take the coil from Fig. 5.11. If we change the material—that is, the magnetic permeability—but keep everything else the same, we notice a proportional change in the voltage. This behaviour is obvious, because we have seen that the strength of the magnetic flux B is proportional to the magnetic permeability and that this flux is the mechanism of action for the induced reverse current.

Since the flux B is due to a superposition of the fluxes of the adjacent conductor, it is proportional to the number of windings n. But we also expect the voltage change to be proportional to the number of windings. So we have two mechanisms, both of which have an effect on the voltage: the proportional increase of the voltage by the flux B and the proportional increase of the voltage by the number of windings. The experiment confirms this finding. If we only change the number of windings, we see that the voltage is proportional to n^2.

If we do further experiments, we also find a dependence on the geometry of the coil core. The larger the cross-sectional area A of the coil body, the larger the voltage increase. This observation can be explained by the increase in magnetic flux B. The larger the cross-sectional area A, the more field lines can be 'captured' in the material. We also observe that the measured voltage decreases the longer the coil body is. If we double the length of the coil body, we halve the voltage increase. This can also be explained by the nature of the B field, because, as we saw in eqn (5.26), we have already observed this relationship between H and B.

All these dependencies between geometry and material of the coil and the voltage are described by the inductance L. L has the unit Henry $[H]$. The following applies:

$$L = \mu n^2 \frac{A}{l}. \tag{5.29}$$

To determine how much energy is stored in a coil, we need to measure how much power $P = U \cdot I$ flows into the coil. The relationship between the coil voltage U_L and the change in coil current $\frac{\mathrm{d}}{\mathrm{d}t} I_L$ over time gives the energy stored at time T:

$$E_L = \int_{t=0}^{T} I_L \, U_L \, dt = \int_{t=0}^{T} I_L \, L \cdot \frac{\mathrm{d}}{\mathrm{d}t} I_L \, dt = \frac{1}{2} L \, I(T)^2. \tag{5.30}$$

Eqn (5.30) shows that the energy inside a coil depends only on the amount of current. A coil is therefore a current storage device, while the capacitor is a voltage storage. We have seen that in a capacitor the energy is stored by the electric field between the two capacitor plates. By adjusting the geometry or permeability, we could change the capacitance of the capacitor. In a coil, the storage mechanism is done by the magnetic field. Similarly, by adjusting the geometry or permeability, we can adjust the capacity of the coil.

Exercise 5.7 We want to store an energy of $E_L = 13$ Ws in a coil. To do this, a current of $I_L = 3$ A is passed through the coil within one second. The core is made of permalloy. It has a circular cross-section with a diameter of $d = 0.5$ cm. How many turns n are necessary, and how long must the coil be so that this energy can be stored?

Solution: We first determine the inductance L:

$$L = \mu A \frac{n^2}{l} = \mu \pi \frac{d^2}{4} \frac{n^2}{l}.$$

With eqn (5.30) we get:

$$E_L = \frac{1}{2} L\, I(T)^2$$
$$= \frac{1}{2}\, \mu\pi \frac{d^2}{4} \frac{n^2}{l} I(T)^2.$$

We solve for l, since l can have any positive value, but n can only be an integer.

$$l = \frac{1}{2}\, \mu\pi \frac{d^2}{4} \frac{n^2}{E_L} I(T)^2$$
$$= 2.1353 \cdot 10^{-5} \mu n^2.$$

We want to work with a few turns $n = 4$, thus a required length of:

$$l = 0.0342 \text{ m}.$$

Let us now look at the charging and discharging behaviour of a coil. We consider the circuit shown in Fig. 5.12. We have a voltage source $U = U_0$. With a switch S_1, this voltage source is connected to a coil L and a resistor R. If S_1 is open and S_2 is closed, we have a circuit containing only the coil and the resistor. S_1 and S_2 are never closed at the same time. At $S_1 = 1$, the coil is charged. When $S_2 = 1$, the coil is discharged.

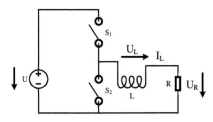

Fig. 5.12 Schematic of a circuit for charging and discharging a coil. The two switches S_1 and S_2 are controlled in such a way that only one is always closed.

We want to charge the coil first. Because of the mesh rule,

$$0 = -U_0 + U_R + U_L. \tag{5.31}$$

We have taken into account that in Fig. 5.12 the voltage source has a different prefix from U_L and U_R. Using Ohm's law and eqn (5.28), this gives:

$$U_0 = R \cdot I + L \frac{\mathrm{d}}{\mathrm{d}t} I. \tag{5.32}$$

We have here a similar differential equation as for the capacitor (eqn (5.18)). We also get comparable solutions for I and U_L:

$$I = \frac{U_0}{R} \left(1 - e^{-\frac{t}{\tau_L}} \right) \tag{5.33}$$

$$U_L = L \frac{\mathrm{d}}{\mathrm{d}t} I = U_0 e^{-\frac{t}{\tau_L}}. \tag{5.34}$$

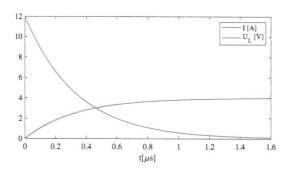

Fig. 5.13 Current I and voltage U_L during the charging process of a coil ($R = 3\,\Omega$, $L = 1\,\mu H$, $U_0 = 12\,V$).

Here, $\tau_R = \frac{L}{R}$ is the characteristic time scale for the charging of a coil. As we can see in Fig. 5.13, the charging process of a coil is similar to that of a capacitor. The only difference is that current and voltage have reversed their roles.

The charging curve can also be interpreted in a different way. A voltage is applied at time $t = 0$. The current to be set is $I = U_0/R$. The coil initially acts against this current. Like a slow water wheel in a current, it does not let the current 'pass' at the beginning; it must first be set in motion, and when this has happened, the desired current of $I = U_0/R$ flows. So we can think of the coil as a buffer that needs to be filled before the current can continue to flow at its full rate. In fact, this is a common application of coils. They serve to dampen current peaks.

Next, consider the discharge process. S_1 is open and S_2 is closed. In this case, the mesh rule gives:

$$0 = U_R + U_L. \tag{5.35}$$

Using Ohm's law and eqn (5.28), the homogeneous differential equation is obtained:

$$0 = R \cdot I + L \frac{d}{dt} I. \tag{5.36}$$

Again, the solution is comparable to that of the capacitor, and we get a comparable solution for I and U_L:

$$I = \frac{U_0}{R} e^{-\frac{t}{\tau_L}} \tag{5.37}$$

$$U_L = L \frac{d}{dt} I = -U_0 e^{-\frac{t}{\tau_L}}. \tag{5.38}$$

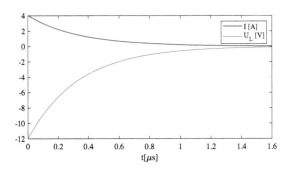

Fig. 5.14 Current I and voltage U_L during the charging process of a coil ($R = 3\,\Omega$, $L = 1\,\mu H$, $U_0 = 12\,V$).

The discharging process is shown in Fig. 5.14. We see that both curves show an exponential decrease. Here, the voltage has a negative sign. This behaviour corresponds to that of the capacitor, but with the roles of current and voltage reversed. If we want to take up the image of the water wheel again, we have the initial situation here that the water wheel turns with the current. Suddenly the current stops, but the wheel continues to turn. The change in current of $I = 0$ initially has no effect: then the water wheel begins to slow down and the current decreases. Since the current acts against the desired current of $I = 0$, logically there is also an increased voltage, which also acts against the actual value.

The circuit in Fig. 5.12 can be used as a buck converter. The basic principle is analogous to the charge pump: by periodically charging and discharging the coil, we generate a time-varying current, which in turn generates a time-varying voltage. Again, we look at the average values $\hat{I}$, $\hat{U}_R$ and consider the current and voltage ripple ΔI, ΔU_R. We modify the circuit by introducing another capacitor C in parallel with resistor R. The resulting circuit diagram is shown in Fig. 5.15. To simplify the mathematical description, we have arranged the voltage arrows for U, U_L, and U_R slightly differently. Fig. 5.16 shows the block diagram for a synchronous buck converter.

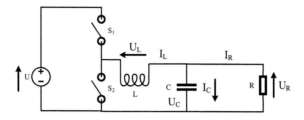

Fig. 5.15 Circuit diagram of a synchronous switched buck converter. The active elements, the switches S_1 and S_2, are used in the same way as in Fig. 5.12.

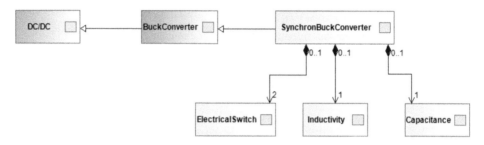

Fig. 5.16 Block diagram of a synchronous buck converter, `SynchronBuckConverter`. It is a specialized realization of a `BuckConverter` and consists of an `Inductivity`, a `Capacitance` and two `ElectricalSwitches`.

The voltage across the inductance is given by:

$$U_{\mathrm{L}} = L \cdot \frac{\Delta I}{\Delta t} \tag{5.39}$$

We use the linearized equation and consider the case where the inductance is charged. From the mesh rule, we find that $0 = U_{\mathrm{L}} + U_{\mathrm{R}} + U$. With $U = U_0$ and eqn (5.39), we get:

$$U_{\mathrm{L}} = U_{\mathrm{R}} = L \cdot \frac{\Delta I}{\Delta t} = L \cdot \frac{\Delta I}{D \cdot T} \Rightarrow \Delta I = \frac{(U_{\mathrm{R}} - U_0)}{L} D \cdot T. \tag{5.40}$$

where D, the duty cycle, describes the ratio of on-time and off-time. $D \cdot T$ corresponds to the time up to which S_1 remains on.

When the coil is discharged—that is, when S_2 is closed—$0 = U_{\mathrm{L}} + U_{\mathrm{R}}$. So we get:

$$U_{\mathrm{L}} = -U_{\mathrm{R}} = L \cdot \frac{\Delta I}{\Delta t} = L \cdot \frac{\Delta I}{(1-D) \cdot T} \Rightarrow \Delta I = \frac{-U_{\mathrm{R}}}{L}(1-D) \cdot T. \tag{5.41}$$

We can combine eqns (5.40) and (5.41) and obtain:

$$\frac{(U_{\mathrm{R}} - U_0)}{L} D \cdot T = \frac{-U_{\mathrm{R}}}{L}(1-D) \cdot L \tag{5.42}$$

$$(U_{\mathrm{R}} - U_0)D = -U_{\mathrm{R}}(1-D) \tag{5.43}$$

$$U_{\mathrm{R}} = D \cdot U_0. \tag{5.44}$$

The voltage at the load resistor depends only on the duty cycle. Both the inductance and the length of the switching period have no influence on the ratio of the input to the output voltage. The situation is different for the current ripple. To calculate this, we can use either eqn (5.40) or eqn (5.41).

Exercise 5.8 We want to realize a synchronous buck converter. This converter shall reduce the voltage from $U_0 = 12$ V to $U_R = 4$ V. The current ripple shall not be greater than 8% of the average load current I_R. The load resistor has a magnitude of $R = 130$ mΩ. The switching frequency we want to use is $f = 25$ kHz.

Solution: Since we already know the output voltage, we can calculate the load current using Ohm's law:

$$I_R = \frac{U_R}{R} = \frac{4\text{ V}}{130\text{ m}\Omega} = 30.77\text{ A}.$$

The required maximum current ripple ΔI_{max} is 8% of the load current—that is,

$$\Delta I_{max} = 8\% I_R = 8\% \cdot 33{,}77\text{ A} = 2.7\text{ A}.$$

To determine the inductance L, we must first determine the duty cycle:

$$D = \frac{U_R}{U_0} = \frac{4\text{ V}}{12\text{ V}} = \frac{1}{3}.$$

For the switching period, $T = \frac{1}{f}$. With eqn (5.41), we now calculate the required inductance:

$$\Delta I = \frac{-U_R}{L}(1 - D) \cdot T$$
$$= \frac{-U_R}{L}(1 - D) \cdot \frac{1}{f}$$
$$L = \frac{-U_R}{\Delta I}(1 - D) \cdot \frac{1}{f}$$
$$= \frac{-4\text{ V}}{2,7\text{ A}}(1 - \frac{1}{3}) \cdot \frac{1}{25\text{ kHz}}$$
$$= \frac{-4\text{ V}}{2,7\text{ A}}\frac{2}{3} \cdot \frac{1}{25\text{ kHz}}$$
$$= 3,95 \cdot 10^{-5}\text{ H} = 0{,}395\ \mu\text{H}.$$

The configuration of the buck converter in Exercise 5.8 shows that for the current ripple, the switching frequency has a influence on the value for the inductance. However, we have not yet considered the voltage ripple. Without a capacitor, the voltage ripple is proportional to the load resistance and the load current. But if we connect a capacitor in parallel to the load resistor, we can also charge it during the charging process of the coil and use it as a buffer.

We assume that the chosen capacitance of the capacitor is so large that the current over the load resistance no longer shows a significant current ripple and the current ripple is compensated by the capacitor—that is,

$$I_C = \Delta I_L.$$

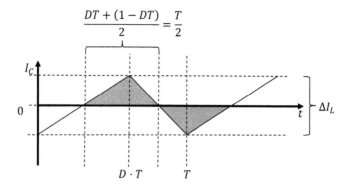

Fig. 5.17 Current waveform across capacitor C in the buck converter. D is the duty cycle. T is the switching period. From $t = 0$ to $T = D \cdot T$, the capacitor is charged; from $t = D \cdot T$ to T the capacitor is discharged.

Now let's look at the current flow at the capacitor. In Fig. 5.17 we have shown this. Since we consider a equilibrium state, the capacitor is charged and discharged within one switching period; the number of charge carriers that are supplied to the capacitor during this process are also discharged again during the discharging process. Therefore, the two areas are equal. The charging time is $T/2$. The height of the triangle is $\frac{\delta I_{\mathrm{L}}}{2}$. Thus, we can determine the number of charge carriers transferred during charging, ΔQ:

$$\Delta Q = \frac{T}{2}\frac{1}{2}\frac{\Delta I_{\mathrm{L}}}{2} = \frac{\Delta I_{\mathrm{L}}}{8f}. \tag{5.45}$$

The relation between the voltage ripple ΔU and the charging numbers transferred during the process is given by the capacitor equation $Q = C\,U$. Thus, we can determine the size of the capacitor directly:

$$C = \frac{\delta Q}{\Delta U} = \frac{\Delta I_{\mathrm{L}}}{8f\Delta U}. \tag{5.46}$$

Exercise 5.9 We have not yet designed the capacitor for the voltage ripple in Exercise 5.8. We want a very small voltage ripple here of at most $\Delta U_{\mathrm{R}} = 0.1$ V. How large is the capacitance of the capacitor that we have to choose?

Solution: From Exercise 5.8 we know that the current ripple is at $\Delta I_{\mathrm{R}} = 2.7$ A. We have a switching frequency of $f = 25$ kHz. The capacitance of the capacitor can be calculated directly:

$$C = \frac{\Delta I_{\mathrm{L}}}{8f\Delta U} = \frac{2,7 \text{ A}}{8 \cdot 25 \text{ kHz} \cdot 0.1 \text{ V}} = 1.35 \cdot 10^{-4} \text{ F} = 0.135 \text{ mF}.$$

How well does the synchronous buck converter meet the requirements for a buck converter? To answer this, let's take a look at these requirements:

- **DC Buck 1:** Just like the charge pump, the synchronous buck converter is able to cover the entire voltage interval $U_2 \in [0, U_1]$.
- **DC Buck 2:** The synchronous buck converter operates through a current storage device, so it meets this requirement as long as enough current is provided by the voltage source.
- **DC Buck 3:** Similar to the charge pump, the efficiency of the synchronous buck converter depends only on the conduction losses and the switching losses. We have a very efficient component here as well.
- **DC Buck 4:** Voltage and current ripple can be well adjusted in the synchronous buck converter by the inductor and the filter capacitor.

The comparison with the various requirements shows that the synchronous buck converter meets all the requirements for a buck converter well.

We have now learned about three components that we can use to implement a buck converter: the capacitor, the coil, and the resistor. However, we have so far neglected one component that we used for both the charge pump and the synchronous buck converter: the switch. It is obvious that with a switching frequency of 25 kHz, a mechanical switch cannot be used anymore. In fact, this is realized by semiconductor devices. In the following section, we want to introduce them.

5.2.4 How to turn alternating current into direct current: The diode as a rectifier

Before we turn to the question of how we actively switch current, let us first look at a semiconductor component that is simpler in its construction: the diode.

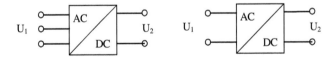

Fig. 5.18 Symbols for rectifier. A rectifier converts AC into DC. Connected to the mains, U_1 is a three-phase AC that is converted into a direct current U_2. However, rectifiers can also be connected to individual AC phases, in which case U_1 is only a single-phase AC current.

So far, we have dealt with the question of how direct current with a certain voltage, U_1, is transformed into direct current with a different voltage, U_2. In our example in Chapter 3, these DC/DC controllers were used to charge the battery of the bus or the buffer storage, or to use the solar power. However, we also needed a component that converted the AC from the grid into DC. In Fig. 5.18 we have shown its symbols. Rectifiers can be designed as three-phase or single-phase. Three-phase rectifiers are typically used when a large amount of power is drawn from the grid. In this case, the AC from all three phases is converted into DC. For lower power, such as charging a mobile, a single-phase rectifier is used. As with the DC/DC converters, there are various realizations of this task. We will look at the simplest realization here: the passive rectifier.

The following requirements apply to a passive rectifier:

pas. Rect. 1: `AS A user, I WANT an alternating current to be converted into a direct current SO THAT I can use an alternating current source to supply a direct current consumer.`

pas. Rect. 2: `AS A user, I WANT the DC current to be as constant as possible and have only a small voltage ripple SO THAT my loads do not have a shorter lifetime due to the voltage fluctuations.`

The first requirement describes the function of a passive rectifier. The second requirement refers to the quality of the generated DC. Since we are changing from a signal that is periodic in time to a signal that is constant in time if possible, it is to be expected that the fluctuations in the DC are visible. Therefore, we must demand that these fluctuations are lower, ideally by giving a quantitative limit for the current ripple ΔI and the voltage ripple ΔU.

There is one aspect here that we have not addressed. We have not required that the DC voltage level be adjustable. In fact, a passive rectifier is not capable of adjusting the voltage level. Other topologies are necessary for this.

Fig. 5.19 shows the system components for single-phase and three-phase passive rectifiers. These consist of a capacitor, as well as a number of diodes. Before we look at the construction of a passive rectifier in more detail, let's take a closer look at the diode, as a new element in our construction kit.

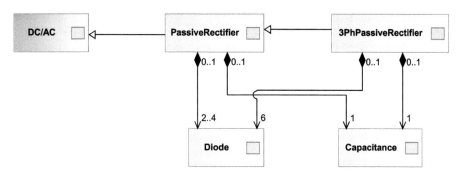

Fig. 5.19 Block diagrams for single-phase and three-phase passive rectifiers. These consist of several diodes and a capacitor.

The structure of a diode is shown in Fig. 5.20. A diode consists of two semiconductor layers that are connected to each other. One layer is n-type. This means that the semiconductor material has been mixed with a material that has more electrons than would be necessary for chemical bonding in the crystal. The excess electrons are mobile.

The other layer of the diode is p-type. This means that here the semiconductor has been mixed with a material that has slightly fewer electrons than could be served to form the crystal. In the p-type material, there are therefore free places, holes, which can be occupied by an electron.

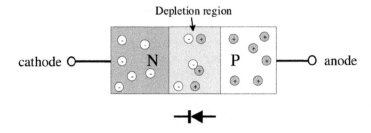

Fig. 5.20 Structure of a diode. A diode consists of an *n*-type and a *p*-type semiconductor that are connected to each other. In the equilibrium state, i.e. when no voltage is applied, an electrically neutral depletion region forms at the interface.

In the equilibrium state—that is, without voltage—electrons and holes are initially evenly distributed in the two semiconductor layers. They move randomly through the crystal, and their mobility depends only on the temperature. If holes and electrons come closer together, they attract each other. If they come close enough, the electron recombines with the hole. Both lose their electrical charge. This process occurs more frequently directly at the contact surface of the two semiconductor layers. Here, an area called the depletion region or space charge zone is formed. It is electrically neutral.

What happens when we apply a voltage to the diode? Suppose we apply the voltage so that there is a positive voltage on the *p*-type side and a negative voltage on the *n* doped side (Fig. 5.21 left). In this case, we are basically strengthening the attraction between the electrons and the holes. The depletion region becomes smaller because electrons are attracted to the *p*-type side and holes to the *n*-type side. Part of it neutralizes in the depletion zone, which becomes narrower, and there will be electrons flowing across the depletion zone, through the *p*-type semiconductor to the anode, thus completing the circuit. So the diode becomes a conductor. The current depends on the voltage. The higher the voltage an the narrower the depletion zone, the more charge carriers overcome it, resulting in higher current flows.

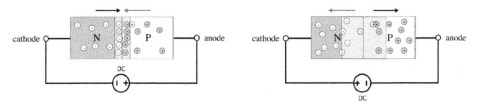

Fig. 5.21 Function of a diode. Depending on the polarity of the applied voltage, the depletion region becomes smaller or larger. If the depletion region increases, no more charge carriers can pass through the diode. The diode blocks the current. If the depletion region is small, the barrier for the charge carriers is significantly smaller and a current can flow through the diode.

Now we reverse the polarity (Fig. 5.21 right). We apply a positive voltage to the n-type side and a negative voltage to the p-type side. The electrons will move away from the p-type side. The holes on the p-type side will do the same. As a result, the depletion zone will increase, creating an ever-greater barrier to any charge carrier that does try to cross the diode. As a result, it becomes less and less likely that a current can flow between the two contacts of the diode. The diode therefore blocks the flow of current.

Fig. 5.22 shows a circuit with a diode. We have an alternating voltage $U_{AC} = U_0 \cdot \sin \omega t$ connected via the diode to a load resistor R. U_0 is the amplitude of the voltage and ω is the frequency of the alternating voltage.

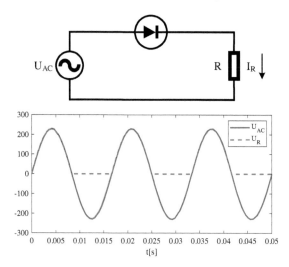

Fig. 5.22 Illustration of how a diode works: the diode is connected to an alternating voltage. A resistor serves as the load. If the voltage is positive, the diode conducts and a current flows. If the voltage is negative, the diode blocks.

As long as the AC voltage is positive, the diode is conductive and a current flows through the resistor. If, on the other hand, the AC voltage is negative, the diode blocks and no voltage is applied to the resistor. The resulting current results from Ohm's law.

The diode thus ensures that the polarity of the voltage at the load resistor can only have one sign, which means that the current can only flow through the conductor in one direction. The oscillating change of sign produced by the alternating voltage does not occur. We observe a pulsed positive voltage whose amplitude corresponds to the amplitude of the alternating voltage. On average, we have a DC voltage with half the amplitude of the AC voltage. Unfortunately, the voltage and current ripples are very high. The voltage ripple has the full height of the amplitude $\Delta U_R = \frac{U_0}{2}$. The current ripple depends on the load resistance $\Delta I_R = \frac{U_0}{2\,R}$.

To reduce this ripple, we add a capacitor. This can supply the load resistor whenever the diode is blocked. The capacitor serves as a buffer. In Fig. 5.23, we have extended the circuit from Fig. 5.22 with the capacitor C. As can be seen over time,

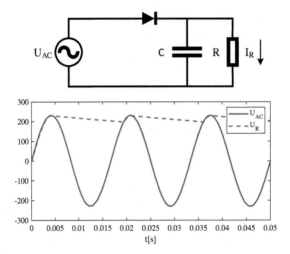

Fig. 5.23 Illustration of a rectifier for a single-phase AC.

the capacitor significantly reduces the current ripple. We are thus able to generate a single-phase DC from a single-phase AC.

Fig. 5.24 shows the circuit diagram for a three-phase rectifier. Each of the three AC phases, u, v, and w, is connected to a diode bridge. The diode bridge ensures that when the voltage is positive, the current can flow through the upper branch, and when the voltage is negative, the current can flow through the lower branch. The resulting voltage at the output is always positive. The gaps visible in Fig. 5.22 are compensated. Because all three phases are rectified, the output voltage U_{DC} has a significantly smaller ripple, which can be additionally smoothed with the help of the capacitor C.

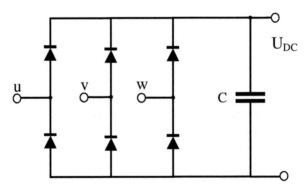

Fig. 5.24 Circuit diagram of a three-phase rectifier. The three AC phases u, v, and w are connected to a diode bridge. Here, too, the capacitor C serves as a buffer to compensate for voltage fluctuations.

The rectification with the help of a diode fulfils the two requirements `pas.-Rect. 1` and `pas.-Rect. 2`. It is simple in its construction and is therefore served in many applications. However, it has two disadvantages that limit its use. Firstly, the DC output voltage is not adjustable. It always results from the amplitude of the AC. If we wanted to charge a battery with a rectifier, we would still need a step-down or step-up regulator so that the voltage would be adapted to the state of charge of the battery.

The second disadvantage is that a diode rectifier only allows power to flow from AC to DC. We could charge a battery with it. But we would not be able to discharge the battery and feed the stored energy back into the grid. There are other solutions for this, which we will describe in Section 5.2.5.

With the diode, we have now integrated the first semiconductor element into our construction kit. We have now learned about circuits with which we can adjust direct voltages and turn AC into DC. In the next section, we will deal with the conversion of AC. In doing so, we will get to know the last two components: the transformer and the transistor.

5.2.5 How to transform AC power

In the previous sections, we first dealt with the transformation of direct current. The rectifier represented a very first component where we were able to bridge between the two types of current: DC and AC. We now want to get to know the next two components: the transformer and the semiconductor switch. We will first start with the transformer and its typical application: the transformation from one AC to another.

AC is described by three properties: the number of its phases, its amplitude, and its frequency. The number of phases refers to how many AC lines are connected together. In everyday life, we mostly encounter AC as the single-phase AC that we get from the socket. We usually have two current-carrying connections. In some countries, a neutral conductor and an earth conductor are also connected to the socket. A single-phase AC voltage is described by $U_{AC} = U_0 \cos(2\pi f t)$. Here, U_0 is the amplitude and f the frequency of the AC voltage.

We encounter an AC in which three single-phase alternating voltages are linked together, in everyday life wherever greater power is required. Electrical machines with a higher power rating are usually operated with three AC phases. The three phases are offset from each other by 120°. The following applies: $U_{AC,n} = U_{0,n} \cos(2\pi f + (n-1)\phi_0$, where $n = 1, 2, 3$ is the number of the respective phase and $\phi_0 = \frac{2\pi}{3}$ is the phase shift.

The electrical grid also works with three phases. This is because until the introduction of renewable energies, the generation of electrical energy was almost exclusively based on the conversion of kinetic energy into electrical energy. Huge electric motors were in use as generators and driven by steam turbines, for example. For the construction of an electric motor, a realization with three phases represents a first realization in which power fluctuations are significantly reduced.

If we want to transform one AC into another, there are cases where only the amplitude has to be changed, but where the frequency remains the same. This transformation is often found in electrical networks. Transformers are used here.

But it can also happen that we have to adjust the frequency as well as the amplitude. This is the case, for example, when an electric motor is to be operated with grid current. Since the speed at which a motor rotates depends on the rotational frequency of the electric field, the frequency must change in comparison to the grid frequency and may also have to be adjusted in time.

5.2.5.1 When the frequency remains the same: The transformer as a simple frequency converter

The electrical grid is roughly divided into three levels. On the top level, the high-voltage level, the large power plants are connected. They transmit their electricity in transmission lines across greater distances. The voltage is in the order of 100 kV to 300 kV. This is called the transmission grid.

The transmission grid is linked to the distribution grid. The distribution grid distributes electrical energy within communities and towns. The distribution grid is divided into the medium-voltage grid, with voltages in the order of 50 kV, and the low-voltage grid, with voltages in the order of 400−500 V.

The frequency is identical in all these grids—in other words, if one wants to transfer power between the different voltage levels, only the amplitude has to be adjusted. The transformer is used for this purpose.

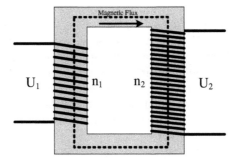

Fig. 5.25 Structure of a transformer. A transformer consists of a ring-shaped core with two wires wound around it.

The transformer is another basic element in our construction kit. Fig. 5.25 shows its construction. It consists of two coils wrapped around the same ferromagnetic core. One coil is wrapped n_1 times around the core. The second coil has been wrapped n_2 times around the core.

What happens if we apply an AC voltage $U_1 = \hat{U} \cos(2\pi\, t)$ in one of the coils? As we have already seen in Fig. 5.11, a magnetic flux builds up inside the coil:

$$B_1 = \mu \frac{n_1\, I_1}{l},$$

where l is the length of the coil, μ the magnetic permeability, n_1 the number of turns, and I_1 the current flowing through the coil. The field lines of B_1 run parallel inside the core. They emerge from the core on one side of the coil and enter the coil again on the other side. Due to the ring-shaped ferromagnetic core, the field lines are guided through the ring and thus through the second coil.

For the second coil, it now looks like no current was originally passed through here, but a magnetic flux flows through it. Since a magnetic flux in a conductor applies a force to the charge carriers, a current now begins to build up in the second coil:

$$B_2 = \mu \frac{n_2 \, I_2}{l}.$$

B_2 is the magnetic flux in the core of the second coil. $B_1{=}B_2$, since it is the same core. l and μ are also identical. Hence, we can combine the two equations and obtain a relationship between the two currents:

$$\mu \frac{n_1 \, I_1}{l} = \mu \frac{n_2 \, I_2}{l} \tag{5.47}$$

$$n_1 \, I_1 = n_2 \, I_2 \tag{5.48}$$

$$I_1 = \frac{n_2}{n_1} I_2. \tag{5.49}$$

Only the ratio of the number of windings, $\frac{n_2}{n_1}$, determines the difference in the currents. So we can adjust the current amplitude by the turns ratio.

What about the voltage? For this, recall the induction law eqn (5.28) and the definition for inductance (5.29). For the voltage on the first coil the following applies:

$$U_1 = L \cdot \frac{d}{d \, t} I_1 = n_1 A \frac{d}{dt} B_1, \tag{5.50}$$

where A is the cross-sectional area of the core. The same applies to the voltage on the second coil:

$$U_2 = L \cdot \frac{d}{d \, t} I_2 = n_2 A \frac{d}{dt} B_2. \tag{5.51}$$

Since the cross-sectional area should be the same and the magnetic flux is also identical, the following holds:

$$\frac{U_1}{n_1} = \frac{U_2}{n_2} = A \frac{d}{dt} B_1 \tag{5.52}$$

$$U_1 = \frac{n_2}{n_1} U_2. \tag{5.53}$$

We see that the voltage also depends only on the ratio of the two numbers of turns.

Let's look at the circuit and the resulting waveform in Fig. 5.26. An AC voltage, U_1, is applied to the primary site. The turn ratio in this example is at 2:1—that is, the n_2 corresponds to half the turn of n_1. As we can see, the voltage amplitude is halved. The phase position is identical. The frequency has remained the same. So we have succeeded in halving the AC input voltage.

Transformers have the great advantage of being relatively simple in their construction and function. The basic construction has hardly changed since the days of Westinghouse, Siemens, and Edison. Improvements have taken place in the area of manufacturing technology and the materials used (Wang *et al.*, 2002; Ookubo, Kousaka, and

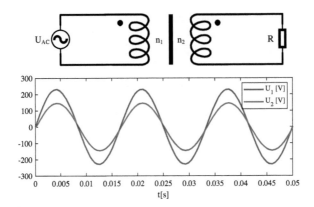

Fig. 5.26 The circuit of a transformer and the corresponding waveform. ($U_1 = 230$ V, $f = 50$ Hz, $\frac{n_1}{n_2} = 2$, $R = 1\ \Omega$).

Ikeda, 2008; Dujic *et al.*, 2012). An advantage of transformers is the galvanic isolation between the primary and secondary sides: as we have seen, the transformer transfers power from the primary to the secondary side via a conversion from an electric field to a magnetic field and back again. This conversion can only work if there is a change in the current (eqn (5.28)). A direct current cannot be transmitted in this way. The two circuits can thus have a different potential without any current flowing between them.

The disadvantage of transformers is their size and the fact that the frequency cannot be adjusted. A change in frequency on the primary side is always transferred to the secondary side. In distribution and transmission grids, this means that a fluctuation in the frequency in the transmission grid is also experienced by the end consumer, even though several transformations take place in between.

To eliminate this disadvantage, one changes to alternative topologies, which also partially contain transformers, but also allow an adjustment of the frequency through the use of switches (Falcones, Mao, and Ayyanar, 2010; She, Huang, and Burgos, 2013).

5.2.5.2 *When everything needs to be adjusted: The inverter*

We now want to turn to the last building block of our power electronics kit: the transistor. We used switches for both the charge pump and the buck converter without taking a closer look at their structure and function. We now want to make up for this.

Fig. 5.27 shows the symbols for an inverter. The inverter converts one AC into another AC, adjusting both the amplitude and the frequency. An inverter consists of a series of diodes, as well as transistors and a capacitor, which is used here mainly for buffering (Fig. 5.28). In this chapter, we will look at the three-phase, unidirectional inverter, one in which the power flow only goes in one direction.

Let's first look at the requirements for such an inverter. We start with the basic use case for an inverter:

uni. Inverter 1: AS A user, I WANT an AC with a variable frequency and amplitude to be converted into an AC with a given frequency and amplitude SO THAT I can meet the requirement of my load for the quality of the AC.

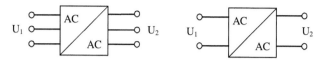

Fig. 5.27 Symbol for an inverter. An inverter converts single-phase or three-phase AC, whereby both the frequency and the amplitude can change.

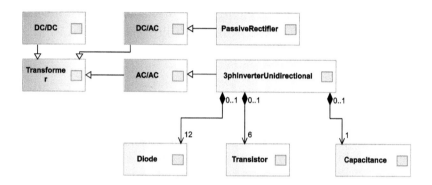

Fig. 5.28 Block diagram of an inverter, which is a deviation of the transformer and a realization of an AC/AC inverter.

uni. Inverter 1 describes the use of an inverter. If we want to use an inverter to drive a motor, we must be able to adjust the frequency and amplitude of the AC that is driving it, regardless of what the mains frequency or mains amplitude is. In industrial applications, the drive must be stable enough so that fluctuations in the mains do not affect my production process. The inverter stabilizes the output voltage and frequency and drives the motor.

Since consumers have different requirements for the properties of the AC, an inverter must be able to meet the requirements for the AC as well.

uni. Inverter 2: AS A user, I WANT the AC current to meet the voltage quality requirements of the load SO THAT damage to the load is avoided and the load operates correctly.

In AC systems, the relationship of the three phases to each other is fixed. An 120-degrees offset between the phases is hardwired into the motors and generators. Therefore, there is no requirement for the inverter to adjust this phase offset. In fact, the most important control methods in electrical drive technology are based on the premise that the three phases have a fixed phase offset. However, the situation is different with regard to amplitude and waveform. Here there may very well be a requirement that these are adapted or varied.

uni. converter 3: AS A user, I WANT to be able to vary the amplitudes and waveforms of the three phases SO THAT I can generate an asymmetrical, three-phase AC.

So far we have focused on the output of the inverter, but of course there are also requirements for the input of the inverter:

uni. Inverter 4: `AS A user, I WANT the inverter to work with the different forms of AC on the input side SO THAT variations in waveform and current amplitude do not greatly affect the output of the inverter.`

These requirements apply to both unidirectional and bidirectional converters. The question of whether or not the power flow should be unidirectional depends on the application. Many drive inverters only work unidirectionally, which has the disadvantage that they cannot recuperate. However, they are simpler in their design, which is why we want to look at them in this chapter. But first we start with the introduction of the semiconductor switch.

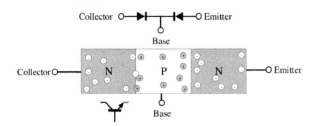

Fig. 5.29 Structure of a bipolar transistor (npn). This consists of two n-type semiconductors with a p-type semiconductor between them. Alternatively, a bipolar transistor can also be constructed as a pnp transistor. In this case, one n-type semiconductor lies between two p-type semiconductors.

The basic structure of a transistor is shown in Fig. 5.29. A p-type semiconductor layer is inserted between two n-type semiconductor layers. Each of the semiconductor layers is contacted separately. The contacts at the n-type layers are called the collector and the emitter. The contact at the middle layer is called the base.

If no voltage is applied, a depletion zone is built up between the two boundary layers due to the thermal movement of the charge carriers. The behaviour is analogous to that of a diode. We can therefore also interpret the transistor as two diodes connected to each other. If the base is isolated, no current can flow through the transistor because one of the two diodes is always connected in the reverse direction.

But what happens if we build up a voltage not only between collector and emitter, but also between base and emitter? In Fig. 5.30 we apply a negative voltage to the p-type semiconductor layer. At the same time, we apply a voltage between the emitter and the collector. Since the holes have a positive charge, a part is removed via the base. Another part recombines with the charge carriers from the n-type semiconductor on the emitter side, as these are now driven to the collector side. As a result, the transistor has now become conductive. The current flowing from the emitter through the collector can even be controlled: the fewer holes are discharged via the base, the lower the number of charge carriers arriving at the collector.

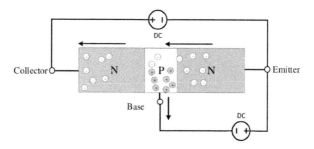

Fig. 5.30 Behaviour of the charge carriers in a transistor when a base-collector voltage is applied. The charge carriers of the *p*-typed semiconductor can now flow away so that the transistor becomes conductive.

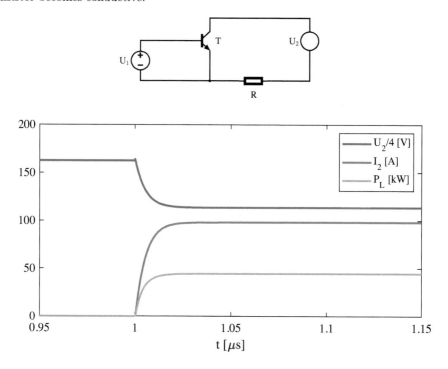

Fig. 5.31 Example of a transistor circuit. The base-collector voltage U_1 is 6 V and is switched on for 2 μs. A voltage of $U_2 = 650$ V is applied between the collector and the emitter. It can be clearly seen that during the switching process the collector–emitter voltage U_2 decreases while the current I_2 begins to flow. Losses P_L occur in this transition phase ($R = 2$ Ω).

With a transistor we can regulate a large current with a small switching voltage. Fig. 5.31 shows an example circuit. Here, with a small base-emitter voltage of $U_1 = 6$ V, we switch a circuit with a voltage supply of $U_2 = 650$ V. As can be seen, the switching behaviour is not ideal. As soon as the transistor is switched, the current increases steadily, while the voltage drops. The losses that occur during this time are called switching losses.

We now have all the building blocks to build an inverter. As already shown in Fig. 5.28, an inverter consists of several diodes and transistors. Analogous to the buck converter, we solve the task of transforming a AC into an AC by generating an alternating current in the time average. That is, we switch the DC voltage on and off so quickly that it appears at the output that an AC is coming out of the inverter.

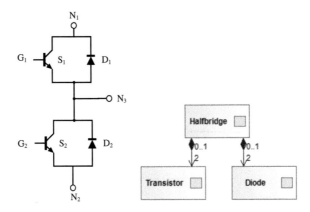

Fig. 5.32 Circuit diagram and block diagram of a **Halfbridge**. It consists of two **Transistors** and two **Diodes**.

The basic element for this circuit is the half-bridge. It is shown in Fig. 5.32 as a circuit diagram and a block diagram. The half-bridge consists of two switches, S_1, S_2, and two diodes, D_1, D_2. It has two inputs, N_1 and N_2, and one output, N_3. The switches are switched via the gate signals, G_1 and G_2. Only one of the two switches is switched on, while the other is switched off.

To understand the basic principle for the construction of an inverter, let us first consider a single-phase inverter. We have shown the circuit in Fig. 5.33. The half-bridge is connected to a DC voltage source. This is split in the middle, so that we have a voltage of $U_{DC}/2$ and $-U_{DC}/2$. The output of the half-bridge is connected to a coil L and the load resistor R. As with the buck converter, we use a coil as a current storage device and filter element. We apply current to the coil with the two switches so that the desired AC is measured at the output of the coil.

The half-bridge allows us to generate three output voltages: 0, $U_{DC}/2$, and $-U_{DC}/2$. The task now is to use the switches S_1 and S_2 so that the current I_L becomes like an AC. One way to control the switches is to measure whether the current I_L is larger or smaller than the desired AC. If the current is greater than the target value, S_2 is switched on. A negative voltage is then applied to the coil and the current will decrease. Conversely, S_1 is switched on if the current I_L is less than the target current. This procedure is illustrated in Algorithm 3.

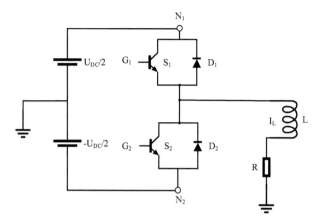

Fig. 5.33 Circuit of a single-phase inverter, using a half-bridge. The intermediate voltage U_{DC} is divided to provide a positive and negative voltage, which is needed to realize the upper and lower halves of the AC current.

Algorithm 3 Simple control for generating an AC using a half-bridge.

---

while inverter is running **do**
    $\hat{I}_L \leftarrow I_0 \cos(2\pi f t)$
    if $I_L > \hat{I}_L + \Delta I_L$ **then**
        $S_1 =\leftarrow 0, S_2 \leftarrow 1$
    end if
    if $I_L < \hat{I}_L - \Delta I_L$ **then**
        $S_1 \leftarrow 1, S_2 =\leftarrow 0$
    end if
end while

To minimize the number of switching operations, a hysteresis band is used. The switching operations only take place when the deviation is greater than ΔI_L. In this way, the number of switching operations is reduced. In addition, the hysteresis band can be used to adjust the quality of the control.

An alternative approach is the so-called pulse-width modulation (PWM). This is very common in the control of inverters and drive converters because it is formulated close to a hardware realization and can therefore be easily implemented.

The procedure described in Algorithm 3 works like a control, including a feedback loop. PWM, on the other hand, has no feedback loop. It does not measure I_L but assumes a certain form of I_L. PWM works by comparing a target signal, in our case a sinusoidal signal, and a carrier signal. Typically, this is a triangular-shaped signal. The carrier signal is periodic. It takes over the function of the FOR loop in Algorithm 3. As long as the target signal is smaller than the carrier signal, the upper switch S_1 is switched on and the lower switch S_2 is switched off. If the target signal becomes larger than the carrier signal, the upper switch S_1 is turned off and the lower switch S_2 is

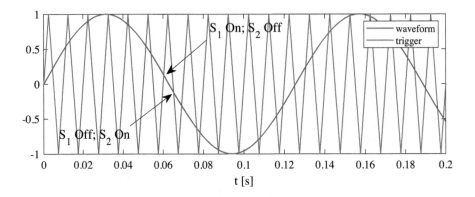

Fig. 5.34 Mode of operation of a PWM with a triangular-shaped carrier. The desired modulation, here a sine, is compared with a carrier signal. If the modulated signal is greater than the value of the carrier, S_1 is switched on and S_2 is switched off. As soon as the value of the modulated signal is below the value of the carrier, S_2 is switched on and S_1 is switched off.

turned on (see Fig. 5.34). Due to the structure of the carrier signal, a trigger can be used here that always changes the switching states when the target signal overlaps with the carrier signal.

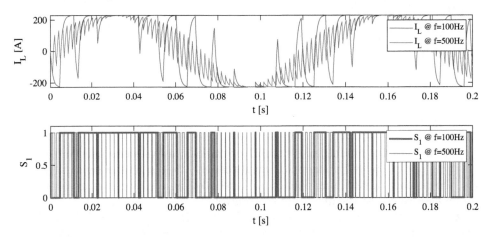

Fig. 5.35 I_L and switching states of S_1 at different PWM frequencies ($f_1 = 100$ Hz, $f_2 = 500$ Hz). The inductance here is $L = 1$ mH.

The frequency of the carrier signal determines the number of switching events per oscillation period. Fig. 5.35 shows the switching states of S_1 and the current curves of I_L at different PWM frequencies. At $f = 100$ Hz, the current curve follows the target signal poorly. One can see here that the coil is almost completely charged and discharged. If the frequency is increased ($f = 500$ Hz), the course of the current I_L corresponds much better to the target signal. There are also considerably more switching events. The coil is also no longer completely charged and discharged. Analogous to the buck converter, the deviation between the target current and the actual current scales with $1/f$—that is to say, we want to halve the inductance for the same deviation by doubling the frequency.

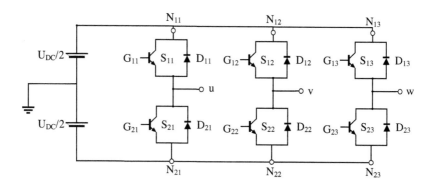

Fig. 5.36 Interconnection of three half-bridges to form a three-phase frequency inverter. Each of the half bridges controls one of the three phases u_2, v_2, and w_2. They share a common DC link. A three-phase rectifier serves as the input. This transforms the three input phases u_1, v_1, and w_1 from AC to DC and thus charges the intermediate circuit.

If we want to build a three-phase inverter, we no longer have to connect one half-bridge, but three. This circuit is shown in Fig. 5.36. Each half-bridge controls one of the three output phases u_2, v_2, and w_2. The DC-link is used by all three half-bridges. The control of the individual phases is realized by individual PWM signals. There is one PWM each, with a given target current, whose switching signals are then sent to the switches.

By connecting the three half-bridges, we have built a DC/AC inverter. However, our goal is to build a unidirectional frequency converter that converts AC to AC. We can achieve this by first rectifying the AC current. Therefore, in Fig. 5.36, a three-phase, passive rectifier is connected upstream. This converts the three AC phases u_1, v_1, and w_1 into DC, which charges the DC link. From this intermediate circuit, the direct current is then converted into the desired target current. An AC/AC inverter that is not based on a transformer therefore always consists of two system components: an AC/DC converter and a DC/AC converter.

Exercise 5.10 Four requirements have been identified for the inverter. Are these requirements met by the solution in Fig. 5.36 or not?

Solution: uni. Inverter 1 required an AC of any frequency and amplitude to be converted into another AC of a different frequency and amplitude. The conversion of the AC into AC is in fact independent of the frequency. The output frequency can also be set as desired. So this part of the requirement is fulfilled.

However, the DC link voltage depends on the amplitude of the input AC current. As a result, the amplitude of the output current is limited. The output voltage of the frequency inverter with a passive rectifier as input stage can depend on the input voltage. If we wanted to change this, a boost converter would have to be integrated. This part of the requirement is therefore only met to a limited extent.

uni. Inverter 2 is a requirement concerning the quality of the output signal. As we have seen in Fig. 5.35, the quality of the output current depends on the pulse-width frequency and the output inductance. These must therefore be selected in such a way that the requirements for the quality of the output current are also met. In principle, this requirement can be met. However, compliance has an effect on the overall design of the frequency inverter.

Since the three output phases are controlled by three independent PWM generators, **uni. Inverter 3** can be fulfilled. Each of the three phases is described by its own objective function. The respective PWM controls the switches according to this target function. This requirement is fulfilled.

uni. Inverter 4 is a requirement for the input stage, but is also fulfilled. However, it may have an impact on the output current characteristics of the inverter. This is because if the inverter requires a stable DC link. This DC link needs to provide sufficient power at a sufficiently high voltage. However, this requirement is met.

With the introduction of the half-bridge and semiconductor switches, we now have all the elements together to convert different forms of electrical energy into each other. There are a variety of circuit topologies that realize different conversion stages. For a concrete realization, it is worthwhile to evaluate at least some of the circuits in question in more detail.

In the next section, we will now look at the question of how kinetic energy can be converted into electrical energy and which system components are needed here.

5.2.6 Electrical drive technology: How motion becomes electrical power and vice versa

So far, we have dealt with the question of how we can convert different forms of electrical power into each other. But in Chapter 2 we have already seen that the transformation of electrical energy into kinetic energy and vice versa is an essential part of storage systems. In this chapter, we want to address the question of which system components we need to implement this transformation.

The core of this transformation are electric drives. Electric drives are divided into two large groups (Fig. 5.37): DC motors and AC motors. The difference is the type of excitation. DC motors are excited by a direct current. AC motors require an alternating current for excitation.

For DC motors, the energy source of the rotor field is used as a further differentiating feature. If, for example, the motor has a rotor equipped with a permanent magnet that generates its own magnetic field, it is referred to as a self-excited DC motor. Alternatively, the rotor magnetic field can also be generated by an electromagnet, in which case it is referred to as an excited DC motor.

With AC motors, a distinction is made as to whether the rotor, the rotating axis of the motor, rotates synchronously with the AC field or asynchronously. In asynchronous motors, the rotational speed of the AC field and the rotor axis differ. In synchronous motors, both are the same. Here, too, there are further subcategories that describe the internal structure of the motors (Kim, 2017).

To understand how an electric motor works, let's look at a realization of a DC motor. We start with a simple experiment. In Fig. 5.38 we have two magnets. The magnetic

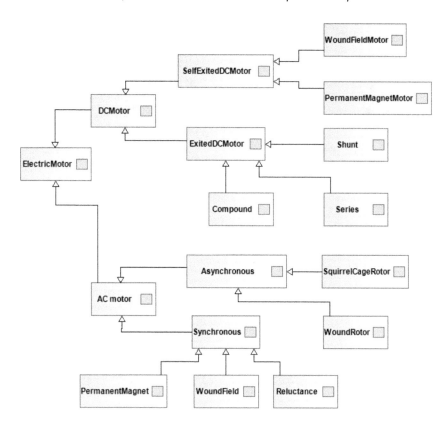

Fig. 5.37 Classification of electric drives. A distinction is made between two main groups: DC and AC drives. BLDC drives are a hybrid group that contain elements of both machines.

flux B is formed between the two bar magnets. It points from the magnetic north pole to the magnetic south pole.

We now insert a rotatable conductor winding with an inductance L between these two bar magnets. If a direct current I_r flows through the conductor loop, we notice that the conductor loop starts to align itself in such a way that the electromagnetic north pole is close to the magnetic south pole and the electromagnetic south pole is close to the magnetic north pole. The force acting on the conductor loop is proportional to the force of the magnetic flux B. However, we also find that we can increase the force if we increase the current I_r. If we double the current, we double the force. We can observe the same if the inductance L of the conductor loop is doubled. The following applies to the force F acting on the conductor loop:

$$F = B \cdot L \cdot I_r. \tag{5.54}$$

We now refine our experiment by measuring the force acting on the conductor loops. We do this by applying a torque to the conductor loop, which ensures that the loop no longer rotates. This torque must be as great as the torque produced by the force. We find that there is an angular dependence. The torque is greatest when the

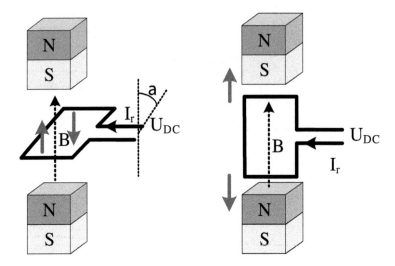

Fig. 5.38 How a DC motor works. A conductor winding is inserted between two magnets. When a current flows through the conductor, the conductor winding aligns itself so that the electric north pole is close to the magnetic south pole and vice versa.

conductor loop is oriented transversely between the two magnets and is smallest when the conductor loop is oriented parallel between the two magnets. If α describes the orientation of the loop, the formula for the torque is:

$$T = B \cdot L \cdot I_r \cos \alpha. \tag{5.55}$$

Exercise 5.11 The angles $\alpha = 0°$ and $\alpha = 180°$ represent equilibrium states. In both positions the torque disappears. How do these two equilibrium states behave with respect to small disturbances? Are both positions stable or unstable equilibrium states?

Solution: To determine the stability of the position of equilibrium, let us first determine the first derivative of the torque as a function of α:

$$\frac{d}{d\alpha}T = B \cdot L \cdot I_r \frac{d}{d\alpha} \cos \alpha = B \cdot L \cdot I_r \sin \alpha.$$

For $\alpha = 0$ the following holds:

$$\frac{d}{d\alpha}T\big|_{\alpha=0ř} = B \cdot L \cdot I_r \sin 0ř = B \cdot L \cdot I_r > 0.$$

The first derivative is greater than zero, i.e. the rest position $\alpha = 0$ is unstable. A small perturbation is amplified. What is the situation with $\alpha = 180°$? Here:

$$\frac{d}{d\alpha}T\big|_{\alpha=180ř} = B \cdot L \cdot I_r \sin 180ř = -B \cdot L \cdot I_r > 0.$$

Here the first derivative is negative, so the equilibrium $\alpha = 180°$ is stable. This means that if we introduce the conductor loop with $\alpha = 0°$, it will remain there until a small disturbance brings it out of the equilibrium position. After that, the conductor loop rotates and settles at $\alpha = 180°$. At this point, even a disturbance can no longer remove it from its rest position.

How can we realize a motor with a conductor loop introduced into a magnetic field? Since the rest position $\alpha = 180°$ is stable, the rotor would turn once by $180°$ and then remain at rest. The reason why this rest position is stable is that at $\alpha = 180°$ the electromagnetic south pole lies directly at the magnetic north pole and reverses. If we want to make this position unstable, all we have to do is change the polarity of the magnet or the electromagnetic field. So the solution is that we change the direction of the current and change the polarity of the DC voltage on the conductor loop $I_r \rightarrow -I_r$, and so $\alpha = 180°$ becomes unstable, while $\alpha = 0°$ becomes stable.

We can realize this switching mechanically, via a commutator, a sliding contact that touches the conductor loop. If the loop rotates, the contact surfaces also rotate and the polarity reverses. In this way, the loop can be permanently set in rotation. But instead of using a mechanical contact, we can also use electronic switches to operate the motor. To do this, we connect our DC motor between two half-bridges.

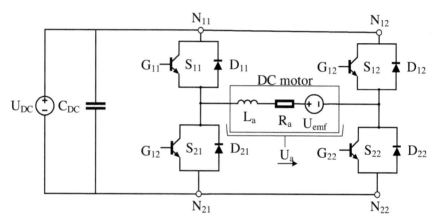

Fig. 5.39 Connection of a DC motor driven via an H-bridge. Either the switches S_{11} and S_{22} or S_{12} and S_{21} are switched on, thus reversing the polarity.

In Fig. 5.39 this circuit is shown. The motor is connected between an H-bridge: one terminal is connected to the left half-bridge, the other to the right half-bridge. If S_{11} and S_{22} are switched on, the following applies:

$$U_a = -U_{DC} = -\left(L_a \frac{d}{dt}I_a + R_a I_a + U_{emf}\right). \qquad (5.56)$$

If, on the other hand, S_{12} and S_{12} are switched on, the following applies:

$$U_a = U_{DC} = L_a \frac{d}{dt}I_a + R_a I_a + U_{emf}. \qquad (5.57)$$

In a DC motor, the conductor loop is called the armature winding, so the subindex a is used in the equation for those quantities that are associated with the conductor loop: the inductance of the conductor loop L_a and the internal resistance L_a, as well as the voltage on the conductor loop U_a and the current flowing through the conductor loop I_a. Since the moving conductor loop always generates a countervoltage via the electromagnetic force, this is also drawn in the circuit diagram as an additional voltage source, U_{emf}.

U_{emf} is called the back electromagnetic field (EMF). We can measure it by applying a torque to the conductor loop and thus simply rotating the conductor loop between the magnets. We find that a voltage is measured at the terminals of the motor. This voltage is proportional to the strength of the magnetic flux of the two magnets B and to the speed of rotation ω_m:

$$U_{emf} = k_{emf} \cdot B \cdot \omega_m. \qquad (5.58)$$

where k_{emf} is the proportionality constant and is called the back EMF constant. Eqn (5.58) also shows us how we can generate electrical energy from a mechanical movement. All we have to do is turn the conductor loop, and we can use the voltage to make a current flow. The motor then works as a generator.

Let's look at the torque eqn (5.55). The torque depends on the angle, so the power we can extract from the motor depends on its rotor position. This is not a good property for an electric motor. Suppose we have a vehicle that wants to start up a slope and is driven by a DC motor that has only one conductor loop. The question of whether we could drive our vehicle up the hill would then depend on how the rotor of the motor is positioned. In the worst case, it would not move at all. To solve this problem, we could add a conductor loop that is twisted by 90°. Then one conductor loop would generate torque when the other conductor loop was in its rest position. This would improve the situation, but we would still have a strong angular dependence. So we add more conductor loops, and in this way, we get an almost constant torque. Therefore, DC motors are usually equipped with extensive armature winding so that the angular dependence is as small as possible.

In addition to DC motors, there is a second group of motors known as the AC motors. These are divided into two subcategories: synchronous and asynchronous motor. Let us look at the induction motor, an asynchronous motor, as another example. In contrast to DC motors, induction motors do not need permanent magnets. This is why they are becoming more widespread in drive technology.

Again, we start with a thought experiment (Fig. 5.40). We take a horseshoe-shaped ferrite core, which we have wrapped with a wire. Between the two poles we insert a rotatable conductor winding. We call the horseshoe the stator. We call the conductor winding the rotor. The current that passes through the wire in the stator is the stator current I_s. If this is constant in time, a magnetic flux B_s builds up between the two poles of the horseshoe. The following then applies to the magnetic field in the stator:

$$B_s = \frac{L}{N \cdot A} I_s, \qquad (5.59)$$

where N is the number of turns of the wire around the ferrite core and A is the cross-sectional area of the ferrite core. The effects of the magnetic flux B_s on the rotor depend only on the material of the rotor if the flux is constant. If the rotor is made of a magnetic material, the conductor winding is aligned horizontally from $\theta = 0$.

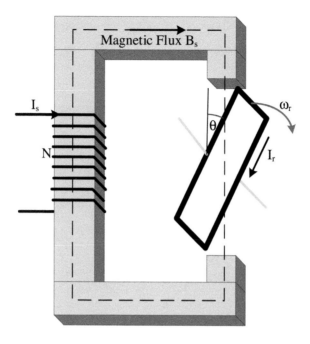

Fig. 5.40 The principle of operation of an induction motor. In this case, the stator current I_s is an AC. This generates a time-varying magnetic flux. A closed conductor winding is introduced into this magnetic flux. The varying magnetic flux induces a current inside the winding, and also generates a magnetic flux. The resulting forces rotate the winding with an angular velocity ω_r.

The behaviour is different if we generate an AC in the stator instead of a direct current. $I_s = \hat{I}_s \cos(2\pi \cdot f \cdot t)$. generates an alternating voltage in the stator, which produces a time-varying magnetic flux B_s:

$$U_s = L\frac{\mathrm{d}}{\mathrm{d}t}I_s = NA\frac{\mathrm{d}}{\mathrm{d}t}B_s. \tag{5.60}$$

Since the conductor winding is in the time-varying magnetic field, a magnetic field B_r, a current I_r, and a voltage are also induced in the conductor winding:

$$U_r = L\frac{\mathrm{d}}{\mathrm{d}t}I_r = NA\frac{\mathrm{d}}{\mathrm{d}t}B_r = NA\frac{\mathrm{d}}{\mathrm{d}t}B_s \sin\theta. \tag{5.61}$$

We now notice that the conductor winding starts to rotate. Since the induced current generates I_r, B_r is no longer aligned equally with B_s and thus a torque is created in total. An induction motor is therefore nothing more than a rotor consisting of closed conductor windings that are magnetically excited via an alternating field.

The induction motor in Fig. 5.40 has a disadvantage: we cannot rotate the exciting magnetic field so that it optimally excites the flux in the rotor. It would be better if we could rotate the magnetic field of the stator. If we had more than one stator field and could combine them, we could realize more than just the one alignment. In Fig. 5.41

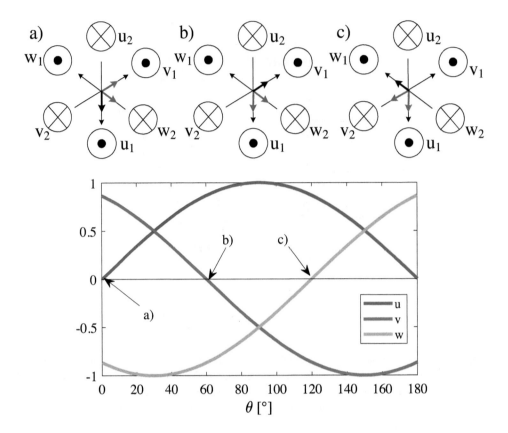

Fig. 5.41 With three pole pairs u, v, w shifted by $120°$ each, it is possible to realize any alignment. a), b), and c) show examples of the resulting alignment of the magnetic field. The grey arrows mark the direction of the magnetic field in u, v, w, and the black arrow marks the sum field.

we have realized a combination of three pole pairs. The pole pairs u, v, and w are shifted by $120°$ each. We can now adjust the magnetic field in each pole pair so that the overlapping magnetic field points in a desired direction. Three examples are given in Fig. 5.41. To align the magnetic field to $\theta = 0°$, we need to set $u = 0$, $v = -w$. For an angle of $\theta = 120°$, $u = -w$ and $v = 0$ are valid. If we want to set $\theta = 240°$, we have to set $w = 0$ and $u = -v$ (Boldea and Nasar, 2016; De Doncker, Pulle, and Veltman, 2020).

We can now generate an AC at each pair of poles. Let these AC be offset in time by 120:

$$B_{us} = \hat{B}_{us} \sin(\omega_s t) \tag{5.62}$$

$$B_{vs} = \hat{B}_{vs} \sin(\omega_s t - 120°) \tag{5.63}$$

$$B_{ws} = \hat{B}_{ws} \sin(\omega_s t - 240°). \tag{5.64}$$

Here, $\hat{B}_{\mathrm{us}}$, $\hat{B}_{\mathrm{vs}}$, and $\hat{B}_{\mathrm{ws}}$ are vectors in the direction of one of the three pole pairs. For the sum of these vectors, the following holds:

$$\hat{B}_{\mathrm{us}} + \hat{B}_{\mathrm{vs}} + \hat{B}_{\mathrm{ws}} = 0. \tag{5.65}$$

The resulting magnetic flux equals the superposition of the three magnetic fluxes:

$$B_{\mathrm{s}} = B_{\mathrm{us}} + B_{\mathrm{vs}} + B_{\mathrm{ws}}. \tag{5.66}$$

We generate a rotating magnetic field through modulation. Let us now assume that the conductor winding is already rotating at a speed of w_r. The axis of the magnetic field B_{s} rotates with a speed w_s. If both velocities are equal, the following happens: since the magnetic field of the loop and the magnetic field of the rotor are identical, there is no longer any induced voltage in the conductor winding. If no voltage is induced, no more current flows and the field in the stator collapses. To avoid this, we need to ensure that the relative speed w_{sr} between rotor and stator is non-zero.

$$w_{\mathrm{sr}} = w_{\mathrm{s}} - w_{\mathrm{r}} = sw_{\mathrm{s}}, \tag{5.67}$$

where s is referred to as slip. If $w_{\mathrm{sr}} > 0$ or $s > 0$, the rotor runs behind the exciting magnetic field. The motor is in motor operation. If $w_{\mathrm{sr}} < 0$ or $s < 0$, the rotor is faster than the exciting magnetic field, which decelerates the motor. The result is that the motor operates in regenerative mode. Kinetic energy is converted into electrical energy.

An induction motor consists of three pole pairs, or $3n$ pole pairs, arranged as in Fig. 5.41. Instead of one conductor winding, the rotor consists of several conductor windings, so that the torque has a lower angular dependence (5.42).

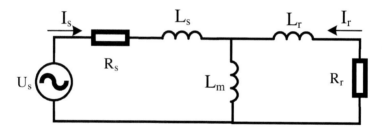

Fig. 5.42 Equivalent circuit diagram of an induction motor. The stator is described by a resistor R_{s} and an inductance L_{s}. The rotor is described by an inductance L_{r} and a resistance R_{r}. The inductance L_{m} describes the leakage inductance of the air gap that exists between the rotor and the stator.

We can interpret the induction motor as two circuits separated by a transformer. The stator circuit consists of an AC voltage source U_{s} to which a resistor R_{s} and an inductor L_{s} are connected to the primary side of a transformer. The rotor circuit consists of a resistor and an inductance L_{r}, which short-circuits the secondary side of the transformer. Unlike a classical transformer, the shorted secondary side, the rotor, can rotate. There is a small air gap. This air gap creates a leakage inductance, L_m, which we have to take into account (Cirrincione, Pucci, and Vitale, 2017; Neapolitan and Nam, 2018).

To represent the behaviour of induction motors, we need to give the rotor resistance a slip dependence. The following equation for the slip results from eqn (5.67):

$$s = \frac{\omega_s - \omega_r}{\omega_s}. \tag{5.68}$$

At $s \to 0$, the rotor and the exciting stator field run the same. The consequence is that no current is induced. This is equivalent to an open circuit or an infinitely large rotor resistance. At $s \to 1$, $\omega_t \to 0$, and the rotor does not move. The transferred energy is therefore completely converted into heat and not into heat plus kinetic energy. We can describe this behaviour by giving the resistance R_r a reciprocal slip dependence—that is,

$$R_r = \frac{\tilde{R}_r}{s}. \tag{5.69}$$

Since we are working with time-varying currents and voltages, we choose the complex notation with $j = \sqrt{-1}$—that is,

$$\mathbf{U}_{s,r} = \hat{U}_{s,r} e^{j\omega_{s,r} t} \tag{5.70}$$

$$\mathbf{I}_{s,r} = \hat{I}_{s,r} e^{j\omega_{s,r} t}. \tag{5.71}$$

If we apply the mesh rule, we get two equations for the voltage for the stator mesh and the rotor mesh:

$$\mathbf{U}_s = (R_s + j\omega_s (L_s + L_m)) \mathbf{I}_s + j\omega_s L_m \mathbf{I}_r \tag{5.72}$$

$$0 = (R_r + j\omega_s (L_r + L_m)) \mathbf{I}_r + j\omega_s L_m \mathbf{I}_s. \tag{5.73}$$

In doing so, we use that $\frac{d}{dt}\mathbf{I}_{s,r} = j\,\omega_{s,r}\mathbf{I}_{s,r}$.

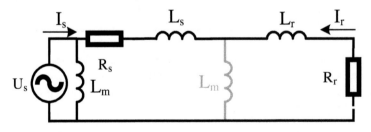

Fig. 5.43 Simplification of the equivalent circuit shown in Fig. 5.42. Since the leakage inductance is small in induction motors, we neglect its effect on the relationship between the motor currents. It is virtually inserted between the voltage source and the windings.

Our goal is to obtain a relationship between electrical power and mechanical power. The only term that consumes active power in our model of an induction motor is the resistor—in other words, the active power transmitted across the air gap P_m is consumed in the resistor. Part of the power is consumed in the form of copper lost, and the rest corresponds to the mechanical power. Since we are working with a three-phase motor, we have to sum up the power over all three phases. The following therefore applies:

$$P_m = 3|\mathbf{I_r}|^2 \frac{R_r}{s} \tag{5.74}$$

$$= 3|\mathbf{I_r}|^2 R_r \frac{s + (1 - s)}{s} \tag{5.75}$$

$$= \underbrace{3|\mathbf{I_r}|^2 R_r}_{\text{rotor copper lost}} + \underbrace{3|\mathbf{I_r}|^2 R_r \frac{1 - s}{s}}_{\text{mechanical power}}. \tag{5.76}$$

In an induction motor, the air gap between the rotor and the stator is kept as small as possible. It must be physically present; otherwise, the rotor cannot rotate and the power transmission via the magnetic field cannot take place, but at the same time one wants to keep the energy that is stored in the leakage inductance as small as possible, since this power cannot be used. So if we are interested in power transmission, we can neglect the leakage inductance, and we can virtually shift it between the voltage source and the stator winding (Fig. 5.43). Thus, for this simplified stator–rotor mesh, the following applies:

$$\mathbf{U_s} = \left(\left(R_s + \frac{R_r}{s} \right) + j\omega_s (L_s + L_r) \right) \mathbf{I_r}. \tag{5.77}$$

Exercise 5.12 With eqn (5.77) we can determine the mechanical power that is transmitted, by neglecting the effects of the air gap. How large is this?
Solution: We first form the absolute value of eqn (5.77):

$$|\mathbf{U_s}|^2 = \left(\left(R_s + \frac{R_r}{s} \right)^2 + \omega_s^2 (L_s + L_r)^2 \right) |\mathbf{I_r}|^2.$$

This results in the magnitude square of the stator current:

$$|\mathbf{I_r}|^2 = \frac{|\mathbf{U_s}|^2}{\left(\left(R_s + \frac{R_r}{s} \right)^2 + \omega_s^2 (L_s + L_r)^2 \right)}.$$

Since we have neglected the losses via the leakage inductance, the total power of the stator is converted into rotor copper lost and mechanical power. The following then applies to the mechanical power:

$$P_{\text{mech}} = 3|\mathbf{I}_{\text{r}}|^2 R_{\text{r}}$$

$$= 3 \frac{|\mathbf{U}_{\text{s}}|^2}{\left(\left(R_{\text{s}} + \frac{R_{\text{r}}}{s}\right)^2 + \omega_{\text{s}}^2 (L_{\text{s}} + L_{\text{r}})^2\right)}.$$

In Exercise 5.12, we have now calculated an expression for the mechanical power generated. In practice, however, we are more interested in the torque. The torque is given by:

$$T = \frac{P_{\text{mech}}}{\omega_r} = \frac{P_{\text{mech}}}{s\omega_s} \tag{5.78}$$

$$= 3 \frac{|\mathbf{U}_{\text{s}}|^2}{s\omega_s \left(\left(R_{\text{s}} + \frac{R_{\text{r}}}{s}\right)^2 + \omega_{\text{s}}^2 (L_{\text{s}} + L_{\text{r}})^2\right)}. \tag{5.79}$$

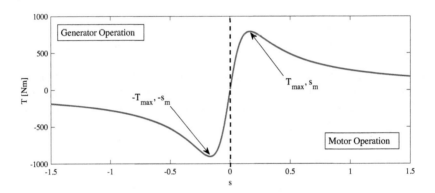

Fig. 5.44 Torque as a function of slip ($\mathbf{U}_{\text{s}} = 650$ V, $\omega_{\text{s}} = 500$, $L_{\text{s}} = 1$ mH, $L_{\text{r}} = 2$ mH, $R_{\text{r}} = 0.25$ Ω, $R_{\text{s}} = 0.1$ Ω).

In Fig. 5.44, the torque is shown as a function of slip. At $s < 0$, we are in regenerative operation. At $s > 0$, on the other hand, we are in motor operation. We see that at $s = 0$, there is no torque. In this case, the stator and rotor fields are parallel, and no current is induced.

Suppose we start with a slip of $s = 0$ and now begin to increase it slowly. We will notice that the torque becomes stronger and stronger. The more slip, the more torque. This corresponds to the linear increase of the torque near $s = 0$. However, if we keep increasing the slip, at some stage we will reach the point where the rotor can no longer follow the field so quickly. At $s = \pm s_{\text{m}}$, the torque reaches its maximum, $T = \pm T_{\text{max}}$.

If we want to lift a load with an induction motor, we have to make sure that we always work in the interval between $\pm s_{\text{m}}$. The reason for this is easy to understand. Suppose we want to lift a load. The weight force generates a torque T_{L} on the rotor. For the load to be lifted, the torque T generated must be greater—that is, $T > T_{\text{L}}$.

If we are between $\pm s_m$, the torque will be greater if we increase the slip. So we will increase the slip until $T > T_L$ is reached.

Now we want to keep the load in a certain position. This is $T = T_L$. If we now increase the weight, we have to increase the slip so that the equilibrium state is reached. If we decrease the weight, we have to make the slip smaller.

Now we increase the weight of the load so that $T_L > T_{max}$. We would increase the slip again because we always do this when the torque of the load is higher, but then we would be beyond s_m. In this case, an increase actually produces a decrease in torque. The consequence is that the difference between T_L and T becomes even greater. We would therefore increase the slip even more and only make the situation worse.

For this reason, it is necessary to stay in the range between $\pm s_m$ when operating the motor. This is a special feature of induction motors. Other types of motors do not have this problem. In the DC motor we discussed earlier, the torque depended solely on the current applied.

Induction motors are widely used. Unlike DC motors, they do not need permanent magnets. Since they are operated via a three-phase AC, they can ideally be connected directly to the mains and operated at the mains frequency. Alternatively, they are driven via an inverter, which allows the exciting AC voltage to be controlled in a very targeted manner.

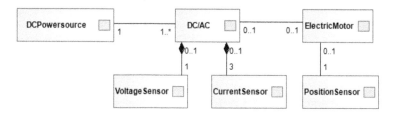

Fig. 5.45 System components for an electric powertrain.

DC motors and induction motors are two examples of the different types of electric motors that can be used to convert electrical energy into mechanical energy. Of course, the motors alone are not enough; other system components are needed. In Fig. 5.45, we have shown the components for an electric powertrain. This consists of a DC power source that supplies the motor with energy, the electric motor, and power electronics. In this example, we are dealing with a DC/AC inverter, as we are working with an AC machine here. In order to operate the motor, the inverter needs three physical quantities: the DC voltage, the current of the three phases u, v, and w, and the position of the motor. The latter is optional, because there are a number of methods that determine the motor position indirectly.

5.3 Description of power transfer units from a systems engineering point of view

In the previous sections we have learned about the different converters and their system components. We want to conclude this chapter by showing how we can describe these components in a simplified way from a system engineering point of view. This simplified description summarizes the essential characteristics of the design of a storage system. At the same time, these characteristics represent the technical requirements that a certain converter must have in order to be used at all for our energy storage system.

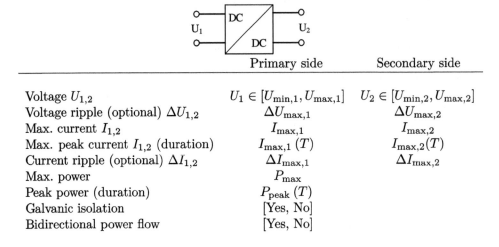

	Primary side	Secondary side
Voltage $U_{1,2}$	$U_1 \in [U_{\min,1}, U_{\max,1}]$	$U_2 \in [U_{\min,2}, U_{\max,2}]$
Voltage ripple (optional) $\Delta U_{1,2}$	$\Delta U_{\max,1}$	$\Delta U_{\max,2}$
Max. current $I_{1,2}$	$I_{\max,1}$	$I_{\max,2}$
Max. peak current $I_{1,2}$ (duration)	$I_{\max,1}(T)$	$I_{\max,2}(T)$
Current ripple (optional) $\Delta I_{1,2}$	$\Delta I_{\max,1}$	$\Delta I_{\max,2}$
Max. power	$P_{\max}$	
Peak power (duration)	$P_{\text{peak}}(T)$	
Galvanic isolation	[Yes, No]	
Bidirectional power flow	[Yes, No]	

Fig. 5.46 Technical specification of a DC/DC converter.

We start with the DC/DC converter (Fig. 5.46). A DC/DC converter converts AC from one voltage level to another. We therefore have to define the allowed voltage interval $[U_{\min}, U_{\max}]$ on the primary as well as on the secondary side. Sometimes a nominal voltage is specified instead of an interval.

It may happen that the components connected to the DC/DC converter are sensitive to high-frequency voltage ripples. In this case, the level of the permitted voltage ripple $\Delta U_{1,2}$ must be specified on the respective connection side.

In addition to the voltage, the maximum current $I_{\max}$ must also be specified. The maximum current represents the maximum continuous current load—that is to say, the unit is designed in such a way that it can be operated with this current over a longer period of time. Additionally, a peak current is also specified—that is, an overload which can only be applied for a short period of time. The peak current is always defined with a time interval, which specifies how long the unit can handle this higher current without damaging itself. Furthermore, it can also be the case that a current ripple $\Delta I_{1,2}$ must be specified.

It may happen that not the maximum current and a peak current are specified, but a maximum power and a peak power. The current required for this can be determined via the relationship $P = U \cdot I$. If information about the maximum current is missing but a maximum power is specified, it must be clarified whether this power is required in the entire voltage range.

Exercise 5.13 A DC/DC converter has a voltage interval of $U_1 \in [200 \text{ V}_{DC}, 300 \text{ V}_{DC}]$ on the primary side. On the secondary side, the voltage interval is $U_2 \in [600 \text{ V}_{DC}, 650 \text{ V}_{DC}]$. A maximum power of $P_{max} = 1$ kW is to be transmitted. The current on the primary side should not exceed 4.5 A. What is the maximum current on the primary and secondary sides? What is the maximum power that can be transmitted at $U_1 = 200 \text{ V}_{DC}$? Up to what voltage can the expected power still be drawn?

Solution: For the secondary side, the maximum current is given by:

$$I_{max,2} = \frac{P_{max}}{U_{min,2}} = \frac{1,000 \text{ W}}{600 \text{ V}_{DC}} = 1.67 \text{ A}.$$

Since the current must increase when the power is constant but the voltage decreases, it is sufficient to use the lower voltage value for this calculation.

Since the maximum current on the primary side is given as 4.5 A, we can determine the maximum power at the smallest voltage:

$$P_{max} \text{ at } U_1 = 200 \text{ V}_{DC} = 200 \text{ V}_{DC} \cdot 4.5 \text{ A} = 900 \text{ W}.$$

The voltage from which we can transfer the full power on the primary side can be calculated as follows:

$$U_{min} = \frac{1,000 \text{ W}}{4.5 \text{ A}} = 222.22 \text{ V}_{DC}.$$

Since both sides must be able to transport the power, these results mean that the DC/DC converter can transmit a power of 900 W at $U_{min,1} = 200 \text{ V}_{DC}$. As the voltage increases, the maximum power increases linearly and reaches the target level of 1,000 W at 222.22 V_{DC}.

There are two other properties that can be specified for a DC/DC converter. One is whether the DC/DC converter provides galvanic isolation between the primary and secondary sides. This is not the case with all DC/DC converters. As a rule, only DC/DC converters that operate via an internal transformer are galvanically isolated. The galvanic isolation is required as a safety measure. It guarantees that no direct current flows unintentionally from the primary side to the secondary side.

Furthermore, it must be specified whether the DC/DC converter enables unidirectional or bidirectional power flow. With unidirectional power flow, the power can only be transported in one direction. With bidirectional power flow, transport is possible in both directions.

For system components that operate with AC, we must pay attention to how current and voltage are specified. Since current and voltage oscillate in an AC source, the transmitted power is usually not equal to the product of their amplitudes. Assume that the AC voltage is sinusoidal:

$$U(t) = \hat{U} \sin \omega t. \tag{5.80}$$

Here $\hat{U}$ is the amplitude of the AC voltage and ω is the frequency of the AC voltage. With the help of Ohm's law, the following applies to the power:

$$P(t) = \left(\hat{U} \sin \omega t \right) \left(\frac{\hat{U}}{R} \sin \omega t \right) = \frac{\hat{U}^2}{R} (\sin \omega t)^2. \tag{5.81}$$

The transmitted power is time-dependent and is transmitted in pulses. Within a period $T = \frac{2\pi}{\omega}$, a power of

$$\bar{P} = \frac{\hat{U}^2}{R} \int_{t=0}^{t=T} \sin \omega t^2 dt = \frac{1}{2} \cdot \frac{\hat{U}^2}{R} = \frac{U_{\text{eff}}^2}{R} \tag{5.82}$$

is transmitted. This corresponds to a voltage reduced by $1/\sqrt{2}$. When describing storage systems that are connected to AC networks, it is common to carry out the power considerations with the help of the effective values.

Exercise 5.14 A power system has an rms voltage of $U_{\text{eff}} = 230\,\text{V}_{\text{AC}}$. What is the amplitude $\hat{U}$ of the AC voltage? A resistor with $R = 19\,\Omega$ is connected to the current source. What is the rms value of the current?
    Solution:
$$\hat{U} = \sqrt{2} \cdot U_{\text{eff}} = \sqrt{2} \cdot 230\text{V}_{\text{AC}} = 325.7\,\text{V}_{\text{AC}}.$$

We use Ohm's law and use the calculated effective value:

$$I_{\text{eff}} = \frac{U_{\text{eff}}}{R} = \frac{230\,\text{V}_{\text{AC}}}{19\,\Omega} = 12.11\,\text{A}.$$

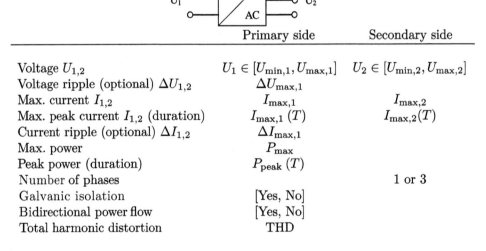

	Primary side	Secondary side
Voltage $U_{1,2}$	$U_1 \in [U_{\text{min},1}, U_{\text{max},1}]$	$U_2 \in [U_{\text{min},2}, U_{\text{max},2}]$
Voltage ripple (optional) $\Delta U_{1,2}$	$\Delta U_{\text{max},1}$	
Max. current $I_{1,2}$	$I_{\text{max},1}$	$I_{\text{max},2}$
Max. peak current $I_{1,2}$ (duration)	$I_{\text{max},1}\,(T)$	$I_{\text{max},2}(T)$
Current ripple (optional) $\Delta I_{1,2}$	$\Delta I_{\text{max},1}$	
Max. power	P_{max}	
Peak power (duration)	$P_{\text{peak}}\,(T)$	
Number of phases		1 or 3
Galvanic isolation	[Yes, No]	
Bidirectional power flow	[Yes, No]	
Total harmonic distortion	THD	

Fig. 5.47 Technical specification of a DC/AC inverter.

Let us first consider the DC/AC inverter (Fig. 5.47). For the DC side, the same requirements apply as for a DC/DC converter. On the AC side, rms values of comparable quantities are used. However, no information about the current or voltage ripple is required on the AC side.

The counterpart to the requirements for current and voltage ripple are requirements for the shape of the AC signal on the AC side. The total harmonic distortion value

(THD) is representative of this. The THD is calculated from the deviation of the effective values. If U_{ref} and I_{ref} are the specified current and voltage curves and U, I are the actual current and voltage curves of the inverter, then the THD results in:

$$\text{THD}_U = \frac{\sqrt{\hat{U} - \hat{U}_{ref}}}{\hat{U}_{ref}} \tag{5.83}$$

$$\text{THD}_I = \frac{\sqrt{\hat{I} - \hat{I}_{ref}}}{\hat{I}_{ref}}. \tag{5.84}$$

For a DC/AC inverter, information can also be provided on the direction of the power flows and the galvanic isolation. In addition, there is the number of phases on the AC side. As a rule, you can choose between one and three phases. Only in motor applications is it sometimes the case that more than three phases are required.

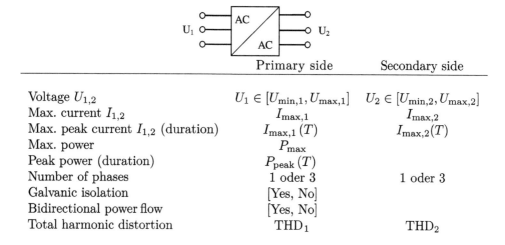

	Primary side	Secondary side
Voltage $U_{1,2}$	$U_1 \in [U_{min,1}, U_{max,1}]$	$U_2 \in [U_{min,2}, U_{max,2}]$
Max. current $I_{1,2}$	$I_{max,1}$	$I_{max,2}$
Max. peak current $I_{1,2}$ (duration)	$I_{max,1}(T)$	$I_{max,2}(T)$
Max. power	P_{max}	
Peak power (duration)	$P_{peak}(T)$	
Number of phases	1 oder 3	1 oder 3
Galvanic isolation	[Yes, No]	
Bidirectional power flow	[Yes, No]	
Total harmonic distortion	THD_1	THD_2

Fig. 5.48 Technical specification of an AC/AC inverter.

The specifications of an AC/AC inverter are similar to the specifications of the AC side of a DC/AC inverter (Fig. 5.48). However, it is possible that the current and voltage quality requirements occur on both the primary and secondary sides.

For the technical description of a motor, current, voltage, and electrical performance data are required (Fig. 5.49). There are no requirements for the voltage quality. The implementation of the requirements for the voltage quality of the feeding generator is done by the drive converter, which is essential for the operation of an electric motor.

In addition, there are requirements for the maximum torque and the maximum speed that the electric motor can operate. These values are based on the mechanical construction of the motor.

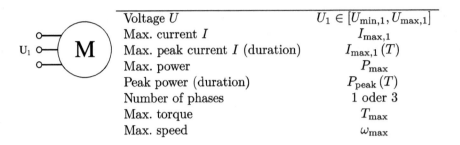

Voltage U	$U_1 \in [U_{\text{min},1}, U_{\text{max},1}]$
Max. current I	$I_{\text{max},1}$
Max. peak current I (duration)	$I_{\text{max},1}(T)$
Max. power	P_{max}
Peak power (duration)	$P_{\text{peak}}(T)$
Number of phases	1 oder 3
Max. torque	T_{max}
Max. speed	ω_{max}

Fig. 5.49 Technical specification of an electrical motor.

5.4 General requirements for storage systems that need electrical components

In Section 4.3, we presented the eight general requirements for a storage system. For storage systems that require electrical components, there are also a number of elementary requirements that will be described below.

One challenge in the realization of storage systems with electrical and electrochemical components are the different time scales within which the different processes take place. For example, the energy management system (EMS), which determines the charging and discharging strategy, works in time scales of seconds or minutes. If the EMS regulates too slowly, losses may be generated, since the residual load cannot be correctly regulated. However, a considerably faster control of the power flow may lead to a more expensive design without a compensating yield, since the energy content at smaller time scales is too low to generate additional yield. Electrochemical processes take place in time scales below seconds to microseconds. The control of power electronics operates at frequencies from 4 kHz to 200 kHz, and yet is slow compared to the time scale of electronic components, some of which operate in the nanosecond range. This leads to the first requirement for electrical storage (ES):

ES 1: AS A development engineer, I WANT to ensure that the system dynamics are controlled in the different time scales SO THAT the system components can react correctly to the technical and physical processes.

This generally formulated requirement is realized in systems by various components. There is not one system component that can act in all time scales simultaneously, so one distributes the responsibility among components that can act on the respective time scales. For example, it is common for the EMS not to take over the control of the power electronics or the battery, but to leave this function to the respective component and only issue instructions and retrieve state variables.

Analogous to the requirement in previous storage technologies, the transport system that transports the power must also be suitable for this. In thermal storage, for example, the pipe system had to withstand the pressure and temperature of the fluid, and the pumps had to be able to build up the required pressure.

In electrical systems, the relevant quantities are current and voltage. The components used must be able to carry the current—that is, they must not overheat or be damaged by the current. They must also be able to withstand the voltage load, i.e. there must be no short circuits or damage to the components. This results in two further requirements:

ES 2: AS A development engineer, I WANT to ensure that the current load on the components is not too high SO THAT there is not too much heating and thus damage to the components.

ES 3: AS A development engineer, I WANT to ensure that the voltage load on the components is not too high SO THAT there is no damage to the components because, for example, short circuits occur.

When designing electrical and electrochemical storage, these two requirements lead to conflicts. For example, the voltage level or current-carrying capacity of electrochemical cells is often not suitable for the planned application. Additional measures have to be taken or system components added to make the design possible.

In Section 5.2.5.2, it was shown that semiconductor elements do not have ideal switching behaviour, which leads to the generation of heat. Keeping these losses low is another requirement:

ES 4: AS A development engineer, I WANT to ensure that switching operations are as low-loss as possible SO THAT switching losses do not become too great.

There are four ways to fulfil ES 4. The simplest option is to avoid switching altogether. This strategy is used for a power electronic switchgear in the megawatt range. Here, the switching operations are reduced as much as possible, since each individual switching operation places a very high load on the components.

The second option is to use semiconductor switches that have lower switching losses. These include semiconductors that work with wide-gap bandwidth materials such as silicon carbite or gallium nitrite. These switches have a much faster switching behaviour and thus have significantly lower switching losses.

The third and fourth options are to perform the switching operation in a state where either no voltage or no current is applied. Various types of circuits allow zero-voltage or zero-current switching.

The electrical and electrochemical storage components cannot always be realized at the same voltage levels. This is due to the physical and chemical properties of the components. Electrochemical cells, for example, have relatively low voltage levels, which can only partially be brought to the voltage at which feeding into the distribution grid is possible through interconnections. This results in a final requirement:

ES 5 AS A User, I WANT to transfer power from one voltage level and between different types of electrical power SO THAT I can combine different loads, sources, and storage components.

This requirement also includes the transformation of DC into AC. For the design of storage systems, ES 5, in addition to ES 2 and ES 3, is another requirement that directly affects the system design. Since each conversion component entails losses, the number of converters in the design should be kept low. On the other hand, this limits the possible combinations of components and their design. Solving this contradiction is the challenge in the design of storage systems.

Electrical systems use a large number of components, whose properties are not to be explained in detail here, but which are nevertheless presented so that an under-

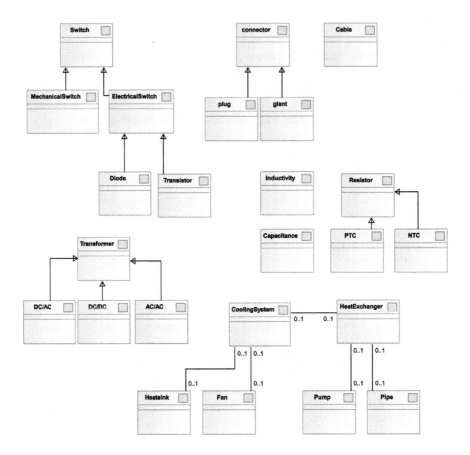

Fig. 5.50 Overview of different system components of electrical storage systems

standing of the system can be obtained. These components are shown in Fig. 5.50. Switches are needed to control the power flows, analogous to the valve and pump in mechanical or thermal storage. Roughly, a distinction can be made between mechanical and electrical switches. The mechanical switches are used for slow switching processes—that is, for switching the system on and off—and often as part of an emergency system ('emergency stop'). The electrical switches are used for the temporally fast and frequent switching operations. As a rule, they are semiconductor elements. The diode and the transistor are examples of other components here.

Capacitances and inductances are also used in electrical circuits. Capacitances—that is, capacitors—have the function of storing electrical power. In addition to the storage function of the storage system, they often also have a storage function resulting from the need for power flow control—that is, they are required for the application in different time scales and are therefore also dimensioned differently. The same applies to the inductors, the coils. These store electrical and magnetic power. In addition, they are also used as transformers. Resistors are used to meet the requirements for current and voltage tolerances. The counterparts to the pipe system are the cables.

Three types of cooling are used in electrical systems: firstly, heat is dissipated from the affected components via thermal conduction. This can be done by heat conduction via metal or via a liquid medium. The second mechanism works via air: the dissipated heat is transported to a heat sink with a large surface area, where it is transferred to the environment. The third variant is liquid cooling, where a suitable liquid is used instead of air.

5.5   Conclusion

In this chapter, we have dealt with the question of which system components we can use to transfer power between the different forms of energy. If we want to transfer power between two DC networks, we need a DC/DC converter. If, on the other hand, we want to convert DC into AC, we need a DC/AC inverter. To transfer power between different AC grids, we need an AC/AC inverter. With this, we have become acquainted with the three components with which we can transfer power between electrical grids. Furthermore, energy storage systems often require a conversion from mechanical to electrical energy. The electric motor is available to us for this purpose. We have learned about its basic functioning with the DC motor and the induction motor.

The required electronic components were introduced for the different types of conversion. The most important system components were:

- **The ohmic resistor:** Realizes a fixed relationship between current and voltage and also serves as a load.

- **The capacitor:** Serves us as a voltage buffer in which it temporarily stores charge carriers.

- **The coil:** Serves as a current storage and filters current peaks from a signal.

- **The transformer:** Can transform AC with one amplitude into AC with another amplitude. The ratio of the two amplitudes is determined by the winding ratio of the primary to the secondary coil.

- **The diode:** A semiconductor element that allows current to flow in only one direction.

- **The transistor:** A semiconductor switch with the help of which we can switch very large currents with a small control voltage.

In addition, we have learned about the most important requirements for electrical storage systems. The requirements ES 1 to ES 6 apply, in addition to the requirements G 1 to G 8 already described.

We now have all the tools together to be able to design and lay out electrical storage systems. Next, we will use these tools to design systems using different storage technologies. We start in the next chapter with mechanical storage technologies. These are based on the use and storage of kinetic or potential energy.

Chapter summary

- Power conversion technologies are needed to transform between different DC and AC voltage levels.
- Buck–boost converters or other DC/DC converters transform DC power between different DC voltage levels.
- Inverter technology is used to transform between different types of AC current.
- To transfer AC power to DC power, active or passive rectifiers are used.
- The transfer from DC power to AC power is done by an inverter.
- The basic elements for power conversions are the resistor, the capacitor, the inductance, the diode, and the transistor.
- Capacitors are voltage storage devices. They are able to store the voltage in terms of separated charge carriers.
- Inductance are current storage devices. They store electrical current by storing the motion of the charge carriers into an magnetic field.
- Power conversion devices use capacitor and inductance as filter and storage component. The power transfer is realized in terms of average values. Therefore, we have to investigate not only the average values of voltage and current, but also their variances, the so-called 'ripple'.
- Semiconductors like diodes and transistors are fast-switching devices.
- A diode allows current to pass in only one direction.
- With a transistor it is possible to switch a large voltage by inducing a small current.
- If the transformation of AC current does not need a change in the frequency, like most of the transformation in the power grid, a transformer can be used. The transformation ratio depends on the difference in the winding number of the wire on the primary and secondary sides.
- A more flexible way to transform AC current is the inverter, which consists of a rectifier and at least six switching cells. These cells are able to generate an AC current with different amplitude and frequency, as is needed for electric drive applications.
- An inverter is able to drive an electric motor to transform electric power into mechanical power or vice versa.
- Different types of electric motors exist. The two main categories are DC and AC motors.

6
Mechanical storage systems

In this chapter we start to investigate storage technologies. We begin with mechanical storage systems. Mechanical storage systems are the oldest forms of energy storage; many machines work with these storage technologies. For example, every mechanical watch has a balance wheel. This is a spring that has to be 'wound up' regularly. The tensioning process corresponds to charging. The discharge process then takes place via the relaxation of the spring.

Another example is the pendulum clock. In addition to the actual pendulum, which periodically converts kinetic energy into potential energy, there is often a counterweight in the clock, or there are weights that drive the clock. These weights must be periodically returned to their initial position. Here, too, energy is stored for a short period and discharged in small portions over a longer period.

Mechanical energy is stored either in the form of potential energy or as kinetic energy. When using kinetic energy, rotational movements are preferable. In the pendulum clock and the balance wheel, on the other hand, potential energy is stored: once through the use of gravity – the pendulum is pulled up and strives to move down again – and once through the use of the restoring force of a spring.

In this chapter, we deal with three forms of mechanical energy: potential energy, which is stored in a body when it is lifted up; the kinetic energy that is in a body when it moves; and the energy that is stored in a body when it is mechanically compressed, as is the case with a spring, for example.

For each of these forms of energy, we will investigate an example application. The chapter then ends with a more extensive example of a hybrid power plant, a power plant in which we want to combine pumped storage with flywheel storage (FWS).

However, we will first start with a consideration of the requirements that apply to mechanical storage systems in general.

6.1 Requirements for mechanical storage systems

In order to design a storage system that works with mechanical storage technologies, we need components that enable the conversion of mechanical energy into electrical energy. This must be possible in both directions. Therefore, the first requirement is:

MS 1: AS A user, I WANT to convert electrical energy into mechanical energy, or convert mechanical energy into electrical energy SO THAT I can store a surplus of electrical energy mechanically and also extract it as electrical energy when discharging.

Energy Storage Systems. Armin U. Schmiegel, Oxford University Press. © Armin U. Schmiegel (2023).
DOI: 10.1093/oso/9780192858009.003.0006

We have already learned about the system component that fulfils this requirement: the electric motor, which operates either in motor or generator mode.

However, the electric motor only fulfils the requirement MS 1 if there is rotational kinetic energy available. An electric motor cannot directly convert translatory motion. Therefore, we need additional system components that convert the stored kinetic or potential energy into rotational energy (Fig. 6.1).

MS 2 AS A user, I WANT to convert kinetic or potential energy into rotational energy SO THAT I can perform a conversion of energy using an electric motor.

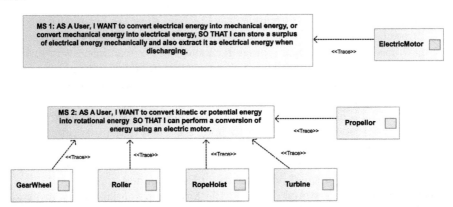

Fig. 6.1 General requirements for the use of mechanical storage systems and system components available for their compliance.

The conversion can be done via gears or rollers. In a crane, the load is moved vertically, but a roller operates the rope hoist. The winding and unwinding of the rope ensures that the linear movement is translated into a rotational movement.

If the medium is liquid or gaseous, turbines or propellers have proven to be efficient system components that convert the linear movement into a rotational movement.

6.2 Energy storage using potential energy part 1: Pumped storage power plants and other concepts

We want to build a simple storage system that can temporarily store electrical power using potential energy. For this we take a crane and a number of containers of different sizes (Fig. 6.2). The containers have a unit height of 2.6 m ≈ 8 1/2 feet. The smallest container has a weight of 30 t, the medium container weighs 60 t, and the largest 90 t.

For lifting and lowering, we use an electric motor equipped with a bidirectional frequency converter so it can take electrical energy from the mains to drive the motor. When the weight is lowered, the motor works in regenerative mode and the generated power can be fed into the grid.

Let's assume that two containers are standing next to each other. We want to stack the two containers on top of each other. During the lifting process, the force of gravity, $F_g = m\,g$, acts on the container ($g = 9.81\,\frac{m}{s^2}$). To calculate how much energy we have

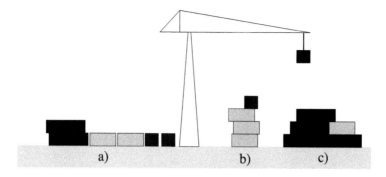

Fig. 6.2 A crane with a number of heavy containers can be used as a storage system.

to apply, we integrate the applied force over the entire lifting path. The potential energy is thus given by:

$$E_{\text{pot}} = \int_{s=0}^{s=h} F_g \, ds = \int_{s=0}^{s=h} m \cdot g \, ds = m \cdot g \cdot h. \tag{6.1}$$

Exercise 6.1 Suppose two containers are standing next to each other on the ground. One of the two containers has mass $m = 3$ t. The crane is connected to the mains: $U_{\text{AC}} = 440$ V. The crane will now to stack the two containers on top of each other. The lifting process should take 4 s. For simplicity, we assume that the electrical power required during the lifting process is the same. What is the electrical power and current required?

Solution: The potential energy is calculated from eqn (6.1):

$$E_{\text{pot}} = m \, g \, h = 30{,}000 \text{ kg } 9.81 \, \frac{\text{m}}{\text{s}^2} \, 2.6 \text{ m} = 765{,}180 \text{ Ws} = 212.55 \text{ Wh}.$$

This energy must be applied within 4 s in the form of electrical energy. Since we assume that the electrical power is applied continuously, this is calculated as:

$$P_{\text{el}} = \frac{E_{\text{pot}}}{4 \text{ s}} = \frac{765{,}180 \text{ Ws}}{4 \text{ s}} = 191{,}295 \text{ W},$$

so we need a power of about 191 kW to move the load in the 4 s. With an effective voltage of $U_{\text{AC}} = 440 \text{ V}_{\text{AC}}$, this corresponds to a current of:

$$I = \frac{P_{\text{el}}}{U} = \frac{191{,}295 \text{ W}}{440 \text{ V}_{\text{AC}}} = 434.476 \text{ A}.$$

The difference in height and the different masses allow us to achieve different storage states. These are, of course, limited by the possible combinations of the containers. In this example, a maximum of six levels can be stacked.

Exercise 6.2 What is the energy capacity of the configurations labelled a), b), and c) in Fig. 6.2? Reminder: $\Delta h = 2.6$ m ≈ 8 1/2 feet, $m_1 = 30$ t, $m_2 = 60$ t, and $m_3 = 90$ t.

Solution: Containers lying on the ground do not contain any potential energy, because we cannot lower the containers any further. For a), this results in an energy content of:

$$E_{\text{a}} = m_3 \cdot g \cdot \Delta h = 765{,}180 \text{ Ws} = 212.55 \text{ Wh}.$$

Suppose we have N containers at one level. m_n is the mass of each container. Then the total mass on this level equals:

$$m_i = \sum_{n=1}^{N} m_n.$$

The energy of the containers on this level is then:

$$E_i = g \cdot i \cdot \Delta h \cdot m_i.$$

Here i is the number of the level. The following therefore applies to the total energy content:

$$E = \sum_{i=1}^{I} E_i = g \cdot \Delta h \left(\sum_{i=1}^{I} i \, m_i \right),$$

where I represents the maximum number of levels.

With this formula it is easy to determine the energy content for b) and c):

$$E_b = g \cdot \Delta h \, (m_2 + 2m_2 + 3m_3)$$
$$= g \cdot 2.6 \text{ m} \, (3 \cdot 60{,}000 \text{ kg} + 3 \cdot 90{,}000 \text{ kg})$$
$$= 1.1478 \cdot 10^7 \text{ Ws} = 3.19 \text{ kWh}$$

$$E_c = g \cdot \Delta h \, ((m_2 + m_3) + 2 \cdot m_3)$$
$$= g \cdot 2.6 \text{ m} \, (150{,}000 \text{ kg} + 180{,}000 \text{ kg})$$
$$= 8.417 \cdot 10^6 \text{ Ws} = 2.34 \text{ kWh}.$$

Exercises 6.1 and 6.2 show us two limitations of this power plant. The storage capacity is limited by the number of containers per level. To be able to store ten kilowatt hours in configuration c), approximately four container rows would be needed. The loading process cannot be completed at any speed, as the containers are also not transported at any speed.

On the other hand, the charging and discharging power can be adjusted by speeding up the lifting process. If a higher charging performance is to be realized, it is basically sufficient to lift the container faster. But we see that there is a clear limitation here as well: we are not able to lift and lower the container as quickly as we want. There are physical limitations caused by the motor and the mechanics of the crane.

These limitations can be solved somewhat by considerably increasing the area and the number of containers and by having several cranes working in parallel.

From Section 2.3.3 we know that for stationary storage applications such as grid support and energy trading we need outputs in the megawatt range and energies in the megawatt hour range. To realize this with a crane storage system, we would have to install many cranes and containers on a large area. But there is another alternative: the pumped hydro storage plant.

Instead of working with containers, we work with a fluid, for example water. We can transport this water back and forth between two basins. We use pumps or turbines to convert the kinetic energy into electrical energy.

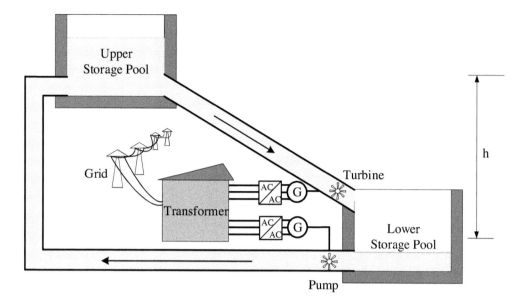

Fig. 6.3 Illustration of a pumped hydro storage plant. Two storage basins are connected to each other via a pipe system. Water can be transferred from the upper basin to the lower basin, thereby driving a turbine. Conversely, water can be pumped into the upper basin. A pump is activated in the process.

Fig. 6.3 shows the structure of a pumped storage plant. Instead of containers, water is used in a pumped hydro storage plant. This is stored in at least two storage basins. The water can flow from the upper storage basin to the lower storage basin via a pipe system. In this pipe system, there is a turbine that drives a generator. The generator is connected to a frequency converter, which is connected to the grid via a transformer. The upper and lower reservoirs are additionally equipped with another pipe system, which contains a pump. This pump is driven by the mains, a frequency converter, and an electric motor, and enables water to be pumped from the lower storage tank to the upper storage tank.

If the storage is to be charged, water is pumped from the lower storage basin to the upper storage basin via a pump. In the process, electrical energy is first converted into kinetic energy—the water flows up the hill. Once the water has reached the upper basin, the kinetic energy has been converted into potential energy.

To discharge, the process takes place in the opposite direction: the water flows down the mountain through the pipe system—that is, potential energy is converted into kinetic energy. This kinetic energy drives a turbine, which converts this energy into electrical energy via a generator.

The storage capacity results from the average height difference Δh between the two storage basins and the storage, the volume V, and the density of the water, ρ_{H_2O}:

$$\kappa = m \cdot g \cdot \Delta h = \rho_{H_2O} \cdot V \cdot g \cdot \Delta h. \tag{6.2}$$

The density of water at a temperature of 20ř is $\rho_{H_2O} = 0.9982067\,\frac{kg}{l}$. One litre of water therefore has approximately the weight of one kilogram.

The energy density of the storage depends on the gravitational constant g, the height difference Δh, and the density of the medium ρ:

$$\rho_{\kappa V} = \frac{\kappa}{V} = \rho_{H_2O} \cdot g \cdot \Delta h. \tag{6.3}$$

Assuming that the difference in elevation between the storage and reservoir basins is significantly greater than the depth of the storage basin, the volumetric flow rate alone determines the rate of loading and unloading:

$$\frac{d}{dt}\kappa = \rho_{H_2O} \cdot g \cdot \delta h \frac{\partial V}{\partial t}. \tag{6.4}$$

Thus, the charging and discharging rate is limited by the volume flow—that is, the pipe and the turbine. An increase in these rates can be achieved by adding more pipes. In order to increase the charging and discharging rate, it is possible, on the one hand, to increase the volume flow of a pipe. However, this also means that the generator, turbine, and pump have to be enlarged. Alternatively, a second, identical pipe, pumps, and a turbine system can be added.

Exercise 6.3 We consider a pumped storage plant that has a difference in height of $\Delta h = 400$ m. How high must the mass flow be so that a power of 1 MW can be withdrawn for a quarter of an hour? (Note: Mass flow means mass per time, i.e. $\dot{m} = \frac{d}{dt}m$).

Solution: To provide one megawatt of power for a quarter of an hour, we need an amount of energy of $\Delta E = \frac{1}{4}$ MWh. We can only realize this amount of energy by mass transport; the amount of water needed for this is:

$$\Delta m = \frac{\Delta E_{pot}}{g \cdot \Delta h}$$

$$= \frac{\frac{1}{4}\,\text{MWh}}{9.81\,\frac{m}{s^2} \cdot 400\,\text{m}}$$

$$= \frac{250,000\,\text{Wh} \cdot 3,600\,\text{s}}{9.81\,\frac{m}{s^2} \cdot 400\,\text{m}}$$

$$= 229.36\,\text{t}.$$

To generate an amount of energy of 1/4 MWh, we would have to pump 229 t of water through the pipes in our power plant. Since the process takes a quarter of an hour, this corresponds to a mass flow of:

$$\frac{\Delta m}{\Delta t} = \frac{229.36\,\text{t}}{15\,\text{min}} = \frac{229.36\,\text{t}}{900\,\text{s}} = 0.255\,\frac{t}{s}.$$

This means that 255 kg of water must be pumped through the pipes per second. The density of water is $\rho_{H_2O} = 0.9982067\,\frac{kg}{l}$, which corresponds to a volume flow of:

$$\frac{\Delta V}{\delta t} = \frac{255\,\text{kg}}{0.9982067\,\frac{kg}{l} \cdot s} = 255.46\,\frac{l}{s}.$$

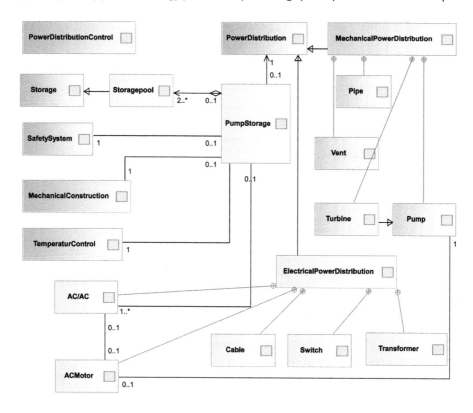

Fig. 6.4 Block diagram of a pumped storage plant. Those components which we have already got to know as basic components are marked in dark grey. The elements of the mechanical power distribution and of the electrical power distribution are marked in colour.

The required system components are shown in Fig. 6.4. The basic components that occur in every storage system and that have no special characteristics here are marked in dark grey. The `Storage` component is realized by at least two `StoragePool`. We can interpret these as specializations of the `Storage` system component. As a rule, the existing topography is used for `PumpStorage` plants. This can lead to several `StoragePool` being connected to each other at different levels of elevation. Therefore, this component is present at least twice. But higher numbers of storage basins are also conceivable.

In the pumped hydro storage plant, we have two `PowerDistribution` systems. On the one hand, we have the distribution of power via the redistribution of water from one basin to another. Here, essentially mechanical power is distributed. This system includes the pipes, valves, pump, and turbine. The second system is the `ElectricalPowerDistribution` system. This is where the `AC/AC`, the frequency converter, the `ACMotor`, which serves as a generator and motor, is located. But we also need `Cable`, `Switch`, and `Transformer` for the grid connection. Between the two systems, the electric motors serve as a connection between the forms of energy.

The use of a `Pump` or `Generator` creates new boundary conditions for the power path. Let's do a simple thought experiment. We have water in the upper storage basin. The connecting pipe between the two basins is filled with water, but the mass flow is zero—that is, there is no flow.

We now open the valves a little bit and a very small mass flow starts, but the turbine will not turn. For the turbine to start turning, the force that the water exerts on the turbine must be great enough. So if the mass flow is too small, in the worst case we would empty the upper storage basin without the turbine turning at all and without generating electrical energy.

For the description of the power flows, this means that the power flows S_G or G_S must have a dependence on the mass flow $\dot{m}_{SG}$ or $\dot{m}_G S$:

$$S_G = \begin{cases} 0, \text{ if } \dot{m}_{SG} < \dot{m}_{SG,\min} \\ [S_{G,\min}, S_{G,\max}] \end{cases} \tag{6.5}$$

$$G_S = \begin{cases} 0, \text{ if } \dot{m}_{GS} < \dot{m}_{GS,\min} \\ [G_{S,\min}, G_{S,\max}]. \end{cases} \tag{6.6}$$

We are working here with two different mass flows, $\dot{m}_{SG}$ and $\dot{m}_{GS}$ respectively, and their boundary conditions, because in Fig. 6.3 we are working with two different pipes and turbines, or pumps in the inflow and outflow of the upper storage basin. These could have different boundary conditions.

Let us extend our thought experiment. We assume that the mass flow is sufficiently large for the turbine to turn and the AC motor to convert the kinetic energy into electrical energy. However, we have not yet opened the valves completely. Now we suddenly open the valves as far as possible. We notice that the mass flow does not immediately reach its maximum value. The inertia of both the water and the turbine ensure that only after some time will the water flow as fast as possible and the turbine turn at a corresponding speed; at this point, it should be possible for the mass flow to immediately reach its maximum value.

So when controlling the power flows, we have to take into account that they cannot be changed arbitrarily quickly, because the components have an internal inertia. This results in additional boundary conditions for the time derivative of the power flows $\frac{d}{dt}S_G$ and $\frac{d}{dt}G_S$:

$$\left|\frac{d}{dt}S_G\right| \le \frac{d}{dt}S_{G,\max} \tag{6.7}$$

$$\left|\frac{d}{dt}G_S\right| \le \frac{d}{dt}G_{S,\max}. \tag{6.8}$$

Exercise 6.4 The boundary conditions in eqn (6.6) are linked to conditions for the mass flow $\dot{m}$. What should be the description attached to the power? (Hint: Stay in the energy and power picture and don't try to understand the physics of a turbine in the flow. This is very interesting and exciting, but not necessary to solve the task.)

Solution: The mechanical power results from the change over time of the stored energy of the pumped storage plant:

$$P_{\text{mech}} = \frac{d}{dt} E_{\text{pot}}.$$

The stored energy can only change by changing the mass flow—that is,

$$\frac{d}{dt} E_{\text{pot}} = \frac{d}{dt} (m \cdot g \cdot h) = \dot{m} \cdot g \cdot h.$$

We have thus directly determined a relationship between mechanical power and mass flow:

$$P_{\text{mech}} = \dot{m} \cdot g \cdot h.$$

The boundary conditions in eqns (6.5) and (6.6) can thus be reformulated as:

$$S_G = \begin{cases} 0, \text{ if } S_G < S_{G,\min} = \dot{m}_{SG,\min} \cdot g \cdot h \\ [S_{G,\min}, S_{G,\max}] \end{cases}$$

$$G_S = \begin{cases} 0, \text{ if } G_S < G_{S,\min} = \dot{m}_{GS,\min} \cdot g \cdot h \\ [G_{S,\min}, G_{S,\max}] . \end{cases}$$

Due to these additional constraints, pumped hydro storage plants are not used for storage operations that require a high dynamic and granular power flow control. They are used where large storage volumes that must hold energy for a long period are needed. These large storage volumes are realized by using storage lakes. The building of separate reservoirs is avoided, but the existing topology is used.

The periods during which the capacity is held in reserve can range from days to weeks to years. Pumped hydro storage plants are typically used for tertiary power provision and for energy trading in MW sizes.

Exercise 6.5 A pumped hydro storage plant shall store 1 MWh. The storage operation shall last 15 min. The maximum pumping power of a turbine is 1,000 kW ($G_{S,\max} = 1,000$ kW), but cannot be faster than 250 kW ($\frac{d}{dt} G_{S,max} \le 250 \frac{kW}{\min}$) per minute. How many pumps must be operated in parallel to store the power?
Solution:

$$S_G = \eta_{GS} \int_0^{T=15 \text{ min}} G_S^t \, dt$$

$$= \eta_{GS} \left(\int_0^{T=4 \text{ min}} 250 \frac{kW}{\min} t \, dt + \int_{T=4 \text{ min}}^{T=15 \text{ min}} 1{,}000 \text{ kW } dt \right)$$

$$= \eta_{GS} \left(\frac{250 \text{ kW}}{2 \text{ min}} 4^2 \text{ min}^2 + 1{,}000 \text{ kW } (15 \text{ min} - 4 \text{ min}) \right)$$

$$= \eta_{GS} 13{,}000 \text{ kWmin} = \eta_{GS} 216 \text{ kWh}.$$

The efficiency of the charging process is 90%, $\eta_{GS} = 90\%$. In this case, six pumps are needed.

6.3 Energy storage by using rotational energy: The flywheel storage

In addition to potential energy, classical mechanics also knows kinetic energy as a form of energy. Kinetic energy can be assigned to objects that move. We know two forms of motion: the translatory movement and the rotatory movement. An object that moves from one place to another makes a translational movement, and we can assign the kinetic energy

$$E_{\text{kin}} = \frac{1}{2}mv^2$$

to it. Here m is the mass of the object and v is the velocity—that is, distance per time.

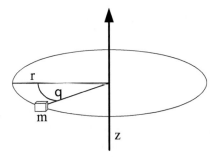

Fig. 6.5 Example of a rotational movement. A mass m moves around an axis z. Its position is determined by the angle θ.

With rotational movement, a body rotates around an axis. Let us consider the example in Fig. 6.5. A body with mass m rotates around the axis z. Its position can be described by the distance to the axis r and the angle θ. Let us first assume that the distance r remains constant. The velocity of the mass is no longer calculated via distance per time, but via angle per time. The angular velocity ω is thus defined via

$$\omega = \frac{\text{d}}{\text{d}t}\theta.$$

Suppose we interrupt the connection of the mass to the axis—that is, the force that causes the mass m to rotate around the axis z—the mass would continue to move along the tangent with a velocity of $v = \omega\, r$. The translational energy would thus be at:

$$E_{\text{kin}} = \frac{1}{2}mv^2 = \frac{1}{2}m\left(r\,\omega\right)^2.$$

We could do this experiment at any angle. Therefore, the rotational energy must be equal to the kinetic energy. So we get an expression for the rotational kinetic energy:

$$E_{\text{rot}} = \frac{1}{2}m\,r^2\,\omega^2 = \frac{1}{2}J\omega^2. \tag{6.9}$$

We used $J = m\,r^2$. J is the moment of inertia of a rotating body. The definition in eqn (6.9) referred to one mass. If we have several mass points m_i, we can add the kinetic energy of the masses. The following is thus valid:

$$E_{\text{rot}} = \frac{1}{2}\left(\sum_i^N m_i r_i^2\right)\omega^2 = \frac{1}{2}\left(\int r^2 \mathrm{d}m\right)\omega^2. \tag{6.10}$$

In eqn (6.10), the moment of inertia is no longer determined via a sum of the masses, but via the integral of all mass points. Alternatively, a representation can be found here in which we integrate over the volume of the body:

$$J = \int_V r^2 \mathrm{d}m = \int_V r^2 \rho(V)\,\mathrm{d}V. \tag{6.11}$$

We have exploited the fact that mass is given by the product of volume and density ρ. The advantage of eqn (6.11) is that we can now integrate over spatial coordinates.

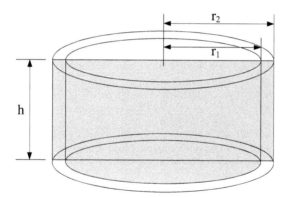

Fig. 6.6 Example of a flywheel. The flywheel is described as a hollow cylinder with radii r_1 and r_2, which has height h.

To store rotational energy, we need a body that we set in rotation. A simple form is the flywheel, which in steam engines overcomes the short dead time in which no force is exerted on the rotating axis by the pistons. The rotational energy depends on the mass, the distance of the masses from the axis of rotation, and the rotational speed. Once the material is fixed, the capacity of a flywheel can be increased in two ways: by increasing the maximum rotational speed and by optimizing the distribution of mass. The more a mass element is moved away from the centre of mass, the higher its contribution to the moment of inertia. We simplify the mathematical consideration by neglecting the influence of the support structure and the axle on the energy storage. Most of the mass of the flywheel is in the outer region of the flywheel, which we will describe as a hollow cylinder with the two radii r_1 and r_2 and the height h (Fig. 6.6).

The moment of inertia is calculated according to the formula in eqn (6.11):

$$J = \int_{h=0}^{h=H} dh \int_{\theta=0}^{\theta=2\pi} d\theta \int_{r=r_1}^{r=r_2} \rho\, r^2\, r\, dr \tag{6.12}$$

$$= 2\pi\, H\, \rho \int_{r=r_1}^{r=r_2} \rho\, r^3\, dr \tag{6.13}$$

$$= 2\pi\, H\, \rho \left[\frac{r^4}{4} \right]_{r=r_1}^{r=r_2} \tag{6.14}$$

$$= \frac{1}{2}\pi\, H\rho \left(r_2^4 - r_1^4 \right). \tag{6.15}$$

The moment of inertia is proportional to the height of the cylinder, but goes to the fourth power with the radius. So if we want to store a lot of energy, we can achieve a great effect by increasing the radius or strengthening the wall thickness.

Exercise 6.6 The mass of a flywheel is contained in an outer metal ring. For simplicity, let this flywheel be a hollow cylinder made of iron ($\rho = 7{,}860\frac{\text{kg}}{\text{m}^3}$). The outer radius is 2 m, the inner radius is 1.5 m. The thickness is 0.5 m. How fast must the wheel rotate to store 1 kWh?
Solution: Using eqn (6.15), we can first determine the moment of inertia:

$$J = \frac{1}{2}\pi\, H\, \rho \left(r_2^4 - r_1^4 \right)$$
$$= \frac{1}{2}\pi\, 0.5\,\text{m}\, 7{,}860\frac{\text{kg}}{\text{m}^3} \left((2\,\text{m})^4 - (1.5\,\text{m})^4 \right)$$
$$= 67{,}519.7\,\text{kgm}^2.$$

Note: The above equation can also be written as $J_{\text{ges}} = J_2 - J_1$, where $J_{1,2}$ represent the moments of inertia of two discs with radius r_1 and r_2. The moments of inertia can therefore be added and subtracted as desired.
To find the rotational velocity needed for the storage of 1 kWh, we use the definition of rotational energy:

$$\omega = \sqrt{2 \cdot \frac{E_{\text{rot}}}{J}} = \sqrt{2\frac{3{,}600{,}000\,\text{Ws}}{67{,}519.7\,\text{kg m}^2}} = 10\frac{\text{rad}}{\text{s}}.$$

This corresponds to a number of revolutions of:

$$n = \frac{\omega}{2\pi} = 1.64\frac{1}{\text{s}}.$$

In order to determine how much power a rotating body has absorbed or released, we must determine the change in rotational energy over time. If the moment of inertia and mass are constant, only the angular velocity must be derived according to time.

$$P_{\text{rot}} = \frac{d}{dt}E_{\text{rot}} = \frac{1}{2}J\frac{d}{dt}w(t)^2 = J\left(\frac{d}{dt}w\right)w = J\dot{w}w. \tag{6.16}$$

The quantity $J\dot{w}$ is the torque M that acts on a body:

$$P_{\text{rot}} = J\dot{w}w = M_{\text{m}}w. \tag{6.17}$$

If we want to store a large amount of energy in an FWS, we can only tune two properties: we can choose a large moment of inertia or work with a high angular velocity. There are mechanical limits to the choice of moment of inertia: we usually only have limited space, we cannot use every material, and the moving parts of the mechanics must also be able to compensate the forces that occur. The angular velocity is also limited.

If we want to absorb or deliver large amounts of power for a short time, we have another degree of freedom. Eqn (6.16) shows that three factors are relevant for high power extraction: a large moment of inertia J, a large change in angular velocity $\dot{w}$, or a large angular velocity w.

Exercise 6.7 We want to implement an FWS for a power application. We need a short-term power of 24 kW for $\Delta t = 1$ min. The FWS shall consist of a hollow cylinder made of iron ($\rho = 7,860\frac{\text{kg}}{\text{m}^3}$, $r_1 = 0.15$ m, $r_2 = 0.25$ m). The braking and acceleration system can apply a maximum torque of $M_{\text{max}} = 200$ Nm.

Solution: To extract 24 kW of power within one minute, we need the following amount of energy:

$$\Delta E_{\text{red}} = 24 \text{ kW} \cdot \frac{1}{60} \text{ h} = 0.4 \text{ kWh}.$$

The moment of inertia, as a function of height, is given for the two bodies to:

$$J_{\text{B1}} = \frac{1}{2}\pi H\rho\left(r_2^4 - r_1^4\right) = 41.97 \text{ kg}^2\text{m} \cdot H. \tag{6.18}$$

From the power equation (6.17) we can now determine the basic speed w_0; for this we assume that the maximum torque may be applied, M_{max}:

$$P = J\dot{w}w = M_{\text{max}}w_0$$

$$24 \text{ kW} = M_{\text{max}}w_0$$

$$w_0 = \frac{24\text{kW}}{200 \text{ Nm}} = 120 \frac{\text{rad}}{s} = 1,145 \text{ rpm}.$$

We therefore need a basic speed of 1,145 rpm in order to be able to extract the required power with the maximum braking torque. We now determine the height of the storage so that the energy to be stored is included.

$$E_{\text{rot}} = \frac{1}{2}Jw_0^2$$

$$0,4 \text{ kWh} = 41.97 \text{ kg}^2\text{m} \cdot H\left(48 \frac{\text{rad}}{s}\right)^2$$

$$H = \frac{0,4 \text{ kWh}}{41.97 \text{ kg}^2\text{m}\left(48 \frac{\text{rad}}{s}\right)^2} = 2.38 \text{ m}.$$

So, to meet this requirement, we need a flywheel consisting of a 2.38 m-high hollow cylinder rotating at a base speed of 1,145 rpm.

If we want to construct a flywheel for a power application, it therefore makes sense to work with a high basic speed in addition to a high moment of inertia. This means that a certain amount of residual energy always remains stored in the flywheel. The rotational speed must therefore not fall below a certain value, ω_{min}.

We will now look at the requirements for a flywheel. The general requirements, the requirements for mechanical and electrical storage systems, naturally also apply here. In addition, there are now specific, technology-dependent aspects.

We had already seen in Exercise 6.7 that with a high moment of inertia and a high rotational speed, we can also call up high power with limited braking and acceleration. Since every braking and acceleration process also places a load on the material, we like to choose a basic speed with a high rotational speed. This is a first additional requirement for FWS.

FWS 1: AS A Design Engineer, I WANT the rotation speed to have a high operating point SO THAT I can achieve a high charging and discharging performance with a smaller torque.

The basic requirements for a storage system include the requirement G 6. G 6 was a requirement for the mechanical design. In an FWS system, there is an extension to this due to the design:

FWS 2: AS A Mechanical Design Engineer, I WANT the axis of rotation to be kept stable SO THAT there is no precession that damages the mechanics.

A stable axle is important for the safety and mechanical stability of the storage. Precession would make controlled charging and discharging much more difficult. Therefore, FWS 2 is an extension of the requirements G 3, G 5, and G 6.

The energy stored in the flywheel is permanently reduced by friction losses. On the one hand, air friction acts on the flywheel, and on the other, there are friction losses in the wheel suspension. Air resistance and friction act as a reducing torque and permanently extract energy from the storage. A flywheel accumulator therefore has high self-discharge due to its design, which should be reduced by suitable measures.

FWS 3 AS A User, I WANT the losses due to friction and air resistance to be kept as low as possible SO THAT the self-discharge generated by these losses is kept as low as possible.

Fig. 6.7 shows the requirements described here and their relationship to the already defined requirements for storage systems. As with the general requirements for mechanical systems, there are also a number of requirements for FWS that extend existing requirements. In order to better recognize these; the respective relationship arrows have been provided with the keyword '«Extend»'.

For the realization of a FWS system, further components are needed (Fig. 6.8). The Rotor—that is, a body that rotates around its own axis—serves as storage in this system. To fulfil requirement FWS 1, an electric motor is necessary. It converts the kinetic energy into electrical energy. In general, this is done with an ACMotor. In addition, a PositionSensor is used to measure the angle of rotation and thus the speed of rotation. It is not necessary for FWS 1 alone, as there are methods with which an electric motor can drive an ACMotor without a PositionSensor. However,

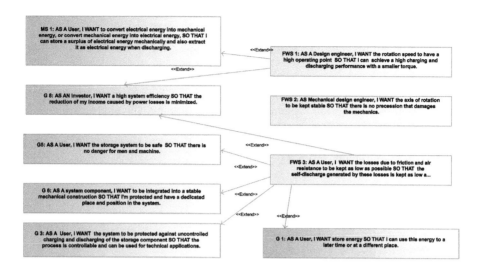

Fig. 6.7 Requirements for a flywheel storage system and the relationships of the requirements to general storage requirements.

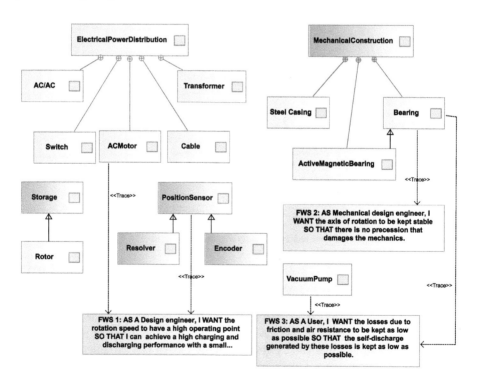

Fig. 6.8 System components of a flywheel storage unit. System components that apply to the specific requirements of a flywheel storage unit (FWS 1 to FWS 3) are connected to these.

a `PositionSensor` is also needed in conjunction with requirement G 3, since the rotational speed is a measure of the charge state of the FWS.

FWS 2 is a requirement for the mechanical design—in particular, the bearing of the rotor. Since the `Bearing` separates the moving part of the storage from the stationary part of the storage, it also has an influence on the friction and thus on the self-discharge of the storage system. Therefore, there is a connection here to requirement FWS 3.

In addition to the friction losses caused by the bearing, the air resistance also creates additional friction. This can be reduced by the rotor being in a vacuum. The vacuum is generated by a `VacuumPump`. This system component is therefore also connected to the requirement FWS 3.

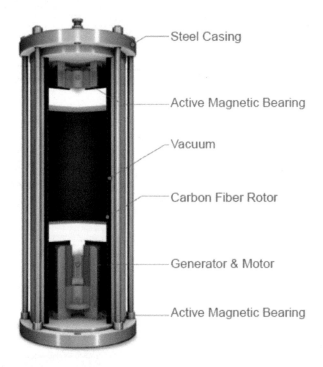

Fig. 6.9 Example of a realization of a flywheel accumulator. (Photo: STORNETIC)

The abstract representation in Fig. 6.8 has been translated into a real system in Fig. 6.9. We see that the rotor is narrow and tall. At first, this seems to contradict the fact that the moment of inertia increases quadratically with the distance to the axis. If we want to increase the storage capacity of a storage—at the same rotational speed—we would only have to increase the diameter of the flywheel mass. This is the reason why in steam engines the flywheels usually have a large diameter and the mass is located at the outer edge. On the other hand, the precession forces become larger when the diameter is large. A small disturbance produces a high force on the bearings, which are thus subjected to a lot of stress. Therefore, a narrow flywheel bearing would be more advantageous for the design of the bearings. The disadvantage of the low moment of inertia can be compensated by a high operating speed, which is in any case preferred for the operation of a flywheel accumulator

Another advantage of this type of design arises from the smaller space requirement for the installation of a flywheel. It is easier to set up a large number of narrow, tall storage units than a few large storage units with less height.

An active magnetic bearing is used here as the axle bearing. With an active magnetic bearing, electromagnets ensure that the axis of rotation remains aligned. The force of these magnets is adjustable. We are therefore able to actively compensate for misalignments during operation.

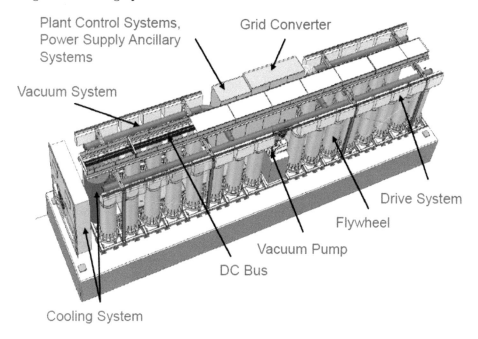

Fig. 6.10 A combination of several flywheel storages allows higher outputs and storage capacities to be realized. (Photo: STORNETIC)

Because of the friction losses, flywheel systems are not suitable for storing energy for a longer period of time. Therefore, flywheels are used as power accumulators (Eckroad, 1999; Strzelecki, 2008; Akhil *et al.*, 2013). As we have seen, we can extract high power from a flywheel accumulator with relatively low torques if the operating point is high enough. Since we can neither build the operating point arbitrarily high, nor build the mass of the flywheels arbitrarily large, we are limited in the maximum available power. However, this can be compensated by aggregating different FWSs into one large storage. In Fig. 6.10, such a system of flywheel accumulators is shown. It consists of four groups of eight flywheels. Each flywheel has its own drive system, but the cooling system and the vacuum pump are executed only once and used by all the storages.

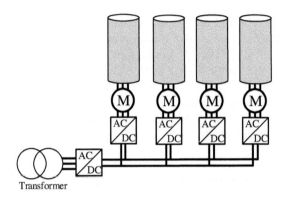

Fig. 6.11 Schematic representation of the flywheel storage system from Fig. 6.10.

The connection to the grid is made here by a central grid inverter. Let us look at Fig. 6.11. Instead of the individual storage units feeding directly into the grid, the flywheels feed and draw electrical energy from a DC bus. Only the central inverter feeding into the grid ensures the power transfer between grid and storage. This design has several advantages: as we already know, inverters need inductances to generate an AC. In a drive inverter, these inductances are contained in the motor, where they also serve to convert electrical energy into kinetic energy. If each of the 32 FWS units contained a grid-feeding inverter, these inverters would be equipped with grid filters. A simple drive converter, on the other hand, does not require filters and is therefore cheaper.

The operation management would also be considerably more complex in the case of a system with 32 grid-feeding inverters, since the characteristics of the current at the output of the entire system are relevant for compliance with the grid code, and in this case the interaction of 32 individual systems must be controlled. The design used here therefore reduces the cost of grid filters and places the responsibility for grid-code compliance on a single system component.

Scaling the FWS to four blocks, each with eight FWS systems, has another advantage. As we know from Chapter 3, each storage system has a constant consumption which reduces its efficiency. If we work permanently with 32 FWS systems, the constant consumption occurs permanently 32 times. It does not matter whether the required power is covered by 32 systems or by eight. By introducing the blocks, the power of the system can be cascaded. As long as one block can store and provide the required power, only one block is put into operation. Although this leads to increased complexity in operation management, it allows a significant reduction in losses (Ried, Schmiegel, and Munzke, 2020).

Exercise 6.8 Both Fig. 6.9 and Fig. 6.10 show a number of system components. How can these system components be assigned to the six basic system components (power distribution, power distribution control, storage, safety system, temperature control, and mechanical construction)? Please extend Tab. 6.1 with the system components and connect each component to the corresponding basic system component.

Table 6.1 Example table for the allocation of the system components of the flywheel accumulators to the basic system components.

	PowerDistribution	Power DistributionControl	Storage	SafetySystem	TemperatureControl	MechanicalConstruction
Systemcomponent A						
Systemcomponent B						
...						

Solution: Tab. 6.2 shows a sensible assignment of the system components. In large part, these assignments are derived directly from the tasks. The DCBus is used to distribute power between the individual flywheel units and to transport the power to the grid, so it is part of the PowerDistribution.

The PlantControlSystem combines various tasks. On the one hand, it controls the power flow between the grid and the FWS. It is therefore part of the PowerDistributionControl. At the same time, however, it also monitors its own system components and is thus part of the SafetySystem and TemperatureControl. These system components therefore contain or realize several basic system components and their tasks.

The question of which components the VacuumPump and the VacuumSystem belong to is somewhat more difficult. They are not part of the PowerDistribution, the SafetySystem, or the MechanicalConstruction. Rather, they are part of the Storage component. Without this system component, the storage would lose a considerable part of its function.

Table 6.2 Example of an assignment of the system components to the basic system components.

	PowerDistribution	Power DistributionControl	Storage	SafetySystem	TemperatureControl	MechanicalConstruction
PlantControlSystem		X		X	X	
PowerSupplyAncillarySystem	X	X				
VacuumSystem			X			
CoolingSystem					X	
DCBus	X					
VacuumPump			X			
Flywheel			X			
DriveSystem	X		X			
GridConverter	X					
SteelCasing						X
ActiveMagneticBearing			X			X
CarbonFibreRotor			X			

In this section, we have dealt with the storage of electrical energy in kinetic energy, or, more precisely, in rotational energy. In doing so, we got to know FWS as a realization. The flywheel accumulator is mainly used as a power accumulator, because movement always generates friction and friction is associated with energy loss. In the next section, we will look at the third form of mechanical energy and how this form can be used to store energy.

6.4 Energy storage using potential energy part 2: The restoring force of a spring

In Section 6.2, we used gravity to store potential energy. We looked at the fact that we have to do work when we lift an object and that we can gain energy when an object falls or is released in a controlled way. In this section, we will look at another form of potential energy: restoring forces. The basic idea here is the same: we do work to change a mechanical state and gain energy when we let the system fall back to its equilibrium state. The simplest example is a restoring force of a spring. In Fig. 6.12, we have a weight placed on a table with a spring attached. The force of the weight is not so great that the spring is completely compressed. Therefore, we apply an additional force F_{charge} until the spring is compressed to the minimum size and lock the table. This process corresponds to the charging process. The restoring force of a spring is

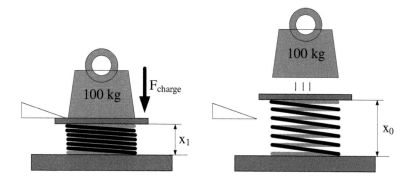

Fig. 6.12 Example of storing energy by compressing a spring. During charging, we compress the spring by adding a force F_{charge}. During the storage time, the spring cannot decompress, because we have locked the spring. If we release the lock, the spring decompresses and accelerates the weight.

proportional to the deflection of the spring Δx and the mechanical properties of the spring D:

$$F(\Delta x) = -D\,\Delta x. \tag{6.19}$$

F is always directed in such a way that the force tries to restore the equilibrium position of the spring. Regardless of whether the spring is stretched, i.e. $\Delta x_0 > 0$, or compressed, $\Delta x < 0$, the force always acts in such a way that the spring returns to the equilibrium position x_0. The energy that we store in the spring results from the integration of the applied force over the path: F is always directed in such a way that the force tries to restore the equilibrium state of the spring. Regardless of whether the spring is stretched, i.e. $\Delta x_0 > 0$, or compressed, $\Delta x < 0$, the force always acts in such a way that the spring returns to the equilibrium position x_0. The energy that we store in the spring results from the integration of the applied force over the path:

$$W_{rest} = \int_{s=x_0=0}^{s=x_1=x} -D\,s\,ds = \frac{1}{2}Dx^2. \tag{6.20}$$

If we release the lock, the spring relaxes. The stored energy is released, and in our example in Fig. 6.12, the potential energy is converted into kinetic energy.

The example shown here has some disadvantages for practical applications. To enable controlled charging and discharging, one would need to be able to vary the spring in discrete units. This is done in clockworks by suitable number wheels. So there is a constructive solution to this problem. But there is another problem. The conversion of kinetic energy into electrical energy is done by electric motors. These always require a rotational movement. The use of an arrester results in discrete, short-term rotational movements. However, it is difficult to convert this into a continuous movement which can then be converted into electrical energy. This raises the question of whether there is another way to use the restoring forces.

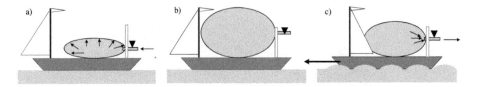

Fig. 6.13 Ship propulsion that uses a compressed air reservoir as a drive. When charged, a), a balloon is inflated. In the charged state, b), it has a significantly higher volume. By opening the valve, c), the storage is discharged and the outflowing gas generates a propulsive force that moves the ship.

Besides the mechanical spring, there are other physical systems that can be stretched or compressed to produce a force. An example is shown in Fig. 6.13. Here we have a ship that has a large balloon on deck. At the end of the balloon is a small valve. When we open it and pump air into it, the balloon expands. The air molecules press against the surface of the balloon. If the pressure of the molecules is high enough, the balloon expands until there is an equilibrium between the air pressure generated by the molecules and the surface tension of the balloon skin. At a certain volume, the balloon is large enough and we close the valve. We have now stored kinetic energy into potential energy in the balloon. We now open the valve. The surface tension and the air pressure inside the balloon drive the air out of the tube. In the process, the air molecules generate a force that propels our ship.

Compared to our spring experiment, this propulsion has the advantage that the valve position can be used to fine-tune the charging and discharging rate. As the gas flows out or in, we have a continuous translatory motion, which we can convert into a rotatory motion with the help of a turbine or a propeller, which can easily be converted into electrical energy with the help of electric motors.

The air balloon is unsuitable as a compressed air storage device for the realization of energy storage systems because the change in volume during charging and discharging creates mechanical stress. This stress ensures that balloons cannot be reinflated too often. Gas containers that have a fixed volume and can store gas under very high pressure are more suitable for a compressed air storage system.

For the description of the charging and discharging process, it is sufficient in the first approximation to use the ideal gas law (Kondepudi *et al.*, 2008):

$$pV = nRT, \tag{6.21}$$

p where is the pressure of the gas, V its volume, and T its temperature in Kelvin. n is the amount of substance in moles, and $R = 8.314 \frac{J}{mol\,K}$ is the general gas constant. The ideal gas law relates the three thermodynamic quantities pressure, temperature, and volume. It describes the behaviour of gases as long as there are no phase transitions or extreme temperature, pressure, or volume conditions.

Let us take a closer look at the storage process. We have a defined amount of gas in a container that is equipped with a piston and connected to a second container via a pipe. In Fig. 6.14 a), we have shown the structure. The gas occupies a volume V_1 in the container, and we can measure a pressure p_1. (Remember: the pressure corresponds

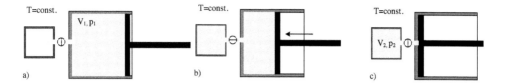

Fig. 6.14 Example of isothermal compression. a) We have a gas in a container that is equipped with a piston. b) After opening a valve to a second, smaller container, we use the piston to press the gas into the second container. c) When this process is complete, there is no more gas in the first container. Initial pressure and initial volume have changed: $p_1 \rightarrow p_2$, $V_1 \rightarrow V_2$.

to the kinetic momentum exerted by the gas molecules on an interface as they move (Feynman, Leighton, and Sands, 1965)).

We now open the valve and use the piston to push the gas into the second container, which has a different volume, V_2. We do this until there is no more gas in the container with the piston. Then we close the valve. The gas is in the volume V_2. We reduce our considerations to the state of the gas in a) and in c), and ignore the transient state during the pumping of the gas in b). In both state a) and in state b), the ideal gas law applies:

$$a) \, p_1 V_1 = nRT \tag{6.22}$$
$$c) \, p_2 V_2 = nRT. \tag{6.23}$$

The right-hand side of the equations is identical, since both the number of molecules and the temperature of the gas should remain the same. So we can combine these equations and get:

$$p_1 V_1 = p_2 V_2 \tag{6.24}$$
$$\frac{p_1}{p_2} = \frac{V_2}{V_1}. \tag{6.25}$$

Exercise 6.9 What is the amount of work done in the isothermal compression process shown in Fig. 6.14?

Solution: In general, work is the integral of force times displacement. The force acting on the piston is the pressure that the gas exerted on the piston during the compression process. The displacement corresponds to the change in volume—that is,

$$W = \int_{V_1}^{V_2} p \, dV.$$

Pressure and volume are linked via the ideal gas law, eqn (6.21). For the pressure the following applies:

$$p = nRT \frac{1}{V}.$$

Thus, we can calculate the mechanical work that takes place during isothermal compression:

$$W = \int_{V_1}^{V_2} nRT \frac{1}{V} \, dV \tag{6.26}$$

$$= nRT \, [\ln V]_{V_1}^{V_2} \tag{6.27}$$

$$= nRT \ln \left(\frac{V_2}{V_1} \right) \tag{6.28}$$

$$= p_1 V_1 \ln \left(\frac{V_2}{V_1} \right). \tag{6.29}$$

The work done and thus also the energy stored depends only on the ratio of the two volumes V_1 and V_2.

The experiment we have described here represents a compressed air storage. Here we take a certain amount of gas and compress this gas to a considerably smaller volume. The work that can be stored is proportional to $\ln \left(\frac{V_1}{V_2} \right)$ and the temperature T. The greater the compression and the ambient temperature, the greater the amount of energy stored.

This type of compression is called isothermal compression. The temperature remains constant during the process. An alternative form of compression is isochoric compression, in which the volume of the gas is kept constant. For the pressure and temperature conditions, the ideal gas law gives:

$$\frac{p_2}{p_1} = \frac{T_2}{T_1}. \tag{6.30}$$

Here $p_{1,2}$ and $T_{1,2}$ are the pressure and temperature values before and after compression. Isothermal compression has the advantage that mechanical work could be used to store energy through the piston movement. In isochoric compression, we have to increase either the temperature or the pressure of the gas without doing any mechanical work in the form of volume change. How could this be realized in a storage system? We have the option of either changing the pressure or the temperature of the gas. Since the volume is constant, it is difficult to increase the pressure of the gas by means of mechanical work. For this reason, the only remaining option is to change the temperature. However, in order to use an isochoric storage for an energy storage system, we need to find a way to add or remove heat from the system and then convert it into another form of energy. As we have already seen in Chapter 2, it is usually very difficult to convert thermal energy back into another form of energy. Therefore, an isochoric storage is not suitable as a compressed air storage.

There is a third method of compressing a gas. This uses a physical quantity newly introduced by thermodynamics: thermal energy or 'heat'.

In thermodynamics, the concept of energy is expanded. In mechanics, a system is described entirely by its coordinates in space and their change over time, and its inherent (rotational) momentum; thermodynamics adds the concept of 'heat' or 'heat energy'. Behind this is the insight that the energy content of a cold and a hot cup of coffee cannot be the same. In order to fully describe an object, it is therefore not sufficient to know only position, velocity, and momentum. Knowledge about the thermal energy content is also needed. Both aspects together result in the inner energy of an object. This results in the first main theorem of thermodynamics:

$$\Delta U = \Delta Q + \Delta W. \tag{6.31}$$

Here ΔU is the change in inner energy, ΔQ is the change in heat, and ΔW is the mechanical work done.

We realized the isothermal compression, shown in Fig. 6.14, by using a thermal bath—that is, the ambient temperature was always constant and the gas could maintain its temperature in exchange with the environment. In the third method, we thermally isolate the container. Since there is no longer any thermal exchange between the gas and the ambient temperature, the amount of heat in the gas is kept constant. This is called adiabatic compression. The following applies to adiabatic compression:

$$\left(\frac{V_2}{V_1}\right)^{\kappa} = \frac{p_2}{p_1} \tag{6.32}$$

$$\frac{T_2}{T_1} = \left(\frac{V_2}{V_1}\right)^{\kappa-1}. \tag{6.33}$$

Here $\kappa = \frac{c_p}{c_v}$ is the adiabatic constant, c_v is the specific heat constant of the gas at constant volume, and c_p the specific heat constant at constant pressure. Both are material constants and depend on the individual gas.

The mechanical work that takes place during an adiabatic compression can be calculated via the adiabatic equation and integration via the volume change. Pressure and volume of the final state result from the adiabatic equation:

$$p = p_1 V_1^{\kappa} V^{-\kappa}. \tag{6.34}$$

Thus, we can determine the mechanical work done during a compression:

$$W = \int_{V_1}^{V_2} p_1 V_1^{\kappa} V^{-\kappa} dV \tag{6.35}$$

$$= p_1 V_1^{\kappa} \frac{1}{1-\kappa} \left[V_2^{1-\kappa} - V_1^{1-\kappa}\right]. \tag{6.36}$$

Exercise 6.10 We want to compare the conducted work in an isothermal compression with an adiabatic compression. To do this, we consider an adiabatic and an isothermal compression of air, which has an adiabatic coefficient of $\kappa = 1.4$. At the beginning, the gas has a pressure of $p_0 = 1$ bar. The volume is reduced from $V_1 = 10$ l to $V_2 = 1$ l. What is the required mechanical work?

Solution: The following is true for isothermal compression:

$$W = p_1 V_1 \ln\left(\frac{V_2}{V_1}\right)$$

$$= 1 \text{ bar} 10 \text{ l} \ln\left(\frac{1 \text{ l}}{1 \text{ l} 0}\right)$$

$$= 101.325 \text{ Pa} 0{,}001 \text{ m}^3(-2,3)$$

$$= -233 \text{ Ws}.$$

Here we have considered that $1 \text{ bar} = 101.325 \text{ Pa}$ and $[1 \text{ Pa}] = 1 \frac{N}{m^2}]$ holds. For adiabatic compression the following holds:

$$W = p_1 V_1^\kappa \frac{1}{1-\kappa} \left[V_2^{1-\kappa} - V_1^{1-\kappa} \right]$$

$$= 1 \text{ bar} (10 \text{ l})^\kappa \frac{1}{1-\kappa} \left[(1 \text{ l})^{1-1,4} - (10 \text{ l})^{1-1,4} \right]$$

$$= -382.98 \text{ Ws}.$$

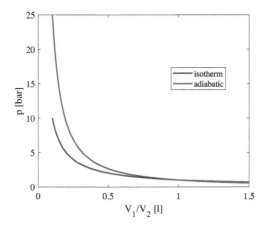

Fig. 6.15 Pressure of an ideal gas, which is compressed from volume V_1 to volume V_2. This compression is done in an isothermal process or in an adiabatic process.

As Exercise 6.10 shows, we need more energy when we want to compress a volume adiabatically compared to isothermal compression. In other words, adiabatic compression allows us to store more energy within the same volume of gas. Fig. 6.15 shows this in the $p(V)$ curve. For the same compression, the pressure build-up is considerably higher in an adiabatic compression than in an isothermal compression. This can be explained from a physical point of view: in an isothermal compression, there is a temperature exchange between the gas in the container and the ambient temperature. As we know, temperature is a physical quantity that can be related to the movement of the atoms of a substance. So when there is a temperature exchange, it means that the kinetic energy of the atoms of the gas is exchanged with the environment. However, the pressure is also derived from the kinetic energy of the atoms of the gas. The pressure corresponds to the average force exerted by the particles at the interface. So if there is an exchange of kinetic energy, the force that the particles can apply at the interface is automatically reduced. The heat exchange leads to a reduction in pressure.

In isothermal compression, we do macroscopic work that causes the kinetic energy of the gas particles to increase. The gas particles now exchange part of their gained kinetic energy with the environment so that the temperature remains constant. The work we have done is therefore lost. If, on the other hand, we perform adiabatic compression, we cut off this exchange. The result is that the pressure increases.

Heat is generated during compression. We know this from everyday life when we use a bicycle pump to inflate a tyre. If we pump fast enough—that is, if we do not give the system time for a temperature exchange with the ambient air—we can feel the pump and valve heating up. Conversely, if the air is let out fast enough, the valve will get cold. We also observe this process in a compressed air storage. Therefore, we add an additional thermal storage component to the compressed air storage. In Fig. 6.16, we have shown the complete energy flow of a compressed air storage. The storage is charged via an electric compressor. The resulting heat is dissipated via a heat exchanger and stored in a heat accumulator. The gas is stored in the actual pressure vessel. During decompression, an electric compressor converts the pressure into kinetic energy and then into electrical energy. During the decompression process, the gas requires heat, which is extracted via the heat storage medium. The heat storage medium cools down and is stored in a second heat storage unit. From there, it can be used again during the next compression process.

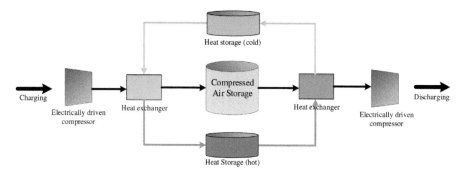

Fig. 6.16 Energy flow of a compressed air storage. To increase the efficiency of the storage, the heat generated during compression is additionally stored and used during decompression to heat the gas.

Additional requirements apply to compressed air storage (CAS), which we will now consider.

CAS 1: AS AN Engineer, I WANT to ensure that the pressure and temperature of the storage medium remain within a defined operating range SO THAT no phase transition occurs.

A substance can have three states of aggregation: solid, liquid, and gaseous. Which of these states of aggregation is obtained depends on the combination of pressure and heat. A pressure cooker, for example, uses this fact and allows food to be cooked at higher temperatures by allowing a higher pressure. What is desirable in a pressure cooker is not necessarily desirable in a compressed air storage, since the described storage process only works with gases (with solids and liquids there are other phenomena). Liquefaction of the gas during charging and discharging would not be desirable. The turbine and the compressor are adjusted to a gas and cannot transform the kinetic energy of a gas and a liquid equally well. This transformation process of the kinetic energy of the flowing gas into electrical energy, and vice versa, leads us to requirement CAS 2 and is an extension of the requirement MS 1:

CAS 2: AS A User, I WANT to guide the outflowing and inflowing gas through a turbine SO THAT I can convert the kinetic energy of the gas into electrical energy, or use electrical energy to compress the gas.

G 8 requires that the stored energy is stored for as long as possible. Self-discharge is to be avoided. We must therefore ask ourselves what effects can cause the stored energy in the storage to decrease over time. In principle, we can attribute it to the three physical variables of mass, pressure, and temperature. In any case, we should avoid stored gas escaping unintentionally.

CAS 3: AS A User , I WANT the stored gas to remain in the storage SO THAT I observe a small self-discharge.

As we have seen, adiabatic compression stores considerably more energy than isothermal compression. The reason is that in an adiabatic compression, no energy can be released to the environment via a heat exchange. If we realize an adiabatic compressed air storage, we have to make sure that no heat exchange can take place between the outside of the storage and the gas. The storage must be thermally well insulated.

CAS 4: AS A Design Engineer, I WANT to ensure that there is no heat exchange between the gas and the storage environment SO THAT the self-discharge remains low.

Fig. 6.17 shows the required system components. The power distribution is realized here by the pipe system. This consists of valves, pipes, temperature, pressure and flow velocity sensors and connects the various subsystems, such as the storage chamber, motor generator, turbine, and compressor. Since heat is generated or required in adiabatic storage, cooling and heating units are necessary. Therefore, a compressed air storage contains at least two heat exchangers, each with at least one thermal storage.

So far, there are only a few compressed air storages in operation. This is in contrast to pumped hydro storage plants, which have been built since the beginning of electrification and which are integrated into the existing landscape. With compressed air storage, either underground caverns or large gas containers have to be developed and used. Both involve an expense that could not be covered by the expected revenues until now.

We have now learned about the three ways of storing mechanical energy: through the use of potential energy, rotational energy, and compression energy. In the next section, we want to build a hybrid power plant that uses two of the technologies described here: the pumped hydro storage plant and the FWS plant.

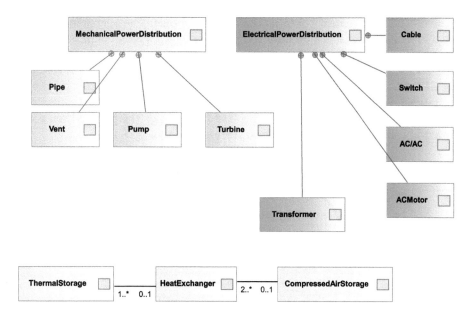

Fig. 6.17 System components of a compressed air storage system.

6.5   Application example: Upgrading a pumped hydro storage plant for the power market

In this section, we want to combine two of the storage technologies we have encountered in such a way that the weakness of one is compensated by the strengths of the other. Our aim is to extend a pumped hydro storage plant, which until now has only been able to operate in the energy market, so that it can also operate in the power market. We want to combine the pumped hydro storage plant with an FWS system so that the entire system can provide the required charging and discharging power sufficiently quickly. When several storage systems are combined, they are referred to as aggregated storage systems. The motivation for this aggregation is the following user requirement:

UR 1: AS AN Operator, I WANT the two storage systems to be addressed as one large storage SO THAT I, as a power market participant, only need to place my power and energy demands on one storage.

The aggregation of the two storages simplifies the control. The market participants do not have to differentiate between one storage and the other. Instead, they are dealing with one system. This simplifies the qualification of the two facilities for the energy and power market.

The task of aggregation is performed by an energy management system (EMS). It coordinates the power flows of the two storages. Market requests are made to the EMS and it decides how the two storages are combined. This can be realized by having one EMS that drives and controls both systems simultaneously, or by ensuring the EMSs that are already present in the pumped hydro storage plant and FWS plant are addressed via a central EMS and carry out disaggregated commands.

The second requirement concerns the actual use case. The aggregation of the two storage systems is necessary because the pumped hydro storage plant alone does not fulfil the requirements for participation in the capacity market:

UR 2 AS A user, I WANT the aggregated storage system to provide a power of 1 MW for one hour within 5 min SO THAT I can participate in the power market with the storage power plant.

The task now is to design the FWS so that it provides the required power during the start-up phase of the pumped hydro storage plant.

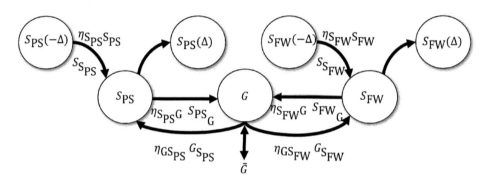

Fig. 6.18 Power flow diagram of the combined storage system. In this example, a pumped storage, S_PS, is combined with a flywheel storage, S_FW. All quantities in this diagram are time-dependent. $S_\text{xx}(\pm\Delta)$ represents the state of the storage xx at time $t \pm \Delta t$.

We start by looking at the power flow diagram. This is shown in Fig. 6.18. We follow the notation from Chapter 3. All quantities are time-dependent, but we avoid a representation of the time argument in the notation. However, the power flow from the past or into the future is abbreviated with the argument Δ. The pumped storage is represented by the node S_PS. The FWS is represented by the node S_FW.

The power flow diagram consists of two storage flows, both connected to the network node G. For participation in the power market, the quantity $\tilde{G}$ is the leading quantity. The network $\tilde{G}$ is both source and sink here. It is the leading quantity specified by the EMS and results from the purchases of energy or the provision of power. The power flows between the storage systems and the grid node must be selected so that $\tilde{G}$ is fulfilled. The associated power flow equation is:

$$\tilde{G} = \eta_{S_\text{PS}G}S_\text{PSG} + \eta_{S_\text{FW}G}S_\text{FWG} - (G_{S_\text{FW}} + G_{S_\text{PS}}). \tag{6.37}$$

Exercise 6.11 Equation 6.37 operates with an '=' relationship, meaning $\tilde{G}$ must be satisfied at all times. Is this a reasonable request? What alternative formulation can be chosen here?

Solution: The formulation that $\tilde{G}$ must be satisfied at all times is required via the market requirement and via UR 2. It is motivated by the fact that $\tilde{G}$ corresponds to the power or energy sold or bought and must, of course, be provided by the storage. However, there is a drawback to this formulation: if $\tilde{G}$ can no longer be covered by the storage systems because the storages are empty or have reached their capacity limit, the equation can no longer be fulfilled. We have to catch this case mathematically in the operation management.

An alternative is to work with a $\leq$ relationship. It is then permitted to take in or deliver less power than is actually demanded. However, the non-fulfilment of the purchase contract has consequences. Penalties become due. These can be taken into account in the yield equation. If $\Delta \tilde{G}$ is the difference and Y_{req} would be the original revenue from the request, then the revenue equation would be set to be extended

$$Y = Y_{\text{req}} - c_{\text{p}} \Delta \tilde{G},$$

where c_{p} represents the penalty fee for non-compliance with the contract.

This ensures that the system of equations is always solvable, and we only need to make sure that we optimize the yield in total.

For the two storage systems, the balance equations for the inflows and outflows of power and storage apply—that is, the flow of power from the past or into the future:

$$0 = \eta_{S_{\text{PS}}S_{\text{PS}}}SS_{\text{PS}}(-\Delta) + \eta_{GS_{\text{PS}}}GS_{\text{PS}} - (S_{\text{PSG}} + SS_{\text{PS}}(\Delta)) \tag{6.38}$$

$$0 = \eta_{S_{\text{FW}}S_{\text{FW}}}SS_{\text{FW}}(-\Delta) + \eta_{GS_{\text{FW}}}GS_{\text{FW}} - (S_{\text{FWG}} + SS_{\text{FW}}(\Delta)). \tag{6.39}$$

We compose the objective function Y_{req} from two aspects: the amount we can realize by buying or selling power, Y_{m}, and the losses incurred by charging and discharging the storage, Y_{loss}.

The power and energy market operates energy exchanges; therefore, there is no fixed purchase or sale price. This is represented by a time-dependent tariff $c_{\text{m}}(t)$. The revenue we achieve in the period $t = 0$ to $t = T$ is then given by:

$$Y_{\text{m}}(T) = \int_{t=0}^{t=T} c_{\text{m}}(t) \left(\eta_{S_{\text{PS}}G}S_{\text{PSG}} + \eta_{S_{\text{FW}}G}S_{\text{FWG}} - (G_{S_{\text{FW}}} + G_{S_{\text{PS}}}) \right) \text{dt}. \tag{6.40}$$

In order to be able to value the losses monetarily, we have to put a price on them. We represent this by the time-dependent tariff c_{loss} and obtain the costs for transfer losses for the period $t = 0...T$:

$$Y_{\text{loss}} = \int_{t=0}^{t=T} c_{\text{loss}}(t) \left[(1 - \eta_{S_{\text{PS}}G})S_{\text{PSG}} + (1 - \eta_{S_{\text{FW}}G})S_{\text{FWG}} + (1 - \eta_{GS_{\text{PS}}})G_{S_{\text{PW}}} \right.$$

$$\left. + (1 - \eta_{GS_{\text{FW}}})G_{S_{\text{FW}}} \right] \text{dt} \tag{6.41}$$

The pumped hydro storage plant is subject to restrictions on changing the charging and discharging power due to the inertia of its mechanical pumps and turbines. This was also the original motivation for combining the two power plants:

$$\left| \frac{\text{d}}{\text{dt}} G_{S_{\text{PS}}} \right|, \left| \frac{\text{d}}{\text{dt}} S_{\text{PSG}} \right| \leq \Delta S_{\text{PS}}^{\text{max}}. \tag{6.42}$$

A similar relationship also applies to the FWS, but with significantly higher values. If the operating point of the FWS is high, a large amount of power can already be taken out or put in with a very small application of force. To reach the limit of the rate of change here, large changes in a very short time are necessary. Since the time scale for the energy market is in the minute range, such changes are not required in this application example. We can therefore ignore the consideration of this boundary condition.

Exercise 6.12 The pumped hydro storage plant has a volume of $30\,\text{m} \times 30\,\text{m} \times 15\,\text{m} = 13{,}500\,\text{m}^3$ and an average difference in height of $\Delta h = 100\,\text{m}$.

a) What is its storage capacity ($\rho = 998\,\frac{\text{kg}}{\text{m}^3}$, $g = 9.81\,\frac{\text{m}}{\text{s}^2}$)?

b) The power plant can change its charging and discharging power by a maximum of $\frac{1}{2}E$ per hour. How long does it take the pumped storage plant to reach the target power of $1\,\text{MW}$? (Remember, the E-Rate relates charging and discharging power to capacity. If a storage with $2\,\text{kWh}$ is discharged with $6\,\text{kW}$ within one hour, this corresponds to an E-Rate of 3.)

Solution:

a) The energy of a pumped hydro storage plant is $E = m\,g\,h$. Therefore, using the above values, the capacity is:

$$\kappa_{PS} = 998\,\frac{\text{kg}}{\text{m}^3}\,13{,}500\,\text{m}^3\,9.81\,\frac{\text{m}}{\text{s}^2}\,100\,\text{m} = 3.671\,\text{MWh}.$$

b) We first determine the maximum rate of change of power:

$$3.671\,\text{MWh} \cdot \frac{1}{2}\,\frac{1}{\text{h}} = 1.8355\,\frac{\text{MW}}{\text{h}}.$$

Within one hour, the pumped hydro storage plant can ramp up from $0\,\text{MW}$ to a capacity of $1.8355\,\text{MW}$. However, we only need a power of $1\,\text{MW}$. For this, we need a time of:

$$\frac{1\,\text{MW}}{1.8355\,\frac{\text{MW}}{\text{h}}} = 0.544\,\text{h} = 32.68\,\text{min}.$$

Within $32.68\,\text{min}$, the pumped hydro storage plant has reached a charging or discharging power of $1\,\text{MW}$.

The calculations in Exercise 6.12 show that we have to bridge a period of $32.68\,\text{min}$ with the help of the FWS. We need to design the FWS so that it can supply the required power and has enough storage capacity during this time. The start-up process is divided into two phases. UC 2 gives us a period of 5 min in which to reach the target power of $1\,\text{MW}$. During this period, both the FWS and the pumped storage start up in parallel. During this phase, the power is ramping up from 0 to $1\,\text{MW}$. After this period, the entire power of $1\,\text{MW}$ is provided. This is initially provided mainly by the FWS. Since the pumped hydro storage plant continues to ramp up, the FWS's output can be reduced proportionally until the pumped storage plant provides the entire output. After approximately $32.68\,\text{min}$, the pumped storage is fully active and the FWS, which was linearly reduced during this time, no longer works.

Exercise 6.13 Fig. 6.19 shows the feed-in profile of the two storages and the total feed-in. In the start-up phase ($t < 5\,\text{min}$), the total power fed in may be less than $1\,\text{MW}$. From $t = 5\,\text{min}$, it must be ensured that the power is $1\,\text{MW}$. The power of $1\,\text{MW}$ should be reached within 5 min of UC 2. After 5 min of start-up, what is the power of the FWS to compensate for the power of the pumped storage not yet reached?

Solution: We know that the rate of change of the output of the pumped hydro storage plant is $1.8355\,\frac{\text{MW}}{\text{h}}$. So we just need to calculate the rate of power change of the pumped storage after 5 min:

$$G_{S_{PS}}(t = 5\,\text{min}) = \frac{1.8355}{60}\,\frac{\text{MW}}{\text{min}}\,5\,\text{min} = 0.153\,\text{MW}.$$

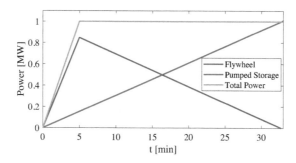

Fig. 6.19 Power profile of the flywheel storage, the pumped storage and the total feed-in. In the start-up phase ($t < 5$ min), the total power fed in may be less than 1 MW. From $t = 5$ min, it must be ensured that the power is 1 MW.

The difference from the target power of 1 MW is then the power required to be provided by the FWS. There is no reason for the FWS to deliver more power than is needed at this time. Therefore, this power is also the maximum power for the FWS.

$$G_{\mathrm{SFW}}^{\max} = 1\,\mathrm{MW} - G_{S_{\mathrm{PS}}}(t = 5\ \mathrm{min}) = 1\ \mathrm{MW} - 1.153\ \mathrm{MW} = 0.847\,\mathrm{MW}.$$

How large must the storage capacity of the FWS be so that the performance profile shown in Fig. 6.19 can be run and the start-up phase of the pumped storage is completely compensated? What is the maximum E-Rate at which the storage is operated?

Solution: The power profile is triangular. Therefore, the integral of the power profile can be determined by the area of the triangle:

$$\kappa_{\mathrm{FW}} = \frac{1}{2} 847\ \mathrm{kW}\ 32.68\ \mathrm{min} = \frac{1}{2} 847\ \mathrm{kW} \frac{32.68}{60}\ \mathrm{h} = 230.66\ \mathrm{kWh}.$$

The E-Rate results from the ratio of the energy content and the maximum or final loading rate:

$$\frac{847\ \mathrm{kW}}{230.66\ \mathrm{kWh}}\mathrm{h} = 3.6721\ \mathrm{E}.$$

The required storage capacity of the FWS is low, but in this exercise the E-rate is high. For the technical realization, one would not use one flywheel unit, but combine several units analogously, as shown in Fig. 6.10. By design, the units provide specifications on their capacity and E-Rates.

Exercise 6.14 Units with a storage capacity of $\kappa = 30\,\mathrm{kWh}$ and a maximum power of $200\,\mathrm{kW}$ are to be used for the FWS. How many units must be used? What is the resulting capacity?

Solution: We determine the number of FWS units required with respect to the power:

$$N = \frac{847\ \mathrm{kW}}{200\ \mathrm{kW}} = 4.235 \approx 5.$$

Here we have to round up the number of FWS, as four can only provide 800 kW of power and thus the target power cannot be reached. The number of FWSs must, of course, be an integer.

These five FWSs have a total capacity of:

$$\kappa_{FW} = 5 \cdot 30 \text{ kWh} = 150 \text{ kWh}.$$

However, we need 230 kWh. We are still missing 80 kWh of storage capacity, which corresponds to three FWS units. For this reason, the total number of storages must be eight instead of five.

As we saw in Exercise 6.14, overcapacity can occur in the design of the storage system. In this case, the FWS can provide much more power than is actually needed to start up the pumped hydro storage plant. The eight FWSs together would be able to feed in a power of 1.6 MW. In order to optimize here, FWS of different designs can be compared with each other. The calculation of the E-Rate can help us here.

Exercise 6.15 From Exercise 6.13, we know that the FWS has an E-Rate of 3.6721 E. What capacity and power would the FWS need to have if we only want to use four FWSs?

Solution: The E-Rate represents a relationship between power and storage capacity. From the default for the number of FWSs, the power rating of the individual storage is:

$$P_{FW} = \frac{847 \text{ kW}}{4} = 211.75 \text{ kW}.$$

Using the E-Rate, we can now determine the required storage capacity of each storage component:

$$\kappa = \frac{211.75 \text{ kW}}{3.6721 \text{ E}} = 57.66 \text{ kWh}.$$

Thus, if we fix the number of FWSs to four, the single FWS must provide a power of 211.75 kW and have a capacity of 57.66 kWh.

Here the importance of the E-Rate for the system design becomes clear. If we determine the E-Rate, we can compare different FWSs. The closer their E-Rate is to our target, the better the storage fits our task and overcapacity does not occur.

6.6 Conclusion

In this chapter, we have learned about mechanical storage systems. These are based on the use of the two forms of energy known to classical mechanics: potential energy and kinetic energy. Kinetic energy is preferably used in FWS. Potential energy is used in pumped hydro storage plants. Another form of potential energy utilization is compressed air storage, where gases are compressed and energy is recovered by expanding the gas.

Pumped storage and compressed air storage power plants are very usable for energy applications. FWS is used as power storage because of its high self-discharge rate.

In the next chapter, we will look at electrical storage technologies.

Chapter summary

- Three energy forms can be used for storing mechanical energies: kinetic energy, potential energy, and deformation energy.
- If we want to use potential energy, we need to lift mass from one level of height to another. This is preferably done with fluids, like in pumped storage systems.
- Pumped storage power plants have a large capacity and a very low self-discharge range. They can be used for seasonal storage of energy.
- Pumped storage power plants are seldom used as power storage.
- If we want to store energy in kinetic energy, rotational kinetic energy is the prefered form. This is because the conversion from electrical energy to kinetic energy and vice versa is done with electric motors, which are generating rotational energy.
- Flywheels are the established way of storing kinetic energy. Flywheels are used as power storage systems, because they allow fast and dynamic charging and discharging.
- Flywheels have a high self-discharge rate, because of friction. Therefore, they are not used for storing larger amounts of energy over a long time.
- If we want to use deformation energy to store energy, we prefer to use gas.
- We can compress gas in a isothermic process; in this case, the temperature of the gas is constant. We can compress gas in an isochore process; in this case, the volume of the gas stays the same. We can also compress gas in an isobar process; in this case, the pressure stays the same. Out of these three processes, for our purposes the isothermic process is the preferred choice.
- We can even improve the storage capacity, if we realize an adiabatic process, where the thermal energy is constant during the compression process. In this case, we are able to store more energy by the same rate of compression.

7
Electrical storage systems

7.1 Introduction

In this chapter, we want to investigate electrical storage systems. Our everyday life is hardly imaginable without electrical storage systems. They are part of every electronic circuit. Here they serve as energy or power storage or are used as filters. In the right combination, they form oscillating circuits and enable the emission of modulated electromagnetic waves. The time scales of their charging and discharging cycles are in the range of milliseconds and below. The capacities of these storages are therefore very small.

Most of the storage technologies described in this book are used to store energy in the form of electric current. However, electrical storage devices themselves are only used for storage in a few applications. They are always present in transfer technologies and circuits. Large electrical storage systems are rarely found. This is due to their high cost.

For mechanical storage systems, we have two basic principles to choose from. First, we could store energy by changing the position of mass—that is, potential energy. Alternatively, we could store energy by setting a mass in motion—that is, kinetic energy.

We also have these two possibilities with electrical storage. The equivalent of using potential energy is the capacitor, which we already got to know as a electric component in Chapter 5. Here, energy is stored by transporting charge carriers from one side of the capacitor to the other side and storing them. In this chapter, we will look at an other variant of the capacitor, the double-layer capacitor.

The second way to store energy electrically is via the coil. In Chapter 5, we could show that the coil can be seen as a current storage device. If we interpret electricity as moving charge carriers, then current storage is the equivalent of flywheel storage.

In this chapter, we will take a closer look at both storage technologies.We start with the less common type of storage, the current storage, and then we turn to the double-layer capacitor. We conclude the chapter with an application example for double-layer capacitors. With the help of double-layer capacitors, we want to store the released potential energy of a lift and make the system capable of recuperation.

7.2 Storage of electrical current

In this section, we look at the possibility of storing electrical current. As with mechanical storage systems, storing motion is difficult. Two issues arise: linear motion takes up a lot of space, as mass moves away from a point, and motion inevitably creates friction

Energy Storage Systems. Armin U. Schmiegel, Oxford University Press. © Armin U. Schmiegel (2023).
DOI: 10.1093/oso/9780192858009.003.0007

and this creates heat and energy loss. In mechanical storage systems, we have solved both challenges with a flywheel storage system. Instead of a translational motion, we used a rotational motion, which we also used for the conversion of motion into electric current. Friction remained a problem, but we were able to reduce it with the help of vacuum pumps and magnetic bearings.

7.2.1   Basic mechanism of current storage

What about electric current? Electric current is the change in charge density at a location per time, or, in other words, moving charge carriers. Current must therefore always flow in a loop. The storage capacity of this current loop, i.e. how much current we can store in this current loop, results from the inductance of the current loop L and the current I:

$$\kappa = \frac{1}{2}LI^2.$$

So we have to increase the inductance of the current loop to get a high storage capacity. Alternatively, we can also increase the current, but every circuit has conduction losses that are proportional to the resistance R and I^2.

Let us consider a simple circuit for storing current. Fig. 7.1 shows this circuit. We have a load resistor R_{Load}, which represents the load that we want to supply. Switch S_1 allows us to connect the load to a power supply U_{DC}. Switch S_2 allows us to connect a current storage device. This consists of a coil L and a resistor R. The resistor R represents the conduction losses of the circuit.

Let us first consider the charging process. In this case, S_1 is closed and S_2 is connected to the voltage source. We do not need to consider the left mesh, which contains the load resistance, if we want to consider the charging process of the storage. Here, on a time scale of $\tau_{\text{L}} = \frac{L}{R_{\text{S}}}$, the current I_{R} is just building up:

$$I_{\text{R}} = \frac{U_{\text{DC}}}{R_{\text{S}}}\left(1 - e^{-\frac{t}{\tau_{\text{L}}}}\right). \tag{7.1}$$

We have an increase of the current to reach $\hat{I}_{\text{R}} = \frac{U_{\text{DC}}}{R}$. At $t = \tau_{\text{L}}$, $I_{\text{R}} \approx 60\%\hat{I}_{\text{R}}$.

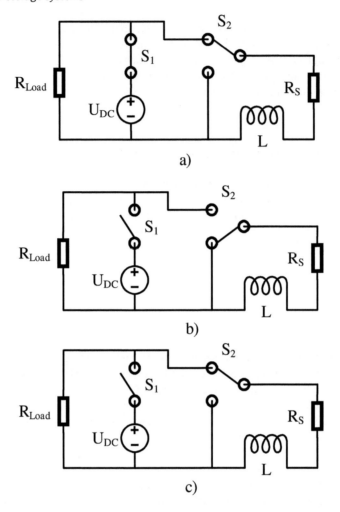

a)

b)

c)

Fig. 7.1 A circuit to store electrical current with an inductance. a) shows the charging process, b) the storage process, and c) the discharging process.

Exercise 7.1 The circuit shown in Fig. 7.1 has the following parameters: $U_{DC} = 100$ V, $R_{Load} = 100$ Ω, $R_S = 12$ mΩ, $L = 342$ mH. What is the size of $\hat{I}_R$? How long does the charging process take to reach a current of 95% from $\hat{I}_R$?

Solution: We are looking for t', at which $I_R = 95\% \cdot \hat{I}_R$. We need to put this value into eqn (7.1):

$$0.95 \cdot \hat{I}_R = \hat{I}_R \left(1 - e^{-\frac{t'}{\tau_L}}\right)$$

$$0.95 = 1 - e^{-\frac{t'}{\tau_L}}$$

$$e^{-\frac{t'}{\tau_L}} = 0.05$$

$$-\frac{t'}{\tau_{\mathrm{L}}} = \log 0.05$$
$$t' = -\tau_{\mathrm{L}} \cdot \log 0.05$$
$$t' = 85 \text{ s}.$$

As soon as we disconnect the storage circuit from the voltage source (Fig. 7.1 b), we have a closed current loop. Since there is electrical current stored in this loop, current continues to flow. So the coil starts to discharge:

$$I_{\mathrm{R}} = \hat{I}_{\mathrm{R}} e^{-\frac{t}{\tau_{\mathrm{L}}}}. \tag{7.2}$$

The discharge process also takes place on a time scale of τ_{L}.

Exercise 7.2 We have charged the storage from Exercise 7.1 to $I_{\mathrm{L}} = 84\% \hat{I}_{\mathrm{R}}$. How long does it take for the storage to self-discharge through the internal resistor R_{S} to reduce the state of charge to $10\% \hat{I}_{\mathrm{R}}$?

Solution: The calculation is analogous to Exercise 7.1. In this case, we use eqn (7.2). However, we have to reduce the amplitude $\hat{I}_{\mathrm{R}}$ proportionally:

$$0.1\hat{I}_{\mathrm{R}} = 0.84\hat{I}_{\mathrm{R}} e^{-\frac{t'}{\tau_{\mathrm{L}}}}$$
$$\frac{0.1}{0.85} = e^{-\frac{t'}{\tau_{\mathrm{L}}}}$$
$$\tau_{\mathrm{L}} \cdot \log \frac{0.1}{0.85} = -t'$$
$$59.2641 \text{ s} = t'.$$

To use the stored current, we need to connect the storage circuit to the load resistor via switch S_2. Since the internal and load resistors are now connected in series, the total resistance is $R_{\mathrm{sum}} = R_{\mathrm{Load}} + R_{\mathrm{S}}$. The discharge of the coil now takes place on the time scale $\hat{\tau}_{\mathrm{L}} = \frac{L}{R}$:

$$I_{\mathrm{Load}} = \hat{I}_{\mathrm{R}} e^{-\frac{t}{\hat{\tau}_{\mathrm{L}}}}.$$

Exercise 7.3 We want to use the storage from Exercise 7.1 to cover the load. At the time of connection, it has a current of $73\% \hat{I}_{\mathrm{R}}$. How long does it take the coil to discharge to $2\% \hat{I}_{\mathrm{R}}$?

Solution: The calculation is analogous to Exercise 7.2. However, in this case the time scale changes because a larger resistance R_{sum} is applied during the discharging process. With $\tau_{\mathrm{Load}} = \frac{L}{R_{\mathrm{sum}}}$, the result is thus:

$$0.02\hat{I}_{\mathrm{R}} = 0.73\hat{I}_{\mathrm{R}} e^{-\frac{t'}{\tau_{\mathrm{Load}}}}$$
$$-t' = \tau_{\mathrm{Load}} \cdot \log \frac{0.02}{0.73}$$
$$t' = 0.0123 \text{ s}.$$

7.2.2 Requirements for a current storage device

Since in the storage process self-discharge takes place on a time scale of τ_L, a current storage device in the described form is not suitable for storage over a longer period of time. The storage capacity of such a storage is also limited. The energy inside a coil is $\frac{1}{2}LI^2$. To increase the capacity, we can increase the inductance of the coil L. According to eqn (5.29), we achieve this by increasing the cross-section of the ferrite core or by increasing the number of windings. However, both lead to a lengthening of the wire, which in turn increases the conduction losses. The resistance R therefore increases.

Alternatively, we can increase the current. This would even be the better alternative, since it is quadratically included in the energy determination. Unfortunately, the cross-section of the winding wire limits the current. If we want to increase the current, the cross-section of the wire must also be increased. But this also increases the resistance. Because of the electrical resistance, we are limited in storage capacity and in reducing the self-discharge rate. Therefore, in practice, electricity storage is only used for storing electricity for a very small period of time. This leads us to the first requirement for current storage (CS).

CS 1: AS A design engineer, I WANT to keep the internal resistance as low as possible SO THAT the self-discharge of the power storage device is low.

For most electronic and power electronic circuits, the storage capacity of coils is sufficient. However, classic coils cannot be used as energy storage systems. The energy content is too low for this, and due to self-discharge they are also not suitable as power buffers. However, this looks different if we find a technology that reduces the resistance of the coil to zero.

7.2.3 Superconductivity in a nutshell

So far, we have accepted electrical resistance as a property of materials. We have even seen that resistance can be useful when we want to prevent high discharge currents in a capacitor or an inductor. This is because in both cases the amplitude at the time of switching on is $\frac{U_0}{R}$ in each case. We have also established that the conduction losses are $P_L = RI^2$. However, we have not talked about the origin of the electrical resistance and how a conductor can lose its resistance.

In order to be able to answer this question qualitatively, we unfortunately have to leave the classical physics behind and include quantum physics in our considerations. In quantum mechanics, we distinguish between two groups of particles: fermions and bosons (Bransden and Joachain, 1989). Fermions have a half-integer spin $\frac{n}{2}\hbar$—that is, their quantum mechanical intrinsic rotation, the spin, is never zero. Fermions have the property that two fermions cannot occupy the same quantum mechanical state. Bosons, on the other hand, have an integer spin $n\hbar$. Unlike fermions, bosons can occupy the same quantum-mechanical state. What does this mean? We want to illustrate the difference with a simple example. We want to organize a party, but we are inviting humans that have the properties of fermions and bosons. If we are now inviting only fermions to the party, we will realize that each person looks different, stands in a different place, listens to a different type of music, and strictly avoids coming too close to others. On the other hand, if we invite only bosons to our party, everybody looks the same and listens to the same music, and we can hardly distinguish between these people, since they all are occupying the same place.

Bosons like to move together, while fermions prefer to move slightly differently. Electrons are fermions; if they are moving in a conductor, they avoid moving the same way and staying at the same positions, and this interaction causes friction.

Now have a look at a conductor. A conductor consists of an atomic lattice or crystal. In a conductor, there are electrons that are free to move within the lattice. Since these electrons all have a similar energy, one also likes to speak of a conduction band, where the term band means a range of energy in which these electrons are located. Electric current. or more precisely, electrons that are located in this conduction band, consists of a moving charge carrier. Electrons have a negative charge and belong to the group of fermions. If we look at the freely moving electrons in an electrical conductor, they all have a different state. If we apply an electric field E to the conductor, a force $F = -e \cdot E$ acts on all the electrons. The electrons start to follow the field lines. Electric current flows. But since the electrons have a unique state, they cannot all move in the same direction at the same speed. Collisions occur. These are emitted as lattice vibrations in the conductor material as heat. If we switch off the field, the current stops immediately, because the electrons give up all their kinetic energy to the crystal lattice of the conductor.

The lattice vibrations bump electrons and are vibrated by electrons. This creates friction and thus causes energy loss. But if we now cool the conductor down further and further, we observe that the resistance becomes smaller and smaller. For some substances, it remains constant above a certain temperature. This is because vibrational states are still possible or necessary in the atomic lattice. With other materials, we notice that from a temperature T_c, the resistance suddenly disappears. At this temperature, the conductor becomes a superconductor.

What happens at T_c? When the atomic lattices oscillate less and less, another effect becomes visible; we have illustrated it in Fig. 7.2. When the atoms in the lattice remain at rest, a single electron causes the atomic shells of the atoms around it to align and polarize. They form a positively charged shell around the electron. This positive charge attracts other electrons. These also generate a positive charge via the atomic shells of the atoms surrounding them, which attract the first electron. In this way, the two electrons interact. This interaction is a long-distance effect and has an effect over up to 100 atoms in a lattice.

This interaction does not only happen between two electrons, but all electrons that are within the range of this long-range effect interact with each other. This leads to additional movements. But as soon as two electrons have opposite momentum and spin, something new happens. These two electrons form a Cooper pair. They remain coherent, and since they have integer spin, together they now belong to the group of bosons.

As we have already discussed, bosons have the possibility of assuming the same quantum state. More and more Cooper pairs can now form, and all of them can be in the same state. Once the pair formation is complete, there are Cooper pairs in the conduction band of the conductor that react completely uniformly when an electric

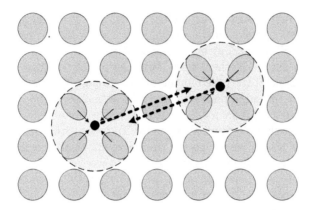

Fig. 7.2 Formation of a Cooper pair in an atomic lattice. An electron generates a polarization of the atoms surrounding it. This positive charge attracts other electrons, which in turn create a polarization that also attracts the original electron.

field is applied to them. There is no more friction and thus no electrical resistance (Fossheim and Sudbø, 2004; Kruchinin, 2021).

To reach this state, three physical quantities must be below a material-dependent threshold. We have already mentioned the critical temperature T_c. In addition, there is also a dependence on the current density j_c and the external magnetic field H_c acting on the conductor.

The fact that the current density must be below a critical value j_c can be explained by the formation of Cooper pairs. A high temperature creates additional oscillations in the atomic lattice, so that Cooper pairs cannot form. An electric field that is too strong, which would produce a higher current density, is also a obstacle. Due to the higher energy, the long-range effect between the polarized atomic shells and the electrons cannot form, or is disturbed, which leads to the Cooper pairs no longer being able to form.

To understand the influence of an external magnetic field, consider the experiment shown in Fig. 7.3. The temperature of the conductor is initially still above the critical temperature, T_c. The conductor does not have current flowing through it, but is placed in a magnetic field. Above T_c, we can observe that the field lines completely penetrate the conductor, as shown in Fig. 7.3 a). As soon as the transition temperature is reached and the conductor becomes superconducting, we observe that the magnetic field lines are forced out of the conductor. The magnetic field lines penetrate only a few nanometres into the conductor. They run along the inside of the outer surface of the conductor. This is because the magnetic field induces a current in the conductor. This generates an oppositely directed field. In a normal conductor, the internal resistance ensures that this opposing field is compensated, and the external field prevails. In a superconductor, there is no resistance and the opposing field completely compensates for the magnetic field or redirects it.

If we repeat the experiment with different field strengths, we find that in addition to the critical temperature, there is also a critical magnetic field strength H_c, but this effect no longer occurs. We can easily deduce the reason: the stronger the field

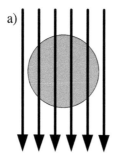 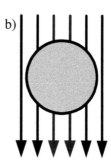

Fig. 7.3 Illustration of the Meissner–Ochsenfeld effect. A magnetic field flows through a conductor. a) If the temperature is above the critical temperature T_c, the magnetic field lines pass through the conductor. b) As soon as the conductor becomes superconducting, the field lines are forced outwards and run along the surface of the conductor.

strength, the more field lines are forced into the surface of the conductor. There, the energy density grows and the compensating counter-field generates higher and higher currents. At some point, this density is too high and the compensation no longer takes place.

We have described here the superconductor of the first type. In a superconductor of the first kind, there is a hard transition between superconducting and normal conducting at H_c. If we want to construct a current storage device with a superconductor of the first type, we must ensure that the operating point is operated well below H_c.

CS 2: AS A User, I WANT to select the operating point in such a way that a hard transition between superconducting and normal conducting cannot occur SO THAT it is ensured that the stored energy is not suddenly lost abruptly through the electrical resistance.

CS 2 is a safety-relevant requirement. If the operating point is too close to H_c, a disturbance of the magnetic field can cause the superconductor to become a normal conductor. In this case, the stored electrical energy is consumed by the electrical resistance, which produces heat and can damage the storage.

An alternative to the superconductor of the first type is the superconductor of the second type. Here there are two critical field strengths, H_{c1} and H_{c2}. If the field is weaker than H_{c1}, the conductor is superconducting. If the field is stronger than H_{c2}, the conductor is conducting. In the intermediate range, there is a mixed state in which not all Cooper pairs have dissolved. This means that there is no longer an abrupt change, so that the heat development can also be better controlled by the electrical resistance.

In addition to the superconductors of the first and second types, which all operate at very low temperatures $< 50°$K, there are the high-temperature superconductors. These are superconducting at $92-138°$K and are already used for many applications. The interesting observation is that these high-temperature superconductors, which are made of ceramics, become superconducting at a temperature for which the Cooper pair formation model described above actually no longer applies. This is because at a temperature above $50°$K, the atomic lattice fluctuations are so strong that the interaction necessary for the formation of the Cooper pairs is no longer visible. Experiments have shown, however, that Cooper pairs also exist in high-temperature superconductors.

The fact that superconductance shows no resistance to a direct current (DC) does not mean that this is also the case for alternating current (AC). In fact, if an AC flows through a superconductor, we will measure a resistance, which depends on the frequency of the AC.

7.2.4   Example of the realization of a superconducting magnetic energy storage (SMES)

With the help of a superconducting material, we are able to build a current storage device that has a significantly lower self-discharge rate than a coil made of normal material. In this section, we will look at the realization of an SMES.

The storage capacity of a superconducting coil is determined by two quantities. Since the current density must be below the critical value of j_c, the current cannot become arbitrarily large. There is therefore a maximum current I_{max} with which the storage can be charged. In order to increase the storage capacity, the inductance can also be increased; this can be achieved by increasing the number of turns or by changing the geometry. However, there is a limitation here given by the critical field strength H_c. These limitations refer to a coil. However, it is still possible to increase the storage capacity by using several connected coils.

To realize a current storage device, we use the circuit shown in Fig. 7.4. Between two switches, S_1 and S_2, we have a SMES. In addition, we have two diodes, D_1 and D_2, connected to the coil, which we need for final charging and during storage.

To charge the coil, S_1 and S_2 are switched on. We can influence the charging current I_L by periodically switching S_1 on and off. Here d_1 is the duty cycle:

$$d_1 = \frac{T - t_{on}}{t_{on}}.$$

The duty cycle is the ratio of the switch-on time, t_{on}, and the switch-off time in relation to a switching period, T. By adjusting the duty cycle, not only can we adjust the charging current, but we can also influence the voltage, U_{DC}. The following applies:

$$C\frac{d}{dt}U_{DC} = I_{DC} - I_L \cdot d_1. \tag{7.3}$$

When discharging, S_1 remains open, and we open S_2 with a duty cycle of d_2. The following then applies to the voltage U_{DC}:

$$C\frac{d}{dt}U_{DC} = I_{DC} + I_L \cdot (1 - d_2). \tag{7.4}$$

During both charging and discharging, a state appears in which the current flows 'in a circle' through the diode, the coil and the switch S_2. This is the state in which the energy is stored.

With a superconducting coil, we have no resistance to limit the current amplitude when charging or discharging. At the same time, we know that there is a critical current density, j_c, that is necessary to maintain superconductivity. We are therefore forced to be able to regulate the charging and discharging current.

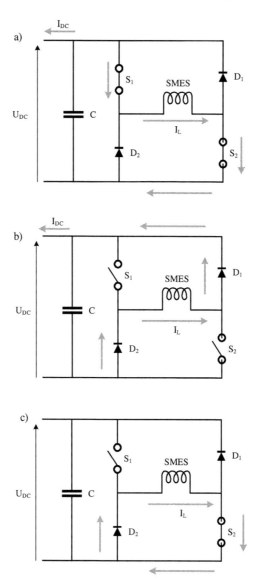

Fig. 7.4 Example of a circuit for an SMES. The system consists of two switches and two diodes, as well as a superconducting coil. a) If S_1 and S_2 are switched on, the coil is charged. b) For discharging, the two switches are opened. c) To store the energy, S_1 is opened and S_2 remains closed.

CS 3: AS A User, I WANT to regulate the charging and discharging current SO THAT there is no damage to components and the superconductivity is maintained.

We can comply with CS 3 by various measures. Since charging and discharging also involve components that are normal conductors, there is a natural limitation of the current amplitude to $\frac{U_{\text{DC}}}{R_{\text{ges}}}$, where R_{ges} represents the resistance of the entire charging and discharging circuit. This represents a passive safety measure. Furthermore, the choice of duty cycle gives us the possibility to control the current. However, this only affects the duration of the current load, not the amplitude of the load. By choosing a certain duty cycle, we can only say how long the coil is to be charged or discharged for. We cannot directly influence the current peak at the beginning of switching on. Therefore, additional protective circuits are used, for example a coil, to reduce current peaks. These only affect the charging and discharging circuit, not the storage circuit.

The system components and their responsibilities for compliance with requirements CS 1 to CS 3 are shown in Fig. 7.5. The basic components `PowerDistributionControl` and `TemperatureControl` have additional tasks in this system. The `TemperatureControl` must also ensure that the temperature of the superconductor remains below T_{c} (CS 2).

In this system, the `PowerDistributionControl` has the task of ensuring that the current remains below the critical current density (CS 2) and additionally regulates the charging and discharging currents, as well as the power transfer over the duty cycle (CS 3).

In order to additionally compensate for the current peaks during charging and discharging, we have a normal `Inductance` as a system component in the system, which is jointly responsible for CS 3.

7.2.5 Conclusion

To store electricity effectively, we have to struggle with similar problems, as in the storage of kinetic energy in classical mechanics. Since motion always means friction, the main task is to reduce friction. In electrical systems, this means reducing the electrical resistance. We can do this with the help of superconductors. There are various materials that can become superconducting. It turns out that three physical parameters must be below a critical value. First, the temperature must be low enough: $T < T_{\text{c}}$. Furthermore, the magnetic field strength must not exceed a certain value, $H < H_{\text{c}}$, and the current density within the superconductor must not exceed a certain value, $j < j_{\text{c}}$.

These three boundary conditions limit the use of superconducting storage systems. While compliance with the critical temperature is only a question of system engineering, the restrictions on the current and the magnetic field have an influence on the storage capacity of the superconductor. Like flywheel storage, superconducting electricity storage systems are more suitable as power storage systems rather than energy storage systems (Ise, Kita, and Taguchi, 2005; Li et al., 2015; Li et al., 2017).

In the next section, we will look at the second way to store electrical energy electrically: the voltage storage. We have already learned about the component needed for this: the capacitor.

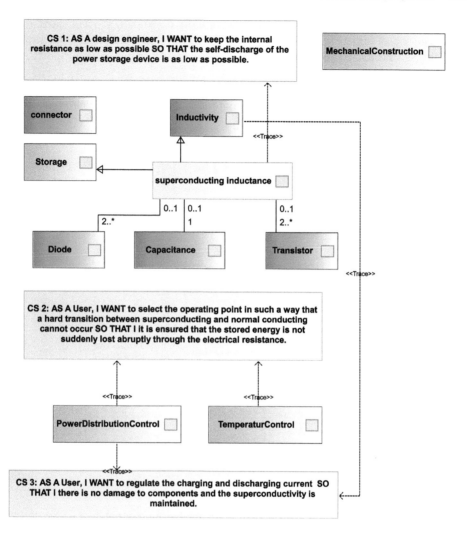

Fig. 7.5 System components of a superconducting magnetic energy storage. The key component is the superconducting inductance, which is a specialized form of a storage component and an inductance.

7.3 Voltage storage systems

The basic idea of the pumped hydro storage plant in Section 6.2 was to store energy by lifting material higher. To do this, gravity had to be overcome and potential energy was in the material, which could be recovered by controlled release. A similar approach can be taken in the storage of electrical energy. Instead of mass particles, charge carriers are separated from each other here and distributed spatially. The system component that works according to this principle is the capacitor, which we got to know in Section 5.2.2. In this chapter, we want to take a closer look at the possibility of using capacitors as energy storage systems. We will first look at the classical capacitor. Then we will show

how the capacity of a capacitor can be additionally expanded. We will learn about the construction of supercapacitors (or supercaps) and electrochemical capacitors. Both use effects that arise when the dielectric has electrically conductive properties.

7.3.1 Aggregation of capacitors: How capacitor banks are designed

As we saw in Section 5.2.2, the energy content of a capacitor is calculated to:

$$\kappa = \frac{1}{2}C\left(U_1^2 - U_0^2\right),\tag{7.5}$$

where U_1 is the upper operating voltage and U_0 is the lower operating voltage. In eqn (7.5), we assume that we operate the capacitor in a certain voltage range and do not discharge it completely to 0 V.

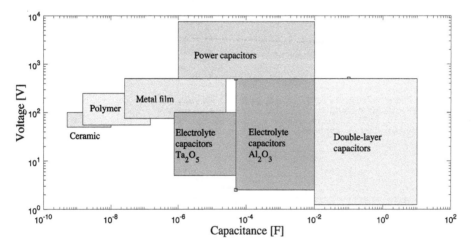

Fig. 7.6 Capacitance and voltage of different types of capacitors (Kurzweil and Dietlmeier, 2018).

We have only two possibilities to increase the energy content of a single condenser. We can increase the voltage or increase the capacity. Fig. 7.6 shows the capacities and voltage range of different technologies. The classical capacitors differ only in their dielectric: ceramic, polymer, metal film, and electrolyte. As we can see, both the maximum voltage range and the capacitance are limited for each technology. The dielectric strength remains limited to a few hundred volts. The capacity reaches a maximum of a two-digit millicoulomb value.

Exercise 7.4 Classical capacitors are distinguished by their dielectrics. How large is the energy content of the following capacitors if you are allowed to discharge the capacitor to 10% of the maximum voltage?

- Ceramic $U_{max} = 80$ V, $C = 12$ nF
- Metal film $U_{max} = 230$ V, $C = 24$ μF
- Electrolyte $U_{max} = 320$ V, $C = 3$ mF

Solution: The energy content is calculated according to eqn (7.5). This results in the following for the ceramic capacitor:

$$\kappa = \frac{1}{2}C\left(U_1^2 - U_0^2\right)$$
$$= \frac{1}{2}12 \cdot 10^{-9}\text{ F}\left((80\text{ V})^2 - (8\text{ V})^2\right)$$
$$= 3{,}802 \cdot 10^{-5}\text{ Ws}.$$

The energy content of the metal film capacitor is:

$$\kappa = \frac{1}{2}24 \cdot 10^{-6}\text{ F}\left((230\text{ V})^2 - (23\text{ V})^2\right)$$
$$= 0.628 \cdot 10^{-4}\text{ Ws}.$$

The largest energy content is found in the electrolytic capacitor:

$$\kappa = \frac{1}{2}3 \cdot 10^{-3}\text{ F}\left((320\text{ V})^2 - (32\text{ V})^2\right)$$
$$= 0.042\text{ Ws}.$$

The results from Exercise 7.4 confirm that energy storage is not a suitable application for capacitors. The energy content is very small for all three technologies and always remains below one watt second for the individual capacitor.

$$C_g = C_1 + C_2 \qquad\qquad \frac{1}{C_g} = \frac{1}{C_1} + \frac{1}{C_2}$$

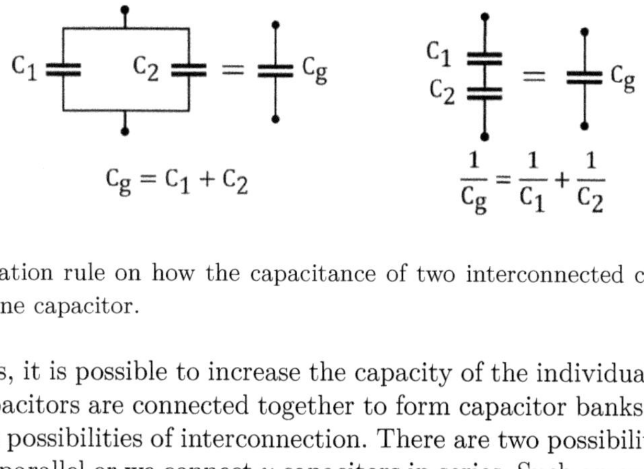

Fig. 7.7 Calculation rule on how the capacitance of two interconnected capacitors can be represented by one capacitor.

Nevertheless, it is possible to increase the capacity of the individual capacitors. To do this, the capacitors are connected together to form capacitor banks. In Fig. 7.7, we have shown the possibilities of interconnection. There are two possibilities: we connect x capacitors in parallel or we connect y capacitors in series. Such an overall connection is then called the xpys-connection. To understand the rules of calculation, let us look at the simple case of a 2p1s and 1p2s.

In an xp1s circuit, the capacitance values add up as follows:

$$C_g = \sum_{i=1}^{x} C_i. \qquad\qquad (7.6)$$

This is because the upper and lower capacitor plates of the individual capacitors are at the same electrical potential. As we know, the capacitance is proportional to the area. But if both capacitor plates are at the same potential, it makes no difference whether they belong to two capacitors or to one capacitor whose plates have the total area of both capacitors. Therefore, we can mentally add the areas of the capacitor plates to one large plate, which is equivalent to adding the individual capacitances for the total capacitance.

In a 1p2s circuit, the reciprocal values add up as such:

$$\frac{1}{C_g} = \sum_{i=1}^{y} \frac{1}{C_i}. \tag{7.7}$$

To understand this behaviour, we make clear that the total voltage U_g over the entire capacitor chain must be equal to the sum of the voltages of the individual capacitors U_i:

$$U_g = \sum_{i=1}^{y} U_i. \tag{7.8}$$

We know that the capacitance is determined from the ratio of the charge by the voltage: $C = \frac{Q}{U}$ with $U = \frac{Q}{C}$ gives eqn (7.7).

Exercise 7.5 Two capacitors are given: $C_1 = 15 \ \mu F$ and $C_2 = 23 \ \mu F$. Both have a maximum voltage of $U_{max} = 14$ V. What is the capacitance Cg and the energy content κ for a parallel and a series connection?

Solution: For the parallel connection, the capacitance is:

$$C_g = C_1 + C_2 = 15 \ \mu F + 23 \ \mu F = 38 \ \mu F.$$

The energy content of the capacitor at U_{max} is thus calculated as:

$$\kappa = \frac{1}{2} C U_{max}^2 = \frac{1}{2} 38 \ \mu F \ (14 \ V)^2 = 0.0037 \ Ws.$$

For the serial connection, eqn (7.7) results:

$$C_g = \frac{C_1 C_2}{C_1 + C_2} = \frac{15 \ \mu F \ 23 \ \mu F}{15 \ \mu F + 23 \ \mu F} = 9.0789 \ \mu F.$$

With a serial connection, the voltages of the individual capacitors add up. Therefore, we have to calculate with twice the voltage here. The energy content of the series-connected capacitors thus results in

$$\kappa = \frac{1}{2} C U_{max}^2 = \frac{1}{2} 9{,}0789 \ \mu F \ (2 \cdot 14 \ V)^2 = 0.0037 \ Ws.$$

As the example shows, the serial connection reduces the capacity C_g, but the voltage increase compensates for this effect. However, the current limitation remains—that is, capacitors connected in series can only carry as much current as is allowed to flow through the individual capacitor. To allow a higher current at the same voltage, capacitors must be connected in parallel. The current which then flows at the connection point of the capacitor bank is the sum of the individual currents—that is,

$$I_g = \sum_{i=1}^{x} I_i. \tag{7.9}$$

If a capacitor bank is used as energy or power storage, it is connected to a DC bus. The application usually requires a minimum energy content, κ_{min}, to be present. As we know from Section 5.3, a DC bus is described by at least three quantities: the allowed voltage interval $[U_{min}, U_{max}]$, as well as the maximum current I_{max} to be carried over this bus. To optimize the cost of the capacitor bank, we need to minimize the total number of capacitors $N = x \cdot y$. At the same time, however, the mentioned constraints must be fulfilled. We are dealing with an optimization problem here:

$$\min N(x, y) = x \cdot y \tag{7.10}$$

with respect to

$$\sum_{i=1}^{y} U_i \leq U_{max} \tag{7.11}$$

$$\sum_{i=1}^{y} U_i \geq U_{min} \tag{7.12}$$

$$\sum_{i=1}^{x} I_i \geq I_{max} \tag{7.13}$$

$$\frac{1}{2} x C \left(\sum_{i=1}^{y} U_i \right)^2 \geq \kappa_{min}. \tag{7.14}$$

Exercise 7.6 A capacitor bank shall have a storage capacity of at least $\kappa = 0.01$ Ws. The voltage interval shall be $U \in [19\,\text{V}, 30\,\text{V}]$. The maximum current shall be at least 10 A.
   The bank shall be composed of capacitors of one type. This has a capacitance of $C = 34\ \mu\text{F}$ and a maximum voltage of $U_{cell,max} = 7$ V, and can carry a maximum current of $I_{cell,max} = 3$ A.
   What is a reasonable configuration?
   Solution: In the first step, we determine the number of capacitors needed for a string to reach the voltage:

$$y = \frac{U_{max}}{U_{cell,max}} = \frac{30\ \text{V}}{7\ \text{V}} = 4.2857 \approx 4.$$

In this case, we want to round down, since we could achieve a maximum capacitor voltage of $U_{max} = 5 \cdot 7\,\text{V} = 35$ V at $y = 5$. (This is a design decision in this case. One could also argue that being able to increase the voltage to above 30 V gives the system more safety.)

Next, we determine the number of strings we need to connect in parallel so that we can transmit the required current of 30 A:

$$x = \frac{I_{\text{max}}}{I_{\text{cell,max}}} = \frac{10 \text{ A}}{3 \text{A}} = 3\frac{1}{3} \approx 4.$$

In this case, we actually have to round up, since three parallel strings will only transfer 9 A.

The last step is to check if we can store enough energy. To do this, we must first determine how high the capacitance of the string is:

$$\frac{1}{C_g} = \sum_{i=1}^{y} \frac{1}{C} = \frac{4}{C} \to C_g = \frac{C}{4}.$$

The energy content is then calculated to:

$$\kappa_y = \frac{1}{2} C_g \left((30 \text{ V})^2 - (19 \text{ V})^2 \right) = 0.0023 \text{ Ws}.$$

We now calculate the number of strings we need to reach the content of κ:

$$\kappa_x = \frac{\kappa}{\kappa_y} = \frac{0.01 \text{ Ws}}{0.0023 \text{ Ws}} = 4.36 \approx 5.$$

So, to be able to store the amount of energy we need, we need at least five strands. This is one more than we originally determined for x, since increasing this increases the maximum current-carrying capacity. The final configuration of our capacitor bank is a 4s5p circuit and requires 20 capacitors.

By interconnecting individual capacitors to form a capacitor bank, we are able to construct a storage that can also be designed for voltages and currents for which the single capacitor is not suitable. We are also able to increase the energy content. In the next section, however, we want to get to know other effects with which we can increase the capacitance of a capacitor.

7.3.2   Double-layer and pseudocapacitance: How supercaps become super

Capacitors store energy via the electric field. We have already seen that we have three adjusting screws for increasing the capacitance: we can choose a suitable dielectric, reduce the plate spacing, or increase the area. But the illustration in Fig. 7.6 shows us that there are limits in both dielectric strength and capacitance that cannot be overcome with this classical design. However, we can overcome these limits by using two additional effects. They are shown in Fig. 7.6 as the capacity of a double layer and the so-called pseudocapacitance, both effects that occur when the dielectric is no longer a solid but a liquid conductive substance, an electrolyte.

Suppose a container is filled with an electrolyte (Fig. 7.8). An electrolyte is a conductive liquid. Unlike an electrical conductor, the charge carrier transport does not take place through conduction of the electrons themselves, but through ionic transport. This means that there is a substance in the liquid that can accept electrons and also later release them.

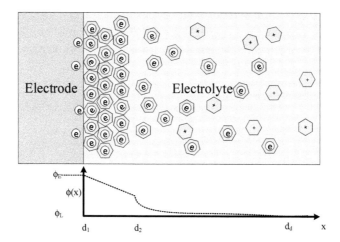

Fig. 7.8 Construction of a double layer at a phase boundary between an electrode and an electrolyte.

The electrolyte in the container has an electrical potential $\phi = \phi_L$. An electrical conductor with a potential $\phi = \phi_E$ is placed in this container. The potentials differ, $\phi_L \neq \phi_E$.

Since there are mobile charge carriers in the electrolyte, they are attracted by the potential difference. They begin to collect on the surface of the electrode. A thin layer of ions forms there. This layer ensures that the potential behind the layer decreases a little: $\phi = \phi_E - \Delta\phi$. The thickness of this first layer d_1 is below the nanometre scale.

At the boundary layer between this ion film and the electrolyte, further charge carriers are now exchanged with the ions of the electrolyte. Another layer is formed. The potential is even lower than that of the previous layer: $\phi = \phi_E - 2\Delta\phi$. In this way, layer after layer builds up and the potential decreases linearly.

Above a layer thickness d_2, the potential difference between the created layer and the electrolyte is too small to build up a new complete layer. Instead, individual ions diffuse to the layer, pick up charge carriers, and distribute themselves in the electrolyte. In this area, the potential now decreases not linearly, but exponentially, until it is equal to the electrolyte potential. This range can extend very far into the electrolyte.

Up to d_2, one speaks of the Helmholtz layer. A distinction is made between the rigid Helmholtz layer up to d_1 and the inner Helmholtz layer up to d_2. The resulting capacitance is analogous to the capacitance of a capacitor; therefore, the total capacitance at the electrode surface is calculated from the capacitances of the rigid C_1- and inner C_2 Helmholtz layer and the capacitance of the diffuse double layer:

$$\frac{1}{C} = \frac{1}{C_L} + \frac{1}{C_H} + \frac{1}{C_d} = \frac{1}{A\epsilon_0}\left(\frac{d_1}{\epsilon_1} + \frac{d_2}{\epsilon_2} + \frac{d_d}{\epsilon_d}\right). \tag{7.15}$$

The rigid and inner Helmholtz layers have a thickness below the nanometre scale of $d_1 \approx 0.05$ nm and $d_2 \approx 0.2$ nm respectively. The diffuse bilayer has a nanometre-scale thickness of $d_d \approx 5$ nm.

The thickness of these layers depends on the mobility and size of the ions in the electrolyte. Large ions increase the thickness of the layers and decrease the capacity. Less mobile ions reduce the thickness of the diffuse layer and increase the capacity.

Exercise 7.7 The permittivity for a double layer would be $\epsilon_1 = 6$, $\epsilon_2 = 30$, and $\epsilon_d = 80$. What is the total capacity per square centimetre $\frac{C}{A}$?
    Solution: Eqn (7.15) gives the total capacitance:

$$\frac{1}{C} = \frac{1}{A\epsilon_0} \left(\frac{d_1}{\epsilon_1} + \frac{d_2}{\epsilon_2} + \frac{d_d}{\epsilon_d} \right)$$

$$= \frac{1}{A\epsilon_0} \left(\frac{0.05 \cdot 10^{-9} \text{m}}{6} + \frac{0.2 \cdot 10^{-9} \text{m}}{30} + \frac{5 \cdot 10^{-9} \text{ m}}{80} \right)$$

$$= \frac{1}{A\epsilon_0} 7.75 \cdot 10^{-11} \frac{1}{\text{F}} = \frac{1}{A} 8.7529 \frac{1}{\text{F}}$$

$$\Rightarrow \frac{C}{A} = 11 \frac{\mu\text{F}}{\text{cm}^2}.$$

The capacitance of a capacitor depends on the distance between the conductors and the area of the conductor plate. The very thin double layer therefore has an extremely small spacing compared to classic capacitors. However, the capacitance of this can be increased even further by significantly increasing the surface area of the conductor. This is done by coating the metallic conductor with a porous granulate of carbon, for example. Activated carbon has a surface area of up to $2{,}000 \frac{\text{m}^2}{\text{g}}$. This relatively large surface area, wetted with an electrolyte, leads to the high capacitance values that double-layer capacitors achieve.

Exercise 7.8 The total capacitance of a double layer is $\frac{C}{A} = 11 \frac{\mu\text{F}}{\text{cm}^2}$. What would be the capacitance with respect to the electrode material if it had a specific surface area of $2{,}000 \frac{\text{m}^2}{\text{g}}$?
    Solution:

$$\frac{C}{m} = 2{,}000 \frac{\text{m}^2}{\text{g}} \cdot \frac{11 \ \mu\text{F}}{0.01 \ \text{m}^2} = 220 \frac{\text{F}}{\text{g}}. \tag{7.16}$$

We can consider the formation of a double layer as a rearrangement effect. Due to the potential difference between ϕ_L and ϕ_E, the charge carriers have to rearrange themselves, and from a distance d_2 onwards, the effect of thermodynamics is added, so that the bonding is no longer strong enough to resist the thermal motion of the charge carriers, and diffusion occurs. However, there is another effect that can occur in a boundary layer between electrolyte and electrode.

Consider an electrolyte in which an electrode is immersed and in which the double layer has already formed (Fig. 7.9). On the surface of the electrode is the substance B. There are atoms of the substance A in the electrolyte. These also reach the surface of the electrode. The following reaction can now occur there:

$$A + B - e \rightleftharpoons AB.$$

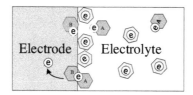

Fig. 7.9 Representation of a redox reaction in a double layer. The electrode or the surface of the electrode consists of the substance B. The substance A is in the electrolyte. Together they form the substance AB, whereby one electron is emitted.

The two substances A and B combine to form a substance AB and emits an electron in the process. This is an ionic, reversible reaction. Only one electron is donated in this reaction, and both atoms share a common electron. In this chemical reaction, there is another potential change, which can be interpreted as an additional capacity. Since this is chemical and not electrical in nature, it is referred to as pseudocapacitance.

Both effects, pseudocapacitance and double-layer capacitance, occur together. Through suitable selection, one effect can be more pronounced than the other, which leads to the fact that one speaks of double-layer capacitors when one effect dominates, and of pseudocapacitors when the other effect dominates (Halper and Ellenbogen, 2006). One also likes to summarize the totality of these capacitors as supercaps.

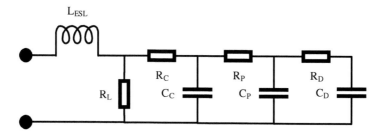

Fig. 7.10 Equivalent circuit diagram of a supercap. The different capacitance components are represented by separate RC elements. The frequency-dependent resistance of the capacitor is represented by an inductance I_{ESL}.

The charging and discharging behaviour of these supercaps differs from the behaviour of classical capacitors. This is because the charge can be exchanged between the different capacitances. In addition, the dynamic of the different capacitances is different. In Fig. 7.10 an equivalent circuit is shown, which can be used to illustrate the behaviour. The frequency-dependent losses are represented in this circuit diagram by the inductance L_{ESL}. Since a capacitor has a weak self-discharge, we have introduced the resistance R_L, which forms the leakage currents. R_L is very large, so that these currents are also very small. C_C represents the capacitance that the capacitor has without the pseudocapacitance C_P and the double-layer capacitance C_D. Since losses can also occur during the exchange between the capacitance components, an additional loss resistance, R_C, R_P, R_D, was introduced in each case.

We have seen that the dynamics of a classical capacitor take place on the time scale $\tau = RC$. At $t = \tau$, the capacitor is about 60% charged or discharged. We can therefore determine with a simple estimate how large the chosen capacitance would have to be in order to store energy over a certain period of time. A supercap has three time scales, τ_C, τ_P, τ_D. To estimate the resulting total capacitance, we can add the three capacitances and neglect the influence of the loss resistances. However, the three time scales have an influence on the behaviour of the supercap, which we have to consider in the system design.

We can illustrate the behaviour with the following thought experiment. Assume we have two capacitors with different time scales $\tau_1 \ll \tau_2$. The two capacitors are connected in parallel and are charged. The capacitor C_1, with its time scale τ_1, will dominate the dynamics of the charging process at the beginning. It will be charged very quickly. The second capacitor C_2 will be charged more slowly. Only at a later time will it dominate the dynamics of the charging process. Suppose we interrupt the charging process before C_2 has been fully charged. In this case, there is an exchange of charge carriers between C_1 and C_2, with C_1 giving off charge. We observe a drop in the total voltage.

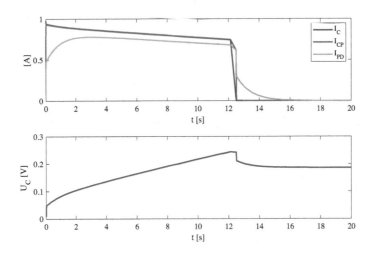

Fig. 7.11 Charging process of a supercap. Up to the time $t = 12$ s, the capacitor is connected to a voltage source. After that, the terminals are opened. U_C represents the voltage at the capacitor. To be able to see the charge distribution between the layers, different currents are shown. I_C represents the current into the capacitor, I_{CP} is the current across the resistor R_P and I_{PD} is the current across the resistor R_D. The parameters were chosen in such a way that the dynamics can be easily seen ($C_C = 0.1$ F, $R_C = 0.01$ Ω, $C_P = 5$ F, $R_P = 0.1$ Ω, $C_D = 50$ F, $R_D = 0.1$ Ω).

Let us now consider Fig. 7.11. We have performed the same experiment with a supercap. We have chosen the parameters of the supercap so that the effect is easy to observe ($C_C = 0.1$ F, $R_C = 0.01$ Ω, $C_P = 5$ F, $R_P = 0.1$ Ω, $C_D = 50$ F, $R_D = 0.1$ Ω).

Up to the time $t = 12$ s, a voltage is present at the supercap. After that, the terminals of the capacitor are open. With a classical capacitor, we would expect the total voltage U_C to follow the terminal voltage exponentially and remain constant after the terminals are opened. However, here we cannot observe this. The voltage shows different curvatures: at the beginning it rises very quickly, and afterwards more slowly. We would have expected this effect, since $\tau_C = 0.01$ s is clearly smaller than $\tau_P = 0.5$ s and $\tau_D = 5$ s. After switching off, there is a voltage dip and a decay of the voltage to a constant level.

We can better interpret the cause if we look at the currents between the layers. I_C corresponds to the current into the capacitor. I_{CP} represents the current across the resistor R_P. This essentially follows I_C, which is not surprising, because it is dominated by τ_C. I_{PD}, the current across the resistor R_D, however, deviates from the dynamics. It is initially much smaller, which we can interpret as C_P being charged first. From $t = 12$ s it also collapses, which is not surprising, because the external voltage supply is switched off. But we can clearly see a now following equalizing current between the two capacitors.

We see that the behaviour of supercaps is partly different from that of classical capacitors. This leads to the fact that we have to consider some additional requirements when realizing a capacitor bank consisting of supercaps, which we will now look at.

7.3.3 Requirements for capacitor banks with supercaps

In this section, we will look at the requirements for capacitor banks consisting of supercaps. Even though supercaps have a much higher capacitance and a much higher power, they are usually also used in *xpys* circuits as capacitor banks. For the component design, the requirement is that the user wants to see this capacitor bank as one large capacitor. The complexity behind this circuit should be reduced.

`SC 1: AS A User, I WANT to work with only one large capacitor SO THAT I don't have to deal with the internal wiring, safety measures, and temperature management.` In order to fulfil this requirement, additional requirements for the design are derived, which we will consider in the following. They comprise three aspects: the charge transport between the layers, the handling of manufacturing variances, and the lifetime of supercaps.

We start with the charge transport between the layers. As we have already seen, the three types of capacitance change the dynamics of a supercap compared to a classical capacitor. In the example shown in Fig. 7.11, the effect was simply a redistribution of charges between the capacitances and a reduction in the clamping voltage. However, the different capacitances can also lead to current or voltage peaks. Therefore, measures must be taken to compensate for these.

`SC 2: AS AN engineer, I WANT to realize a suitable protective circuit or charge and discharge management SO THAT no current or voltage spikes damage components outside the capacitor bank.`

Filter capacitors and inductors can be used for the protective circuit. Both components serve to buffer high voltage or current rises. An alternative or additional measure is a charge controller that monitors the charging and discharging of the capacitor bank and regulates the charging and discharging currents so that the peaks do not occur.

Table 7.1 Technical data of various supercaps.

Capacitance [F]	Capacitance tolerance	Rated voltage [V]	Cont. peak current [A]
1	(−0%/+100%)	2.7	0.9
3	(−10%/+30%)	3.9	4.71
5	(−0%/+100%)	2.7	3.21
30	(−10%/+30%)	2.7	9.2
40	(−10%/+30%)	2.7	25
100	(−20%/+20%)	3.0	8.3
200	(−10%/+30%)	2.7	96.43
400	(−10%/+30%)	2.7	180
600	(−20%/+20%)	3.0	20
3,000	(−10%/+30%)	2.7	2,168

Let us look at the technical data of supercaps from different manufacturers (Tab. 7.1). The nominal voltage of the supercaps is comparable. Since they are based on comparable chemical substances and these define the nominal voltage, no major variations are to be expected here. The maximum continuous current depends on the area of the capacitor. The smaller the surface area, the smaller the maximum continuous current, because the charge carrier exchange takes place via the charge carrier transport from the surface to the 'other side' of the capacitor. Therefore, the maximum continuous current also correlates with the capacitance of the capacitor.

We see, that the supercaps have high capacitance tolerance values. These depend on the manufacturer and the total capacitance of the supercap. This is due to the variance in the manufacturing process. Roughly speaking, the substrate is applied to a metal foil, and then a separator is placed between two of these metal foils to ensure that there is no short circuit. These three layers are rolled together, packed into a case and filled with an electrolyte. The capacitance of the capacitor then depends on the quality of the applied layer and the composition of the electrolyte. For electrolytic capacitors, which have a similar construction to supercaps but without a coating on the metal foils, the capacitance tolerance is ±20%. With film capacitors, even ±5% is possible. This leads us to the next requirement. If we want to design a capacitor bank, we have to consider the dispersion in the capacitance of the capacitors.

SC 3: AS AN engineer, I WANT to combine the single capacitors in a smart way SO THAT I take into account the capacitance tolerance while reducing the overall numbers of capacitors.

This can be done in two ways. We can design the capacitor bank to include an additional 10–20% of capacitance. So we assume that each supercap has 10–20% less capacitance than nominally specified. Alternatively, we measure each supercap and combine suitable supercaps into a bank. Both options have a price, which should be evaluated before making a decision.

There are two aspects to the interconnection of double-layer capacitors that have a greater impact than with classical capacitors due to the manufacturing tolerances, large energies, and power ratings. In a serial connection, the charging and discharging currents are the same for all capacitors, while the voltages add up. The charging current decreases exponentially during the charging process. The time scale is defined by the product of load resistance and capacitance. If the capacitances are the same, all capacitors of a string are charged equally, since the load resistance is the same. However, if one of the capacitors has a lower capacitance, the charging process of all capacitors connected in series is terminated, as the total current must be the same for all capacitors. This reduces the storage capacity. In a capacitor bank in a string of N capacitors, if one capacitor has 20% less capacitance, the total storage capacity is not reduced proportionally by $\frac{20\%}{N}$, but in total by 20%.

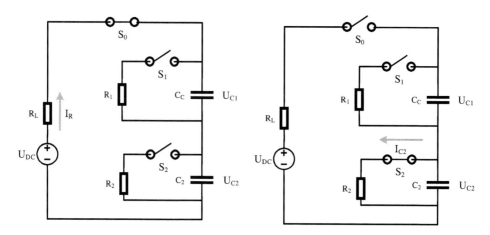

Fig. 7.12 Connection of two capacitors with passive balancing. If the switches S_1 and S_2 are open and S_0 is closed, the capacitors can be charged. If the second capacitor is to be discharged, S_2 is closed and S_0 is opened at the same time.

In order not to solve this problem by sorting and selecting, supercap banks are equipped with a balancing circuit (Ibanez, 2017). The basic idea is to balance the state of charge of the capacitors. Fig. 7.12 shows the circuit for passive balancing for two supercaps. Each of the capacitors C_1 and C_2 is connected to a resistor, R_1 and R_2, which are connected in parallel. The switches S_1 and S_2 can switch on the resistor if required. In addition, there is a switch S_0 that can disconnect the supercapstring from the voltage source.

In Algorithm 4 we have presented the logic for balancing. The decision of how the switches are positioned results from the voltage of the capacitors U_{C_1}, U_{C_2}. To understand the logic, consider Fig. 7.13. We have shown two cases here: one with balancing switched on ('w Balancing') and one without balancing ('w/o Balancing'). C_2 has only 80% of the capacitance of C_1. We can follow the expected course of the capacitor voltage very well here. C_2 goes into saturation, so that C_1 must also go into saturation, and the current I_R can no longer flow across a fully charged capacitor. Now we consider Algorithm 4. The normal state is that S_1 and S_2 are open and only S_0 is switched on. Now, as soon as one of the two voltage measurements on the capacitors measures a voltage above U_{max}, the S_0 or S_1 is closed and the capacitor is discharged. This also stops the charging process, so S_0 must therefore also be opened.

Algorithm 4 Simple balancing algorithm.

$S_0 \leftarrow 1$
$S_1 \leftarrow 0$
$S_2 \leftarrow 0$
while System is running **do**
  if $U_{C_1} > U_{\max}$ AND BalancingFlag $=$ false **then**
    $S_0 \leftarrow 0, S_2 \leftarrow 0, S_1 \leftarrow 1$
    BalancingFlag $\leftarrow$ true
  end if
  if $U_{C_2} > U_{\max}$ AND BalancingFlag $=$ false **then**
    $S_0 \leftarrow 0, S_1 \leftarrow 0, S_2 \rightarrow 1$
    BalancingFlag $\leftarrow$ true
  end if
  if $U_{C_1} < U_{\max} - \Delta U$ AND BalancingFlag $=$ true **then**
    $S_0 \leftarrow 1, S_1 \leftarrow 0, S_2 \leftarrow 0$
    BalancingFlag $\leftarrow$ false
  end if
  if $U_{C_2} < U_{\max} - \Delta U$ AND BalancingFlag $=$ true **then**
    $S_0 \leftarrow 1, S_1 \leftarrow 0, S_2 \leftarrow 0$
    BalancingFlag $\leftarrow$ false
  end if
end while

We see this process in Fig. 7.13. At $t = 1$ µs, the voltage U_{C_2} is greater than the maximum value of $U_{\max} = 6$ V. Now the capacitor C_2 is discharged for a short time, which can be seen from the rapid drop in voltage. As can be seen Algorithm 4, the switch is not closed immediately when the voltage drops below $U_{\max}$, but only when the voltage drops below $U_{\max} - \Delta U$. In this way, rapid switching on and off is avoided. We see that when balancing, C_1 also reaches a higher voltage. We can therefore use the full energy content with the help of the balancing procedure.

This balancing process allows us to compensate for differences within a string. But the manufacturing differences also have an effect on parallel strings. Since the charging current can be different for capacitors connected in parallel, the capacitors are always charged, regardless of their capacity. When one capacitor has reached its maximum state of charge, no more charging current flows, while the charging current continues to flow in the other capacitors. However, the capacitors then have a different open-circuit voltage at the end of the charging process. Due to the parallel connection, there are therefore equalizing currents between the capacitors when the open-circuit voltage differs. These balancing currents can be very high and lead to damage to the capacitor bank or other components of the storage system. This results in a further requirement:

SC 4: AS AN engineer, I WANT to ensure that in a supercap bank the string voltages are not too different SO THAT it is ensured that there are no excessive balancing currents flowing between the strings.

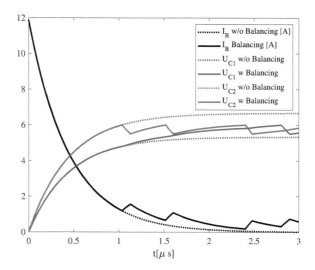

Fig. 7.13 Voltage curve of the charging process of two capacitors connected in series ($C_1 = 1$ μF, $C_2 = 0.8$ μF). As soon as the voltage of C_2 has reached the final value of approx. 6.6 V, the charging process of C_1 is also finished. By balancing, C_2 is discharged as soon as the voltage of one of the two capacitors is greater than 6 V. Through this balancing, it becomes possible to reach a common state of charge ($R = 1$ Ω, $U_{\mathrm{DC}} = 12$ V, $U_{\max} = 6$ V, $\Delta U = 0.5$ V).

A similar effect can be achieved with an interconnected resistor. One can use a positive temperature coefficient (PTC) resistor. With a PTC resistor, the resistance value increases with the temperature. If balancing currents flow, it heats up, which increases its resistance, so that the equalizing currents in turn decrease.

Another possibility is to use a balancing system across strings. This selectively switches strings on or off and reduces the voltage difference by active or passive balancing within the switched-off string, so that the balancing currents are lower.

Double-layer capacitors have a higher power density than energy density. Double-layer capacitors can be charged and discharged with a power that is hundreds to thousands of times greater than their energy content. The capacitance values can still be in the range of several thousand farads. However, if we look at the time scale $\tau = RC$, on which the charging and discharging process takes place, we see that the energy to be stored is comparatively small.

Exercise 7.9 A capacitor has a nominal voltage of $U = 3$ V and a capacitance of $C = 2{,}000$ F. The charging time scale is $\tau = 45$ s. What is the storage capacity and what is the charging current? What power does this correspond to at the time of charging? What is the E-Rate?

Solution: In this example, the capacitor is completely discharged. The energy content is then given by:

$$\kappa = \frac{1}{2}CU^2 = \frac{1}{2}2{,}000 \text{ F} \cdot (3\text{V})^2 = 9{,}000 \text{ Ws} = 2.5 \text{ Wh}.$$

To find the charging current, we need to find the resistance of the circuit. Considering the time scale τ, we calculate the resistance R:

$$\tau = RC \Rightarrow R = \frac{45 \text{ s}}{2{,}000 \text{ F}} = 0.0225 \ \Omega.$$

The maximum current at $t = 0$ is calculated from the ratio of voltage and resistance:

$$I(t = 0) = \frac{U}{R} = \frac{3 \text{ V}}{0.0225 \ \Omega} = 133 \text{ A}.$$

The current and voltage are then used to calculate the power:

$$P = U \cdot I = 3 \text{ V} \cdot 133 \text{ A} = 399 \text{ W}.$$

The E-Rate is the ratio of power and storage capacity related to one hour. In this case, the E-Rate is:

$$E_1 = \frac{P}{\kappa} = \frac{399 \text{ W}}{2.5 \text{ Wh}} 1 \text{ h} = 159.6.$$

As we saw in Exercise 7.9, the energy content of a supercap is small. However, the E-Rate at which the capacitor can be discharged is large. Supercaps are therefore very well suited as power storage devices.

In the case of mechanical storage systems, we have not made any statements about the lifetime of the storage system. Yet G 7 is one of the basic requirements for storage systems. This is because the systems discussed are usually designed to last for 10 to 20 years. Pumped storage plants can have significantly longer life cycles. Regular service intervals and repairs can be maintained here, since the moving parts of a mechanical storage system are also exposed to mechanical stresses and may need to be replaced.

Since capacitors have no mechanical moving parts, their service life should actually be long. However, the service life depends on the used dielectric. One factor in ageing is how strongly the dielectric reacts to heat, which is generated during charging and discharging, but can also be supplied from ambient. With film capacitors, heating leads to small length contractions, which cause mechanical stress. In electrolytic capacitors, the stress leads to a decomposition of the electrolyte. The service life of electrolytic capacitors is approximately 5,000 h under standard conditions—that is, 25 ̊C. A film capacitor, on the other hand, reaches up to 300,000 h. How temperature and charging and discharging currents affect the service life is specified by the manufacturers either in the form of ageing models or in the form of characteristic curves. It is customary to speak not of the number of cycles, but of an AC load at different frequencies.

In the case of a capacitor, the service life depends on how the dielectric reacts to charging and discharging as well as to the ambient temperature. With double-layer capacitors, there are two areas in which ageing effects occur: analogous to the electrolytic capacitor, heat and charging and discharging affect the electrolyte. It can decompose or even oxidize due to high heat. In addition, the porous electrode material can also decompose due to thermal stress. This leads to a reduction of the surface area, which results in a reduction of the capacity. Under standard conditions, double-layer capacitors achieve lifetimes of several decades. However, this long service life is countered by the fact that double-layer capacitors are exposed to high currents and strong temperature fluctuations when uncooled in applications as power storage devices. Thus, a double-layer capacitor achieves a service life of up to 500,000 h—that is, 60 years—only if the charging and discharging currents are at 30% of the nominal current. If the charging current remains in the nominal range of the capacitor, the service life is reduced to 90,000 h, which corresponds to around 10 years. Reducing the current load by one third leads to a fivefold increase in service life. This effect can be used in system design. Even a slight increase in capacity or the parallel connection of strings reduces the current and thus leads to a longer system life. However, it should be noted that this also has an influence on the manufacturing costs, so that costs and benefits must be weighed up here.

Fig. 7.14 shows the system components that are necessary for the realization of a voltage storage with the help of supercaps. Since the **supercap** combines the properties of a double-layer capacitor **DoublelayerCap** and a pseudocapacitor **PseudolayerCap**,

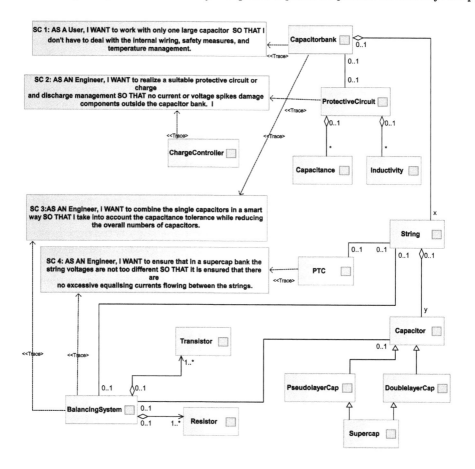

Fig. 7.14 System components for the realization of a voltage storage system.

it is shown here via a double relationship. In the aggregation of the `CapacitorBank`, we have entered the *xpys*-connection in the cardinality. The `BalancingSystem` component is associated with both `capacitors` and capacitor `String`. This relates to two different tasks that the balancing system has. The link to the capacitor serves to comply with SC 3. This is focused on acting within a string. The link to the capacitor strings serves to comply with SC 4.

Now that we have got to know the supercap as a voltage storage device, we want to deal with an application example in the next section. In doing so, we want to apply the various steps that are necessary for the design of a storage system and that we learned about in Chapter 3.

7.4 Application example: Recuperation of a passenger lift

In the previous sections of this chapter, we have looked at storage systems that can store electrical energy in the form of current or voltage. We have seen that these storages are usually power storages because their energy density is small, or the charging and discharging rates are much larger than one compared to the stored energy. We now want to design a storage system that uses supercaps. This is an extension for a passenger lift.

In passenger lifts, electrical energy is used to raise or slowly lower a load. In the process, kinetic energy is converted into potential energy. Basically, this corresponds to a mechanical storage power station. The basic idea is shown in Fig. 7.15. A lift has a connection to the electrical grid through which the inverter draws its power. An electric motor and our storage system are connected to this inverter. The motor can raise and lower the lift cabin via a winch. A counterweight reduces the required power. Without the storage or when the storage is empty, the inverter draws the required electrical power P_{el} from the grid. With the help of the inverter and the electric motor, the electric power is converted into mechanical power P_{mech} and the lift cab is lifted. If the persons are now to be lowered again, the potential energy is converted into kinetic energy. Since the lift is not allowed to become arbitrarily fast, this kinetic energy is converted into electrical power by the electrical brake of the motor. We store this power in our storage. This way, when the lift cabin is lifted again, we no longer need to draw power from the grid.

7.4.1 Requirements analysis

Let us first look at the requirements associated with such a system. The first requirement concerns the basic use case of a lift. The purpose of a lift is for us to move vertically. Generally, we use a lift when we do not want to climb stairs. The requirement is therefore:

RL 1: AS A User, I WANT to be transported vertically using the lift SO THAT I do not have to climb stairs.

The strength of an electric drive is that once the electric field has the necessary energy and the right modulation, the torque is established. This is extremely fast and thus allows very strong accelerations. Of course, these are not suitable for the user. Therefore, the requirement here is:

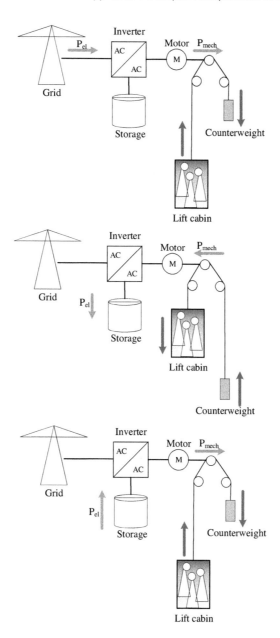

Fig. 7.15 Use of a storage for reducing the energy consumption of a passenger lift. When the storage system is empty, power is taken from the grid and converted into mechanical energy to lift the loads. When the loads are lowered, the mechanical energy of the lift is stored as electrical energy in a storage system, which provides the required power when the loads are lifted again.

RL 2: AS A User, I WANT the cabin not to accelerate or decelerate too fast SO THAT I am not damaged, and neither are the things I am carrying.

RL 1 and RL 2 are user requirements that must be met by every lift. We have not yet created a requirement that motivates our use case at all. This comes with the third requirement:

RL 3: AS AN operator, I WANT the energy consumption of the lift to be as low as possible SO THAT I have low operating costs.

The idea, then, is to reduce operating costs. The energy that is released during deceleration and lost in the form of heat in a mechanical brake should now be stored so that this energy can be used the next time the lift is raised. An alternative would be not to store the energy but to feed it into the grid. However, this only makes sense from an economic point of view if this energy could also be sold. However, since the trips of a passenger lift in a building can only be roughly planned, the possible energy to be fed into the grid cannot be planned. We will also see later in this section that the amounts of energy are relatively small, so participation in energy trading will not be possible due to the small quantities.

Exercise 7.10 Requirements RL 1 to RL 3 represent only very few of the requirements for such a system. Before we look at the technical implementation and power flows, let us consider what other requirements there are. In Chapter 4, we got to know a number of actors; in this exercise, we restrict ourselves to the following: product manager, regulator, production, purchase, and service engineer. What are the requirements of these actors for the recuperative passenger lift? For each actor, we want to create two requirements. (First hint: It makes sense to look at the role description of the actors from Chapter 4 again. Second hint: This task only requires imagination. Of course, if you have not taken on one of these roles, you may not know what requirements to have in that role. But you can imagine what one might need. There is no right or wrong here. Don't be afraid of the white sheet of paper.)

Solution: The product manager sees the recuperation elevator as part of his product portfolio—that is, as only one of a number of other products. Therefore, his requirements reflect the relationships between this product and the other products in his portfolio.

RL 4: AS A product manager, I WANT an elevator system to be able to be upgraded by adding storage SO THAT old elevator systems can be upgraded or I can offer the same lift with or without storage system.

RL 5: AS A product manager, I WANT to make the storage capacity variable SO THAT I can offer different storage options to the customer.

By 'regulator', we mean a person who has defined norms and standards, or who demands them.

RL 6: AS a regulator, I WANT the energy consumption of the lift to be reduced by 20-30% SO THAT the CO_2 footprint becomes smaller.

RL 7: AS a regulator, I WANT measures to be taken to prevent uncontrolled discharging when storage is used on passenger vehicles SO THAT there is no fire in the elevator system and persons are not put at risk.

We have designated as Production those actors who are involved in the manufacture of the product or its subcomponents. Their requirements naturally focus on the question: what does the manufacture of this product mean for my processes? What tools do I need? Do my employees have the right skills?

RL 8: AS a production, I WANT to be able to assemble the storage system on the same production line where the other lift components are assembled SO THAT I don't have to build additional production lines.

RL 9: AS a production, I WANT to assemble the equipment with the normal tool set SO THAT I don't need any additional tools.

The actor Purchase is responsible for procuring the material, which is then assembled by the production. In the development phase, they procure the material for the development. The focus here is on speed. In the production phase, quantities and price are important. Therefore, their requirements are commercial in nature, but can have a backlash on the development.

RL 10: AS A purchase, I WANT the supercaps to be able to be used from two different manufacturers SO THAT in cases of supply shortage I will always have an alternative for procurement as well.

RL 11: AS A purchase, I WANT the development to design the memory bank in such a way that we don't have to build in excess capacity SO THAT the product cost stays low, or we can build more memory banks with fewer supercaps.

RL 10 and RL 11 are development requirements that need to be considered at an early stage of development. We will therefore also use two different supercaps for the design in this chapter.

The last actor we want to consider is the service engineer. These are the people who work on site at the customer's premises during commissioning, in the event of a fault, upgrade, or dismantling. Since passenger lifts have a long life span, service engineers have contact with relatively old products that still need to be maintained and repaired.

RL 12: AS A service engineer, I WANT an indicator of supercap lifetime SO THAT at regular intervals, with a few simple steps, I get to know if the supercaps need to be replaced.

RL 13: AS A service engineer, I WANT to replace parts of the supercap bank SO THAT I only need to transport and repair small supercap modules.

In Exercise 7.10, we have compiled the requirements of the different actors. However, we should be aware that there are additional requirements that apply. These are, firstly, requirements G 1 to G 8, which apply to every storage system. Since the lift converts electrical energy into mechanical energy, requirements MS 1 and MS 2 must be fulfilled. We are working with an electrical system, so ES 1 to ES 5 have to be fulfilled and—since we want to work with supercaps—requirements SC 1 to SC 4 are also relevant. In total, we have to consider 32 requirements.

7.4.2 System design and power flow diagram

After collecting the requirements, we think about the technical implementation (Jabbour and Mademlis, 2017; Kubade, Umathe, and Tutakne, 2017). In Fig. 7.16, the overall system is shown. As we have seen in Fig. 7.15, the basic system consists of a frequency converter and an electric drive that operates a hoist to which the cabin of the elevator is attached. A frequency inverter consists of three system components: a rectifier, a DC link, and a drive inverter. The task of the inverter is to drive the electric motor, i.e. to generate a magnetic field in the motor that sets the rotor of the motor in motion in a controlled manner. If the motor operates in generator mode, the inverter converts the AC of the generator into DC and charges the DC link. The DC intermediate circuit, consisting of a capacitor bank, serves here as a short-term buffer storage to compensate for fluctuations that the rectifier and inverter generate through their switching operations.

The task of the rectifier is to convert the AC from the grid into DC. If this rectifier is operated unidirectionally—that is, if the power is only drawn from the grid—the rectifier usually consists of three diode rectifiers.

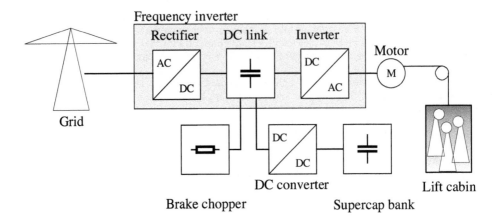

Fig. 7.16 Schematic representation of the system components of a passenger lift equipped with a capacitor bank for recuperating braking power.

When the cabin is raised, the motor power determines the speed and acceleration of the lift. If no braking were applied when lowering the lift, all the potential energy would be converted into kinetic energy. This is not in the interest of the passenger and would violate the requirements RL 2. For this reason, there is a braking system that is used to control acceleration and speed.

One part of the braking system is the mechanical brake. This reduces speed by increasing mechanical friction. It converts kinetic energy into heat. Alternatively, braking can be done with the help of the motor. This can work in regenerative mode when lowering. It converts the kinetic energy back into electrical energy. This can be consumed in three ways. Firstly, it can be fed back into the grid. For this purpose, the rectifier must be designed bidirectionally and be able feed energy into the grid. If electricity is consumed in the building at the same time, the grid consumption of the entire building can be reduced.

Alternatively, a brake chopper can be connected to the DC link. This converts the electrical energy into heat. Compared to a bidirectional rectifier, which consists of three half-bridges and additional inductances for mains filtering, a brake chopper is simpler and less expensive. Fig. 7.17 shows the system components of a brake chopper. It consists of a series of resistors and electrical switches that connect the resistors to the DC link. The brake chopper controller is configured to switch on above a certain voltage and switch off below a certain voltage. In order to control the power drawn from the DC link, the resistor is not connected continuously to the DC link, but is switched on and off periodically. The duty cycle—that is, the ratio of switch-on and switch-off time—then regulates the power drawn.

Both solutions, brake chopper or mechanical brake, have the disadvantage that the braking power is not used,or only partially used, but essentially 'burned'. Therefore, we extend the system with a supercap bank as power storage. We have placed a DC converter between the DC link and the supercap, because we don't know at the moment what the voltage level will be between the DC link and the supercap bank.

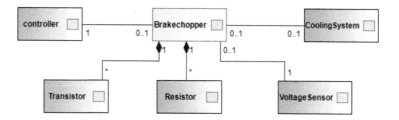

Fig. 7.17 System components of a brake chopper, consisting of a series of resistors that can be switched with the help of transistors. The switching on and off is pulsed so that the braking power can be controlled. A controller is needed for this. Since the power is dissipated in heat, cooling is necessary.

However, it is very likely that we will need such a component. This is because the passive rectifier generates a voltage of $U_{\mathrm{DC}} = \sqrt{3}U_{\mathrm{AC}}$. So, with a mains amplitude of $\hat{U}_{\mathrm{AC}} = 400$ V_{AC}, the DC link has a voltage of $U_{\mathrm{DC}} = \sqrt{3}\,400$ $V_{\mathrm{AC}} = 692$ V_{DC}. If we look at the technical data of different supercaps in Tab. 7.1, we see that a single supercap has a rated voltage between 2.7 V and 3.9 V. So we would have to connect a large number of supercaps together just to get to the DC link voltage. The result may be oversizing. Therefore, we assume that we need a DC converter.

7.4.3 The powerflow diagram of the recuperation lift

After we have familiarized ourselves with the function and the components, let us draw the power flow diagram. We have four power nodes in this system: $G(t)$ or G represents the power flows to and from the grid; $L(t)$, or L represents the power demand of the lift; and the storage is represented by the node $S(t)$ or S for short. In addition, we also use the brake chopper $R(t)$, R. This allows us to make the storage smaller, for example to save costs. We may also get into a situation where the storage is full and we still have to brake. In both cases, it makes more sense to use a brake chopper than a mechanical brake.

The power flow diagram is shown in Fig. 7.18. We want to set up the equations for the power flows and consider the possible connections for this. Power can be transferred from the grid G to all three nodes. However, it makes no sense to transfer power from the grid to the braking resistor, so we do not consider this path. So from the grid G, power G_{S} can be transferred to the storage with an efficiency η_{GS}. Furthermore, the lift can also be operated directly from the grid—that is, we have a power flow G_{L} with a transmission efficiency of η_{GL}. If we have a bidirectional rectifier, the power S_{G} from the storage could be transferred to the grid with efficiency η_{SG}. The same applies to the load—that is, the recuperation power would be fed in directly. We therefore still have a transmission path L_{G} with the efficiency η_{LG}. As a balance variable, we again introduce $\tilde{G}$, which offsets the purchased and the fed-in power. The power flow equation for the grid node is thus:

$$\tilde{G} = \eta_{\mathrm{SG}}S_{\mathrm{G}} + \eta_{\mathrm{LG}}L_{\mathrm{G}} - (G_{\mathrm{S}} + G_{\mathrm{L}}). \tag{7.17}$$

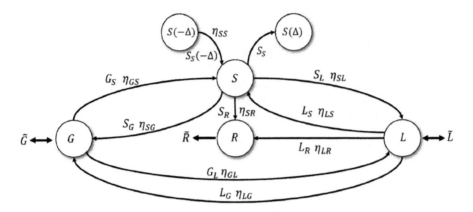

Fig. 7.18 Power flow diagram of an elevator system equipped with a double-layer capacitor as power storage S and a braking resistor R. This corresponds to the system shown in Fig. 7.16.

The maximum power transfer rate is limited by the maximum power provided by the converters and the grid connection point.

The braking resistor is to consume power from the load or the storage. There is a power flow from the storage to the braking resistor S_R and from the load to the braking resistor L_R:

$$\tilde{R} = \eta_{SR} S_R + \eta_{LR} L_R. \tag{7.18}$$

The load L can transfer power through the two power transfers S_L and G_L. If the lift recuperates, this power can be stored, and then the power flow L_S with efficiency η_{LS} is stored. It can also be fed into the grid L_G, if we have a bidirectional inverter. If neither the grid nor the storage is available as a power sink, we can use the brake chopper L_R:

$$\tilde{L} = \eta_{SL} S_L + \eta_{GL} G_L - (L_S + L_G + L_R). \tag{7.19}$$

Here $\tilde{L}$ is the currently required or generated power of the lift system. The time series $\tilde{L}$ is the load profile of our application. Negative values correspond to a surplus of power—that is, it is decelerated. Positive values correspond to a need for power.

We charge the storage S via the grid with the power G_S or via the recuperation of the load with the power L_S. The storage unit can feed its power into the grid S_G or cover the load S_L. It is also possible to discharge the storage unit via the brake chopper S_R:

$$0 = \eta_{GS} G_S + \eta_{LS} L_S + \eta_{SS} S_S(-\Delta) - (S_S(\Delta) + S_G + S_L + S_R). \tag{7.20}$$

In this example, costs are incurred by purchasing power from the grid at a purchase price of c_{gc}:

$$Y(T) = \int_{t=0}^{T} c_{gc} \left(G_L + G_S\right) dt. \tag{7.21}$$

Table 7.2 Overview of the power flow equations of the model of Fig. 7.18.

Node	Equation
Grid	$\tilde{G} = \eta_{\mathrm{SG}} S_{\mathrm{G}} + \eta_{\mathrm{LG}} L_{\mathrm{G}} - (G_{\mathrm{S}} + G_{\mathrm{L}})$
Brake chopper	$\tilde{R} = \eta_{\mathrm{SR}} S_{\mathrm{R}} + \eta_{\mathrm{LR}} L_{\mathrm{R}}$
Load	$\tilde{L} = \eta_{\mathrm{SL}} S_{\mathrm{L}} + \eta_{\mathrm{GL}} G_{\mathrm{L}} - (L_{\mathrm{S}} + L_{\mathrm{G}} + L_{\mathrm{R}})$
Storage	$0 = \eta_{\mathrm{GS}} G_{\mathrm{S}} + \eta_{\mathrm{LS}} L_{\mathrm{S}} + \eta_{\mathrm{SS}} S_{\mathrm{S}}(-\Delta) - (S_{\mathrm{S}}(\Delta) + S_{\mathrm{G}} + S_{\mathrm{L}} + S_{\mathrm{R}})$
Yield	$Y(T) = \int_{t=0}^{T} c_{\mathrm{gc}} (G_{\mathrm{L}} + G_{\mathrm{S}}) \, \mathrm{d}t$

Table 7.3 Overview of the monetary valuation of the power flows of the model Fig. 7.18.

Powerflow	Value of powerflow
$L_{\mathrm{S}}, S_{\mathrm{L}}, S_{\mathrm{S}}$	$-c_{\mathrm{gc}}$
$S_{\mathrm{G}}, L_{\mathrm{G}}, L_{\mathrm{R}}$	0
$G_{\mathrm{L}}, G_{\mathrm{S}}$	c_{gc}

In order to derive a simple control of the power flows, the paths are to be monetarily evaluated. The aim is then to select the power flow in a time slice Δt in such a way that the power flow with the highest value is used.

When power is transferred from the grid into the system, costs arise that are proportional to the electricity tariff c_{gc}.

The paths L_{S} and S_{L} get the value of the saved electricity costs—that is, $-c_{\mathrm{gc}}$. The power paths from the storage and the load to the brake chopper are cost-neutral. The same applies to the power paths to the grid S_{G} and L_{G}, as here it is assumed that no compensation is paid and there is only one lift in the building. We evaluate the power flow into the future S_{S} in the amount of the saved electricity costs, because we assume that we save the purchase from the grid in the future when we use it.

Tab. 7.2 summarizes all equations. The prioritization of the load flows and the monetary valuation are listed in Tab. 7.3.

The evaluation of the power flows shows that power transfer between storage and load is to be preferred from an economic point of view. A transfer to the grid is cost-neutral in terms of the operation cost, but here we have to take into account the higher investment costs. Here, the brake chopper would actually be the cheaper option. The purchase of energy is to be avoided in any case.

7.4.4 Energy management of the recuperating elevators using voltage control

We now want to define a simple algorithm that takes the above considerations into account. In this example, we do not work with a direct energy or power control, as we did with the opportunity charger in Chapter 3, but we control the power flow via the DC link voltage. The preferred source or sink is our storage. When the inverter drives the electric motor, it draws energy from the DC link, which causes the DC link voltage to drop as the DC link capacitors are discharged. As soon as this is the case, the DC/DC converter of the storage unit ensures that the DC link capacitors are charged so that the target DC link voltage of U_{DC} is maintained. In the process, the supercaps are discharged.

After the supercaps have been discharged or if the minimum voltage of the supercap bank has been reached, the DC/DC converter terminates the charging process. If the lifting process still continues, the DC link voltage continues to drop until it reaches a value of $\sqrt{3}U_{AC}$. At this point, power is taken from the grid.

When the motor recuperates, electrical power is transferred to the DC link. In this case, the voltage in the DC link increases. The DC/DC converter discharges the DC link capacitors, reduces the voltage to U_{DC} and charges the supercap bank. If the supercap is fully charged, the DC/DC converter can no longer limit the voltage to U_{DC} and it rises. Now the brake chopper intervenes. As soon as the voltage rises above a value of U_{BC}, the brake chopper switches on and remains active until the DC link voltage U_{DC} is reached again.

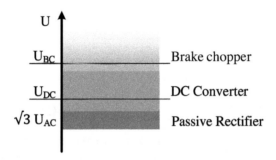

Fig. 7.19 Voltage window for regulating the elevator via voltage control. Different components are responsible for three different voltage ranges. The passive rectifier operates in the lower range, limiting the voltage to $\sqrt{3}U_{AC}$. The width of the voltage band is caused by the fluctuations of U_{AC}. Above a voltage of U_{BC}, the brake chopper reduces the voltage to U_{DC}. The DC converter works in the interval in between, charging or discharging the supercaps.

Fig. 7.19 shows the different voltage ranges and the system component responsible for each. Since U_{AC} varies within the voltage range allowed by the grid code, we have specified a range here. We have presented the algorithm realized with this concept in Algorithm 5.

7.4.5 Design of the supercap bank

We will now look at the design of the supercap bank. Since we want to use the storage as a power storage, we can concentrate on singular transport events in the design. We do not have to consider a time series that runs over a longer period of time.

In Tab. 7.4, we have compiled data for the elevator system (Jabbour and Mademlis, 2017). We first want to address the question of what maximum charging power we might expect.

Algorithm 5 Simple energy management for recuperating elevator.

BrakeChopperFlag ← false
while System is running **do**
    if $U \geq U_{DC}$ AND $U \leq U_{BC}$ AND BrakeChopperFlag $=$ false **then**
        DC/DC Converter voltage control $U \leftarrow U_{DC}$
    end if
    if $U > U_{BC}$ **then**
        BrakeChopper voltage control $U \leftarrow U_{DC}$
        BrakeChopperFlag $=$ true
    end if
    if $U < U_{DC}$ AND BrakeChopperFlag $=$ true **then**
        DC/DC Converter voltage control $U \leftarrow U_{DC}$
        BrakeChopperFlag $=$ false
    end if
end while

Table 7.4 Rated values for the elevator system.

Rated torque	270 Nm
Rated motor voltage	400 V
Rated motor current	10.6 A
$\cos \phi$	$0, 9$
Inverter efficiency	95%
Duty load	630 kg
Cabin weight	750 kg
Counterweight	1.065 kg
Car travel	27 m Braking resistor 75Ω

Exercise 7.11 Tab. 7.4 contains all the data we need to calculate the maximum charging power. What does this consist of?

Solution: This task can be approached in two ways. We can use the maximum distance and the maximum weight. However, in this case we would have to make an assumption about the travel time. Alternatively, we look at the engine data, since we can assume that we do not need to store more power than the amount provided by the engine. The maximum recuperation power is calculated from the rated voltage U_m and the rated current I_m. Since a motor in regenerative operation uses part of the power to build up the magnetic fields, part of the power is used as reactive power. The apparent power is reduced proportionally, which is described by the factor $\cos \phi$. The recuperation power is thus given by:

$$P_{rec} = U_m \cdot I_m \cdot \cos \phi$$
$$= 400 \text{ V} \cdot 10.6 \text{ A} \cdot 0.9 \qquad\qquad = 3{,}816 \text{ W.}$$

3,816 W is the power that the motor provides. However, this is not yet the power with which we can charge the supercap bank. This power is still corrected by the efficiency of the inverter, because P_{rec} is converted into direct current by the inverter. The maximum power P_{el} with which we have to charge the supercap bank is therefore:

$$P_{el} = P_{rec} \cdot \eta = 3.816 \text{ W} \cdot 0.95 = 3.625 \text{ W.}$$

We will design the supercap bank for a maximum power of 3.6 kW. We now need an estimate of the amount of energy that needs to be buffered.

Exercise 7.12 From the data in Tab. 7.4, we can also get an estimate for the maximum amount of energy we need to buffer. What is this?

Solution: The maximum amount of energy is released when the elevator is fully loaded and moves from the top floor to the ground floor. From Tab. 7.4, we know that the maximum travel distance is 27 m. The mass m that is lowered is the sum of the cabin weight and duty load minus the counterweight:

$$m = (630 \text{ kg} + 750 \text{ kg} - 1{,}065 \text{ kg}) = 315 \text{ kg}.$$

The maximum amount of energy results from this potential energy:

$$E = m\,g\,h = 315 \text{ kg } 9.81\,\frac{m}{s^2}\,27 \text{ m} = 83{,}434.05 \text{ Ws} = 23 \text{ Wh}.$$

Table 7.5 Choice of supercaps for the elevator supercap bank.

Label	Capacitance [F]	Capacitance tolerance	Max. voltage [V]	Cont. peak current [A]	Price
Cap 1	800	+50%/−20%	2.7	20	$25
Cap 2	850	+30%/−10%	3.2	40	$75

RL 10 requests the usage of two different types of supercaps. Tab. 7.5 lists our choice. We will discharge the supercaps to a maximum of 40% of the maximum voltage.

Exercise 7.13 How many supercaps do we need to store the desired amount of energy? Consider the spread of capacitance. Please give an upper and a lower value.

Solution: The amount of energy of a supercap is given by the formula:

$$E = \frac{1}{2}C\left(U_2^2 - U_1^2\right).$$

Thus, for Cap 1, the amount of energy is:

$$Cap\ 1: \begin{cases} \frac{1}{2}1{,}200 \text{ F }\left((2.7 \text{ V})^2 - (1.08 \text{ V})^2\right) = 3{,}674.16 \text{ Ws} = 1.02 \text{ Wh} \\ \frac{1}{2}640 \text{ F }\left((2,7 \text{ V})^2 - (1.08 \text{ V})^2\right) = 1{,}959.52 \text{ Ws} = 0.54 \text{ Wh}. \end{cases}$$

Thus, to store the amount of energy of $\kappa = 23$ Wh, we need 23 to 42 capacitor for Cap 1. For Cap 2, the calculation looks like this:

$$Cap\ 1: \begin{cases} \frac{1}{2}1{,}105 \text{ F }\left((3.2 \text{ V})^2 - (1.28 \text{ V})^2\right) = 4{,}752.4 \text{ Ws} = 1.32 \text{ Wh} \\ \frac{1}{2}765 \text{ F }\left((3.2 \text{ V})^2 - (1.28 \text{ V})^2\right) = 3{,}290.1 \text{ Ws} = 0.913 \text{ Wh}. \end{cases}$$

Here, between 18 and 26 supercaps would be needed.

If we now consider the costs, Cap 1 would incur a maximum cost of $42 \cdot 25 = \$1{,}050$ for the procurement of the supercaps. For Cap 2, the costs would be $26 \cdot 75 = \$1{,}950$. So, from an economical point of view and only in terms of the amount of energy, our preference would be *Cap* 1, even though we would have to combine significantly more supercaps in one assembly.

The consideration in Exercise 7.13 only referred to the energy content. This represents the small number of capacitors we need for our task. However, we also need to look at the aspects of power, current, and rated voltage. We know that U_{DC} is $\sqrt{3}\,400$ $V_{AC} = 692$ V_{DC}. However, the maximum voltage of the supercaps is 2.7 V and 3.2 V respectively. So we will need a buck converter, which we learned about in section 5.2. The transformation ratio was given by the duty cycle D—that is, the ratio between switch-on and switch-off time. The lower we have to set the voltage, the longer the switch-on times. But power can only be transmitted during the switch-on times. If the switch-off times are too long—that is, we have to set the voltage too low, we are limited in the transmission of power. In addition, the current increases at low voltage.

Exercise 7.14 How would the duty cycle have to be set so that we lower the voltage from 692 V to 2.7 V? What would be the inductance if the current ripple ΔI is 10% of the maximum current that must be applied at a power of 3.6 kW? The switching frequency of the buck converter in this example is 16 kHz.

Solution: The duty cycle D results directly from the ratio of the two voltages:

$$D = \frac{U_1}{U_0} = \frac{2{,}7\text{ V}}{692\text{ V}} = \frac{1}{256}.$$

The switch-off time is therefore 256 times longer than the switch-on time.

With a voltage of 2.7 V and a power of 3.6 kW, the current to be carried is:

$$I = \frac{3.6\text{ kW}}{2.7\text{ V}} = 1{,}333.3\text{ A}.$$

The allowed voltage ripple is $\Delta I = 10\%I = 133$ A. The following still applies to the voltage ripple:

$$\Delta I = \frac{(U_0 - U_1)}{L}D \cdot T = \frac{(U_0 - U_1)}{L}D \cdot \frac{1}{f}.$$

Thus we can find the required inductance:

$$L = \frac{(U_0 - U_1)}{\Delta I}D \cdot \frac{1}{f}$$
$$= \frac{(629\text{ V}_{DC} - 2{,}7\text{ V}_{DC})}{133\text{ A}} \cdot \frac{1}{256} \cdot \frac{1}{16{,}000\text{ Hz}}$$
$$= 1.14\ \mu\text{H}.$$

A duty cycle of $\frac{1}{256}$ is not used in practice. In addition, the high currents of 1,333 A require a large cable cross-section, which means expensive cables or busbars.

For further analysis, we choose a transformation ratio of $D = \frac{1}{4}$—that is, we require a voltage of 173 V. This voltage is the lowest voltage that the buck converter should provide.

With these boundary conditions, we can now design the supercap bank. We proceed as follows:

1. We determine the required number of capacitors to reach the voltage level. With this we have determined y.
2. We determine how many strings we have to connect in parallel to get the required power. With this we have determined x.
3. We check that the energy content of the interconnection can provide the required energy.

Exercise 7.15 What is the desired design if the minimum final charge voltage is not to fall below 173 V, has a power of 3.6 kW, and has a storage capacity of 23 Wh?

Solution: We need to be careful, because we want to discharge the supercaps to 40%. This means that we have to calculate with a reduced voltage to determine the interconnection. For Cap 1, we get

$$N_y = \frac{173 \text{ V}}{1.08 \text{ V}} = 160.185 \approx 161.$$

For Cap 2, on the other hand, we need:

$$N_y = \frac{173 \text{ V}}{1.28 \text{ V}} = 135.15 \approx 136.$$

We have rounded up in both cases, as rounding down would result in falling below the minimum voltage.

Cap 1 has a peak charge current of 20 A. This corresponds to a power of 173 V $\cdot$ 40 A = 3,460 W. The power of one string is not sufficient to carry the required 3.6 kW. Therefore, we have to use two parallel strings: $N_x = 2$. If we want to use Cap 1, we have to choose a 2p161s circuit. We need a total of 322 capacitors, which have a total cost of $8,050.

Cap 2 looks different. Cap 2 has a peak charge current of 40 A. For this capacitor, the power is even 173 V $\cdot$ 40 A = 6,920 W. Here, one string is enough. Here we can use a 1p136s connection. We only need 136 capacitors, which have a total cost of $10,200.

From Exercise 7.13, we know that for the amount of energy to be stored, we need 23 to 42 capacitors for Cap 1 and 18 to 26 supercaps for Cap 2. The realized voltage window is sufficient. It makes more sense to check whether the voltage level can be reduced further. Due to the amount of energy, a voltage of approximately 48 V would be desirable.

So the target interconnection is 2x161y for Cap 1 and 1x136y for Cap 2. Here, the price of the capacitors is lower for Cap 1 than for Cap 2.

7.4.6 System components and requirement traceability

Finally, let's get an overview of the system components used and see which components are responsible for compliance with the various requirements.

There is no universal solution for the system architecture. The way the relationships between the components are chosen is subject to degrees of freedom. In Fig. 7.20, we have shown an architecture. We have divided the elevator system into three sub-components. The cabin system, the drive system, and the energy system.

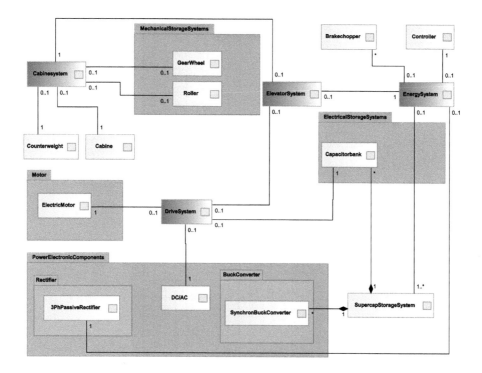

Fig. 7.20 System components of a lift system that can recuperate braking power using a double-layer capacitor.

Cabinsystem contains all the mechanical components of the lift: the Cabin, Counterweight, GearWheel, and Roller. The DriveSystem consists of the drive inverter DC/AC, the DC link, a Capacitorbank, and the ElectricMotor.

The EnergySystem includes the Brakechopper, the 3PhPassiveRectifier, and the SupercapStorageSystem. To meet the requirements of RL 4 and RL 5, we have chosen to make the storage system modular. It can be included in the EnergySystem, but it does not have to be. Furthermore, a supercap storage system always consists of a capacitor bank containing supercaps and a SynchronBuckConverter.

Due to this modular approach, we are able to realize different lift systems and adapt them to the respective customer requirements.

As we saw at the beginning of this section, we have 32 requirements to comply with. In order to have a better overview of the mapping of the requirements to the system components, we use a requirement traceability matrix (RTM). This is shown in Tab. 7.6. In the rows of this table, we have listed the system components shown in Fig. 7.20. The columns represent the requirements. If a system component is responsible or partially responsible for a requirement, a cross has been made. We need to make sure that there is at least one cross in each column.

Let us first look at the Basic requirements. G 1, the use case energy to store, is realized by the Capacitorbank of the SupercapStorageSystem. For the power transfer, G 2, the DriveSystem and the EnergySystem are responsible. A split is made

Table 7.6 Requirement traceability matrix.

	Basic requirements								Mech. Sys.		Electrical systems					Supercaps				Customer requirements												
	G1	G2	G3	G4	G5	G6	G7	G8	MS1	MS2	ES1	ES2	ES3	ES4	ES5	SC1	SC2	SC3	SC4	RL1	RL2	RL3	RL4	RL5	RL6	RL7	RL8	RL9	RL10	RL11	RL12	RL13
Cabinesystem		x																									x	x				
Cabin				x	x	x	x													x												
Counterweight																																
GearWheel																																
Roller																																
DriveSystem					x	x	x		x	x										x							x	x				
ElectricMotor								x				x	x								x				x							
DC/AC								x				x	x	x	x										x							
EnergySystem		x			x	x	x	x			x				x							x			x		x	x				
Controller				x							x																				x	
SupercapStorageSystem					x	x	x	x			x	x	x	x	x								x	x		x	x	x				x
SynchronBuckConverter			x								x	x	x	x	x								x		x				x	x		
Capacitorbank	x		x					x			x	x	x			x	x	x	x										x			
Brakechopper											x	x	x																			

here because both systems are responsible for parts of the power transfer. G 3, the requirement that uncontrolled charging and discharging should be avoided, must be observed by the `Capacitorbank`, because the `Capacitorbank` stores energy. In addition, we also give the `SynchronBuckConverter` a partial responsibility, because the buck converter is able to actively influence the power flow between the supercap bank and the DC link.

In G 4, we require that the state of charge of the storage is known. We assign this function to the controller of the `EnergySystem`.

The safety of the entire system G 5 with regard to the storage system is distributed among the three components `DriveSystem`, `EnergySystem`, and `Supercap-StorageSystem`. The mechanical integrity G 6 and the long service life of the entire system G 7 are also divided among these main components.

Since the efficiency of the whole system is important to the operator, G 8 is kept by all the components that can generate a power loss.

The `DriveSystem` is responsible for converting mechanical power to electrical power, MS 1, and vice versa, MS 2.

ES 1 required that the dynamic be controlled on different time scales. This requirement is sensibly distributed among all electrical system components.

The situation is analogous for ES 2 and ES 3. Each electrical system component must be designed in such a way that the current and voltage load is also maintained.

The requirement that switching losses be minimized, GL 4, is met by the power electronic components `DC/AC` and `SynchronBuckConverter`.

The supercap connections SC 1 to SC 4 can be clearly assigned to the `Supercap-StorageSystem`. However, the DC link capacitors will also have to comply with the SC 1 and SC 2 requirement.

Let us now look at the customer requirements. `Cabinsystem` and `DriveSystem`, including their subcomponents, are responsible for realizing the vertical transport, RL 1. The accelerations are controlled by the drive system. Therefore, the drive system must also comply with the requirement RL 2.

RL 3 was the motivation for adding a storage component to the lift system. Therefore, we assign this requirement to the `EnergySystem`.

We assign the product manager's requirements for a modular, expandable system, RL 4 and RL 5, to the `SupercapStorageSystem`. This is where the design decisions are made that are responsible for compliance with these two requirements. We also assign the regulator's security requirement RL 7 directly to the supercap storage system.

RL 6, the regulator's requirement for an increase in energy efficiency, is distributed among `ElectricMotor`, `DC/AC`, and `EnergySystem`.

Production's requirements, RL 8 and RL 9, apply to all components of the lift system; therefore, these requirements must be met by the `Cabinsystem`, `DriveSystem`, `EnergySystem`, and `SupercapStorageSystem`.

The purchaser's requirements were RL 10 and RL 11, which we have already taken into account in the design process, as the `Capacitorbank` must comply with them.

For the service engineer, we provide an interface with the help of the `Controller` of the `EnergySystem`, which allows them to know the status of the supercaps. RL 12 would then be fulfilled.

The service engineer's requirement to be able to exchange parts of the `Supercap-StorageSystem`, RL 13, must be taken into account by the `SupercapStorageSystem` in its design.

We see that all requirements have been assigned to at least one system component. The storage bank is designed, and we know the power and energy requirements. We now have all the information to start building the system.

7.5   Conclusion

In this chapter, we have learned about electrical storage systems. We have seen that we can store electrical power either via a current storage or a voltage storage. For current storage, inductors or superconducting inductors are suitable. Capacitors or supercaps are suitable for voltage storage.

The chapter ended with a detailed example of an electrical storage system. In the next chapter we will look at the next large group of storage technologies: electrochemical storage systems. These store electrical power via an electrochemical process.

Chapter summary

- We can store electrical power in two ways: we can either use a voltage storage technology or a current storage technology.
- Current storage technology is the electrical equivalent to storing kinetic energy, like using a flywheel. Similar to the flywheel, the current storage technology has the disadvantage of a high self-discharge rate.
- For storing current, we use an inductance. The inductance transfers the electrical energy into magnetic energy and vice versa and stores this energy until no current flows. However, electrical resistance decreases the amount of stored energy.
- Superconductors can be used, to reduce the self-discharge rate of an inductor.
- Voltage storage technology is the electrical equivalent of storing potential energy, i.e. lifting mass from one level of height to another.
- The electrical component that stores voltage is the capacitor. But the storage capacity of capacitors are rather small, and they also have a high self-discharge rate.
- Two additional effects can be used to increase the capacitance of a capacitor: building a double-layer capacitance and a pseudocapacitance.
- Capacitors combining these two effects are called 'supercapacitors, or supercaps' and have capacitance values 10–100 times higher in comparison to the classical capacitor.
- Since supercaps have three different mechanisms to store energy, these three different mechanisms interact with each other. Hence, we need to monitor these interactions if we use supercaps.

- In order to increase the capacitance of capacitors or to increase their charging and discharging power, capacitors are combined into capacitor banks. They are described by the way the capacitors are connected. We speak of a $xpys$ connection. This means y capacitors are connected in series to one string, and x strings are connected in parallel.

- Capacitor banks can be described as one single capacitor.

- Supercaps show larger manufacturing variances. This cause limitations in the assembly of capacitor banks. If we combine y capacitors in series, the capacitor with the lowest capacitance dominates the complete string. This can be compensated if we add a balancing system, which redistributes the charge carrier inside the string.

- Supercaps have an average voltage level of about 2.7 V_{DC}. The y-connection allows this voltage to be increased. However, in most cases the usage of a DC/DC-converter is necessary to integrate supercaps into an energy storage system.

- Both current and voltage storage systems are used as power storage systems. They have the capability to be charged and discharged with very high E-Rates.

8
Electrochemical storage systems

8.1 Introduction

It is impossible to imagine our everyday life without electrochemical storage systems. Only a few people today still wear a mechanical watch whose movement is driven by a mechanical spring, which draws its power from a mechanical storage system. More widespread is the digital watch, which receives its time from a quartz crystal, powered by a small battery.

We have already seen in Chapter 2 that electrochemical storage systems are used everywhere. In mobile applications such as laptops or smartphones, electrochemical storage systems based on lithium ions are generally used. The situation is similar in electromobility, but here solutions using lead acid batteries and high-temperature batteries have also been realized.

The more we rely on renewable energy sources such as solar or wind for energy production, the more we have to deal with the question of how to deal with the problem that the wind does not always blow and the sun does not always shine. As we saw in Chapter 2, we are also working with storage systems in these cases. Here, too, electrochemical storage technologies are strongly represented.

In this chapter, we will take a closer look at these storage technologies. The four most important technologies are examined in more detail in individual sections. Besides describing how they work, we will also look at the characteristics and special requirements that have to be met when using each of them. Although these technologies are all different, they have a lot in common, which is why this chapter begins with a section on the general mechanisms, requirements, and system components that are present in all technologies to varying degrees.

8.2 General considerations on electrochemical storage technologies

While electrical storage devices store energy by spatially redistributing charge carriers and thus creating or modifying an electric field, chemical reactions take place in electrochemical storage devices in which electrons are released and later reabsorbed. We have already learned about the basic reaction in supercapacitors. It was responsible for the formation of the pseudocapacitance in addition to the double-layer capacitance. We start our investigation with a description of this basic reaction; then we will look at the requirements that apply to the use of electrochemical storage technologies.

After that, we want to deal with the topic of ageing. We have touched on the subject of ageing in supercaps in Chapter 7. In this chapter we will look at this topic

Energy Storage Systems. Armin U. Schmiegel, Oxford University Press. © Armin U. Schmiegel (2023).
DOI: 10.1093/oso/9780192858009.003.0008

in more detail, and we will conclude this section with a system design of electrochemical storage systems.

8.2.1 The basic electrochemical reaction

Electrochemical storage technologies are all based on the same basic concept. This is illustrated in Fig. 8.1. We have a cell in which two electrodes, the negatively charged anode and the positively charged cathode, have been placed in an electrolyte. Between the two electrodes is a separator. This is a material that does not allow electrons to pass between the two electrodes. Thus, electrons can only pass from one electrode to the other via the terminals.

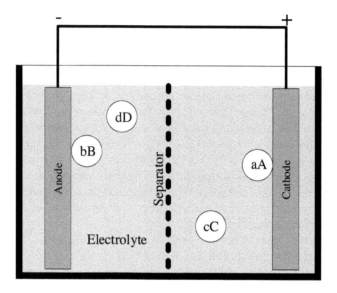

Fig. 8.1 General structure of a battery cell. The smallest system component of an electrochemical storage device.

The structure shown here is called a battery cell. It is the smallest system component of an electrochemical storage device. The basic reaction takes place in this battery cell. It is a redox reaction (Reddy, 2011). It is the same for all electrochemical storage devices. On the anode side, n electrons are released due to a chemical reaction, converting b parts of B into d parts of D. In the process, n electrons are released:

$$bB - ne \rightleftharpoons dD. \tag{8.1}$$

On the cathode side, if there is an electrical interconnection between the anode and cathode, these released electrons are used to convert a parts of A with n electrons to c parts of C:

$$aA + ne \rightleftharpoons cC. \tag{8.2}$$

These two individual reactions can be summarized into one overall reaction:

$$aA + bB \rightleftharpoons cC + dD. \tag{8.3}$$

Now it is also clear why there must be a separator between the two electrodes. The separator blocks the direct flow of electrons. We can interrupt the reactions that take place in eqns (8.1) and (8.2) by opening or closing the electrical contact. Without the separator, the overall reaction would cause A and B substances to convert to C and D in the electrolyte. By introducing the separator and allowing the flow of electrons to be interrupted via the conductor (i.e. a switch), the cell can be kept in an energetic imbalance and thus energy can be stored.

In eqns (8.1)–(8.3), ' $\rightleftharpoons$ ' was used as the reaction symbol in the reaction. This means that the reaction can occur in both directions—that is, it is reversible. We can therefore make the reaction take place in one direction as well as the other by changing the cell voltage. This does not apply to all electrochemical storage technologies. A distinction is made here between the electrochemical storage devices that are reversible in these reactions and those that are not. In the non-reversible case, we speak of primary batteries. Many everyday goods use these primary batteries. Torches, remote controls, toys, and other entertainment devices usually use primary batteries. These are purchased as charged batteries and are then disposed of in the recycling process after discharging. Primary batteries can only be discharged. Therefore, they are not suitable for the realization of energy storage applications. We can therefore only use electrochemical technologies that realize a reversible reaction, because only then can our system be charged and discharged.

If the reaction is reversible, we speak of secondary batteries. These cells can be charged and discharged again and again and thus be used several times. Secondary batteries are used in energy storage applications because these systems go through many charging and discharging cycles.

The fact that a reaction is reversible does not mean that we have no losses. The term reversible only means that the reaction can take place in one direction or the other. We can charge and discharge the battery cell. The losses that occur during a charging and discharging process result in the roundtrip efficiency η_r of an electrochemical reaction. It results from the energy that is transferred into the cell during the charging process E_{in} and the energy that we get out of the cell again during the final charging process E_{out}:

$$\eta_r = \frac{E_{out}}{E_{in}}.$$

Tab. 8.1 (Kairies, 2017) shows the roundtrip efficiencies of different electrochemical cells. The efficiency varies from 70% to 96%. These numbers do not represent the ideal roundtrip efficiency. This is because generation of heat losses is influenced by various aspects. For example, the structure of the active material, the electrical contacting, or the composition of the electrolyte can have an additional influence on η_r.

Table 8.1 Roundtrip efficiencies of different electrochemical storage technologies. For lithium ion batteries and redox flow cells, different chemistries are possible, which are labelled.

Chemistry	η_r
Lead acid	82%
Lithium ion (NMC/LMO)	92%
Lithium ion (LFP)	86%
Lithium ion (Titanate)	96%
Lithium ion (NCA)	92%
High temperature (NAS)	80%
Redox flow (Vanadium)	70%
Redox flow (ZnBr)	70%

Let us take a closer look at the charging process. In eqn (8.1), two things happen. First, an electron is released on the anode and bB becomes dD. We call the energy needed for this the free enthalpy of reaction ΔH. In the experiment, however, we find that this amount of energy is not enough to fully charge the battery. We have to put more energy into it, because the molecules have to overcome spatial and structural obstacles to find each other. The energy needed for this can be measured as a temperature change, ΔT. We call this part the heat of reaction. Both parts together result in the total energy or the enthalpy of reaction ΔG:

$$\underbrace{\Delta G}_{\text{free enthalpy of reaction}} = \underbrace{\Delta H}_{\text{enthalpy of reaction}} - \underbrace{\overbrace{T}^{\text{temperature}}\overbrace{\Delta S}^{\text{entropie}}}_{\text{heat of reaction}}. \tag{8.4}$$

Only the free enthalpy of reaction of the total reaction in eqn (8.3) is also available during the discharge process. The heat of reaction represents a loss that can at best be used thermally.

The voltage level of the electric cell can be calculated from the free enthalpy of reaction: during the charging process, a number of charge carriers are transported from the anode to the cathode. This can be determined by integrating the charging current I over the total period of the charging process. In the case of a complete charge, this charge quantity corresponds to the proportion of charge carriers moved and stored per mole:

$$Q = \int_0^T I \mathrm{d}t = \underbrace{eN_A}_{F} nz. \tag{8.5}$$

n is the number of charge carriers used in eqn (8.3). z embodies the number of moles of the active material of the cell. $N_A = 6.022 \cdot 10^{23} \mathrm{mol}^{-1}$ is the Avogadro constant and indicates the number of particles per mole. The product of the elementary charge and the number of particles per mole is called the Faraday constant, denoted by

$$F = eN_A = 9.6485 \cdot 10^4 \frac{\mathrm{C}}{\mathrm{mol}}. \tag{8.6}$$

After the charging process, two things can be observed. First, the temperature of the electrochemical cell has changed. Secondly, a voltage can be measured between the cathode and the anode. This voltage multiplied by the transported charge gives the electrical energy that is now stored in the cell:

$$\Delta G = E_{el} = Q\Delta U = nFz\Delta U. \tag{8.7}$$

The energy content of a battery cell depends linearly on the potential difference between anode and cathode and the number of transferable electrons. If the storage content of an electrochemical cell is to be increased, according to eqn (8.7) either the amount of active material can be increased, which would correspond to an increase of z, or the cell chemistry can be adjusted, which would correspond to a change of the quantities n or ΔU.

We can also use eqn (8.7) to calculate the potential difference. The free enthalpy of reaction ΔG includes the enthalpy of reaction and the heat of reaction. If these are known for all substances and if one considers the equilibrium state—that is, if no reaction takes place—the following applies

$$\Delta G_0 = (G_A + G_B) - (G_C + G_D). \tag{8.8}$$

The corresponding equilibrium voltage is then given by:

$$U_0 = \frac{\Delta G_0}{nF}. \tag{8.9}$$

However, this voltage can only be measured in the equilibrium state. During the charging or discharging process, an imbalance occurs because the substances occur in different concentrations $(a_A^a, a_B^b, a_C^c, a_D^d)$. The measured stress is described by the Nernst equation and is then (with T as the ambient temperature and R as the ideal gas constant, i.e. $R = 8.314\,\frac{\text{kgm}^2}{\text{s}^2\,\text{molK}}$):

$$\Delta U = \Delta U_0 - \frac{RT}{nF}\ln\frac{a_C^c a_D^d}{a_a^a a_B^b}. \tag{8.10}$$

In the equilibrium state, the concentration of all substances is approximately equal, in which case the term in the logarithm is approximately one, corresponding to a voltage of ΔU_0. In the charging and discharging state, the term in the logarithm is greater or less than one, and there is a decrease or increase in the cell voltage.

Electrochemical storage systems are more subject to the laws of thermodynamics than mechanical and electrical storage systems because, as shown in eqn (8.10), the reactions also have a direct temperature dependence. There are states in electrochemical storage systems in which stochastic phenomena must be taken into account.

For example, in an electrochemical cell, the reaction in eqn (8.3) occurs continuously when electron flow is enabled. Assuming the circuit was closed, as shown in Fig. 8.1, and the battery was not charged, a reaction releasing an electron would occur now and then due to thermal fluctuations at the anode. The measured current in this case would be:

$$I_A \sim e^{-\frac{nF}{RT}}. \tag{8.11}$$

This is caused by reactions of individual molecules that have sufficient energy to carry out the reaction in eqn (8.1) due to thermal excitation.

If measurements were made at the cathode, it would be observed that random reactions also occur here, which on average produce a small current in the opposite direction:

$$I_C \sim -e^{-\frac{nF}{RT}}. \tag{8.12}$$

If a voltage U is now applied, it increases or decreases the reaction probability at the anode as well as at the cathode. In addition to the temperature, which increases the process, the voltage can also influence the process. Thus, for the anode:

$$I_A \sim e^{-\frac{nF}{RT}(U-U_0)}. \tag{8.13}$$

The current measured when a voltage is applied results from a superposition of both currents. It should be noted that the reactions on the anode side as well as on the cathode side have different probabilities at the same voltage and temperature—that is, the mixing ratio of these two currents is different under the same boundary conditions. For example, if the anode reaction occurs 10% more often at the same temperature and voltage, this current has a share of 60%, and the share of the cathode current is only 40% of the measured total current. This is taken into account in the following formula by the symmetry factor α. The following therefore applies to the total current:

$$I_{\text{sum}} = I_0 \left[e^{\frac{(1-\alpha)nF}{RT}(U-U_0)} - e^{-\frac{\alpha nF}{RT}(U-U_0)} \right]. \tag{8.14}$$

With large potential differences, the cell reacts as one would expect from a voltage source. There is a linear relationship between current and voltage, and depending on the sign, the anode or cathode current dominates. For smaller potential differences, there is a deviation generated by the reverse current (Fig. 8.2). This deviation is called transition polarization.

We can think of the electrochemical cell as a voltage source whose open-circuit voltage depends on the state of charge (SOC). The shape of the voltage $U_{\text{OCV}}(\text{SOC})$ depends on the respective electrochemistry. However, this aspect alone is insufficient to describe an electrochemical cell. The open-circuit voltage is the voltage we measure when no load is applied—that is, when the terminals are open. But as soon as we have a current I_{cell} flowing through the cell, we can measure a higher voltage during charging and a lower voltage during discharging. This is due to the internal resistance R_i, of the cell. The internal resistance summarizes various aspects of the cell that affect the current flow: the structure of the active material, the contact, the conductivity of the electrolyte, etc.

If we look at the construction of an electrochemical cell in Fig. 8.1, it is actually not that different from that of a supercapacitor. Here, too, we have a conductor that is dipped in an electrolyte. Therefore, a double layer will also form here. Compared to the storage capacity of the active material, this makes a very small contribution to the storage of energy, but it has an influence on the dynamic properties. We describe this influence by a double-layer capacity C_{DL} with a parallel resistance R_{DL}.

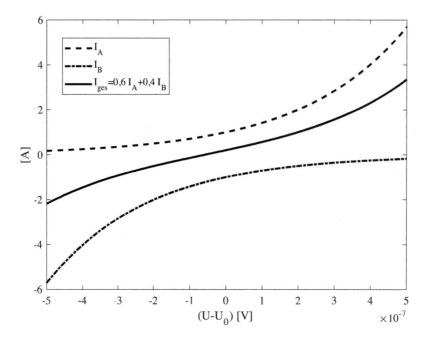

Fig. 8.2 Current at the anode I_A, cathode I_B and the total current I_{sum} with a symmetry factor α at different voltage differential $U - U_0$ at a temperature of $T = 300\,\text{K}$.

If we modulate the charging and discharging current with a frequency-dependent voltage, we can observe that the resulting current undergoes a frequency-dependent modulation. This does not always correspond to the amplitude curve of the modulated voltage. This is the case at very low frequencies. Here we have the situation that the charge carriers in the active material and the electrolyte can follow the excitation slowly. As soon as the frequency increases, other effects occur, so reactions that we can assign to the pseudocapacitance or the double layer become visible. We can describe these effects by additional RC elements R_n, C_n.

We can observe that electrochemical cells slowly lose their stored energy. The reaction in eqn (8.3) takes place although the anode and cathode are not electrically connected. This can occur, for example, through secondary reactions that virtually bypass the separator. But it can also be due to the construction of the cell, which always allows small currents between anode and cathode. Regardless of the cause, it is a self-discharge, which we can represent by a loss resistance R_l.

We have summarized these various effects and equivalent circuits in the equivalent circuit diagram in Fig. 8.3. It is a simple model that allows us to describe the dynamic properties of a cell in a system. We need to be clear on which time scale we want to use to look at the dynamics of the cell. The RC links determine the dynamics at a frequency typically greater than 10 Hz. So if we are interested in the dynamics of the cell on time scales of minutes or days, we do not need the RC members and can simplify the modelling considerably (Magnor *et al.*, 2010).

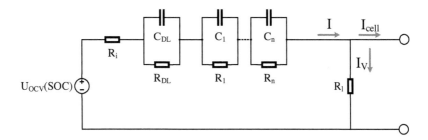

Fig. 8.3 Equivalent circuit diagram of an electrochemical cell. The cell is described as a voltage source. Its voltage level depends on the state of charge, having an internal resistance R_i. The effect of double layers and other electrochemical properties are described in RC-parts. Self-discharge is described by the resistor R_1.

The parameters of the battery model depend on the temperature, the state of charge, and the age of the cell. For a complete modelled description, an extensive determination of all parameters is necessary. In practice, this complexity is reduced by making various experimentally verified simplifications. The same applies to the determination of battery ageing. Here, heuristic models have established themselves, which are adjusted during operation by means of measurement curves.

8.2.2 Requirements and design

Having looked at the fundamentals of electrochemical processes in battery cells in the previous section, in this section we consider the implications of this for the sizing and design of a storage system. In Chapter 5, we had established that the relationship between current and voltage is component-dependent. We also saw that the current and voltage limits are important for the design of a storage system and derived the requirements ES 2 and ES 3.

The general requirement GS 4 demands that we should always know the energy content of a storage system. With previous storage technologies, this was not a particular challenge. With mechanical storage systems, the energy content can usually be measured directly from physical data. If we know the moment of inertia of a flywheel storage system, a measurement of the rotational speed is sufficient to calculate the energy content. In a superconducting coil, the energy content is proportional to the amount of current measured. And in a capacitor, the energy content was proportional to the number of charge carriers. By measuring the voltage, we can determine how much storage capacity is still available. With an electrochemical storage device, the relationship between the state of charge and the measured voltage is more complex.

Let us consider the following experiment. We take two battery cells A and B that use different cell chemistry to store energy. We know that both cells are fully charged. We now measure the open-circuit voltage of each cell and plot this on a graph. Then we discharge both cells a little and repeat this measurement. We do this until no energy can be taken from the cell. The result of this experiment is shown in Fig. 8.4. As we can see, the voltage curve of the two cells differs. The voltage curve seems to depend on the cell chemistry. The shape of the capacitance–voltage curve is much more complex than that of the capacitor.

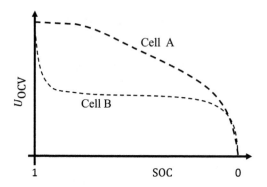

Fig. 8.4 Open-circuit voltage U_{OCV} of two different cell types A and B at different states of charge.

To determine the energy content of the cells, we have to integrate over the changing voltage multiplied by the discharge current:

$$\kappa = \int_{t=0}^{t=T} U(t) \cdot I \, \mathrm{d}t. \tag{8.15}$$

This means that at constant current, the power we draw from the battery changes depending on the state of charge. We had already observed this with the capacitor, where the voltage depends linearly on the state of charge. The difference here is that the voltage is now no longer linear, so that the received power also no longer behaves linearly.

We do not always know the exact shape of the voltage as a function of the state of charge. For most cell chemistries, a simple approximation is sufficient. We use the average voltage and multiply it by the nominal current I_{nom}, with which the cell can be charged or discharged regardless of the state of charge:

$$\kappa \approx \frac{U_{\mathrm{max}} + U_{\mathrm{min}}}{2} I_{\mathrm{nom}}. \tag{8.16}$$

Exercise 8.1 An electrochemical cell has a cell voltage of $U_{\mathrm{OCV}} = 3.9$ V$_{\mathrm{DC}}$ when fully charged. In the discharged state, the voltage is $U_{\mathrm{OCV}} = 2.7$ V$_{\mathrm{DC}}$. The nominal charging current is $I = 1.1$ A. What is the energy content of the cell?

Solution: Since we have no information about the shape of the open-circuit voltage during the final charge, we use the approximation from eqn (8.16):

$$\kappa = \frac{U_{\mathrm{max}} + U_{\mathrm{min}}}{2} I_{\mathrm{nom}} = \frac{3.9 \text{ V}_{\mathrm{DC}} + 2.7 \text{ V}_{\mathrm{DC}}}{2} 1.1 \text{ A} = 3.63 \text{ Ws}.$$

The shape of the open-circuit voltage curve as a function of the state of charge has consequences for the system design. While the B battery has the advantage that the voltage is approximately constant over a wide capacity curve, the electrical components of the system must be able to serve a wide voltage window when using A.

ECS 1: AS AN Engineer, I WANT to ensure that the electrical components of the storage system are suitable for the voltage over the entire capacity curve of the electrochemical cell SO THAT the requirements ES 2 and ES 3 are always met.

The different voltage curve has further consequences. The power is calculated from the product of current and voltage. Therefore, the same charging and discharging power may not be available to the system at every state of charge, as the power electronics or other electrical components cannot carry an increased current. This results in another requirement for electrochemical storage systems:

ECS 2: AS AN Engineer, I WANT to design the current-carrying capacity of the components so that the required power can be provided at any state of charge SO THAT the storage system can also meet the power requirements.

In Chapter 3, we learned about the requirement G 4. It requires that the state of charge of a storage system is known so that the user can adapt his loading and final loading behaviour. In short, there are two methods to determine the state of charge of a storage system. Firstly, the state of charge can—if possible—be measured directly. This would be possible for a battery of type A, for example. The open-circuit voltage is always associated with a state of charge. If we measure the open-circuit voltage, we can determine the state of charge. This procedure is not possible for cell B. Over a wide range of the SOC, we have approximately the same open-circuit voltage. Only shortly before complete charging or discharging can we also assign a state of charge to the open-circuit voltage. In this case we have to use the other method. We determine the state of charge by measuring how much power is loaded into or taken out of the storage. This procedure is called Coulomb counting.

If a generator G is used for charging and a load L is supplied from the storage, we can calculate the SOC using:

$$\text{SOC} = \frac{1}{\kappa} \int_0^T \eta_{\text{GS}} G_{\text{S}} - S_L \; \mathrm{d}t. \tag{8.17}$$

The two procedures can be compared to the task of a bouncer in a bar who has to ensure that there are no more than 50 people inside. The bouncer has two options: he can go into the bar regularly and count the number of people, and close the door for a short time to make sure that nobody leaves the bar. This corresponds to measuring the open-circuit voltage, because we can only measure this when the cell is neither charging nor discharging. The 'door' must therefore also be kept 'closed' in the case of an electrochemical cell. The alternative is to remember how many people enter the bar and how many leave it again. But this method also has its problems. The bouncer would then have to monitor all entrances and exits. He would also have to assume that no person has used an unknown exit or simply disappeared. And he must also assume that he will not miscount. This can be difficult, because he has to keep records throughout the evening. Coulomb counting is similar. In eqn (8.17), the product of voltage and current—that is, two measured quantities—is integrated over time. The measurement error of both quantities adds up. The self-discharge of the cell is not taken into account in the equation. It also eludes direct measurement. The situation is similar for the transfer losses η_{GS}. Since the transfer losses also include losses due to the electrochemical storage process in addition to the losses of the electronics, parts of the power losses elude direct measurement.

We have two other parameters that influence the energy content of a cell and thus the accuracy of eqn (8.17). The amount of energy we can extract from a cell, or with which we can charge a cell, depends on the temperature and the charging or final charging power.

In the experiment, we can observe that a cell can store and release different amounts of energy at different temperatures. This can be explained by the fact that the probability of a reaction taking place depends on the temperature. On the one hand, this is because the active material distributes itself differently in the electrolyte and at the electrode due to its own thermal movement, and on the other hand, it is because with an increased temperature, more energy is available to carry out the basic reaction seen in eqn (8.3) but also unwanted side reactions.

Analogously, the dependence on the charging and discharging rates can be explained. We observe that we can extract more energy from the cell with a slow discharge than with a fast discharge. The reasons for this effect can also be found in the transport processes and the chemical processes of the respective technology. For eqn (8.3), the active material, which is partly in the electrolyte, must reach the electrodes. However, this movement cannot take place arbitrarily fast. Depending on the cell chemistry used, this can lead to unwanted side reactions that hinder further charging or discharging. Conversely, if the rate of charging or discharging is low, we allow enough time for the active material to reach the electrode and react.

Suppose we control the temperature of the battery and limit ourselves in the charging rate. Then the following applies to the error in the state of charge calculation using coulomb counting:

$$\Delta SOC = SOC - SOC_{CC} \tag{8.18}$$

$$= \frac{1}{\kappa} \int_0^T \left(\underbrace{(\eta_{GS} G_S + \eta_{SS} S_S(-\Delta) - (S_L + S_S(\Delta)))}_{\text{powerflow}} - \underbrace{(\eta'_{GS} G_S - S_L)}_{\text{estimates powerflow}} \right) dt. \tag{8.19}$$

Here the SOC is the real state of charge of the cell and SOC_{CC} is the estimated state of charge, determined via Coulomb counting. We see that over time the error in the state of charge computation becomes larger. It grows linearly with time.

In order to satisfy G 4 nevertheless, the requirement for electrochemical storage is:

ECS 3: AS AN Engineer, I WANT to compensate the error in the state of charge estimation process SO THAT I'm able to provide an accurate state of charge value.

How can this requirement be realized? We have two procedures. One can only be used when the battery is not charged or discharged and will give a poor result for some cell chemistries. The other method is intrinsically biased, with its error increasing linearly over time. What would our bouncer do? He can improve his content estimation with two measures. First, he can improve the way he sums up arrivals and departures. For example, he can compare his count with the count of a colleague. Secondly, he can regularly—whenever there is nothing going on, or when the door opens anyway—look into the bar and estimate how full the place really is. He can combine this information to get a better estimate of the number of visitors. Similar strategies are used in the implementation of ECS 3 (He *et al.*, 2013; Wei *et al.*, 2016; Govindarajan, 2017; Sun *et al.*, 2021). Coulomb counting is combined with the measurement of the open-circuit voltage. Additional knowledge, for example the system state or a model of the battery dynamics, is incorporated for improvement. The data are then combined with the help of an adaptive filter (Ma *et al.*, 2020). In this way, as much of the available information as possible is used to comply with ECS 3.

Similar to supercaps, the cell chemistry defines the voltage range in which the battery cell can be used. Increasing the amount of active material changes the energy content, but not the voltage range. For different applications, however, we need different current and voltage ranges. Here, too, we solve the problem by a suitable aggregation of battery cells:

ECS 4: AS AN Engineer, I WANT to be able to combine battery cell in parallel strings with serially connected cells SO THAT I can realize battery modules that provide a suitable current and voltage window.

ECS 4 means that we aggregate cells with a *xpys* connection. The total voltage for such a module is given by the number of cells connected in series:

$$U_{\text{sum}} = \sum_{i=1}^{y} U_i. \tag{8.20}$$

To determine the maximum possible current, the number of strings connected in parallel must be determined:

$$I_{\text{sum}} = \sum_{i=1}^{x} I_i. \tag{8.21}$$

By using a suitable combination of cells connected in parallel and series, configurations can be determined that are suitable for the application in question. However, we are somewhat limited. Since the voltage depends on the state of charge (Fig. 8.4), the voltage window cannot be arbitrary.

Exercise 8.2 An electrochemical cell has a cell voltage of $U_{\text{OCV}} = 3.9$ V_{DC} when fully charged. In the discharged state, the voltage is $U_{\text{OCV}} = 2.7$ V_{DC}. The maximum charging current is $I = 1.2$ A. The capacity is $\kappa = 4$ Ws. The cell is used in a 2p8s circuit. What are the electrical properties for this module?

Solution: To find the minimum and maximum module voltage, we use eqn (8.20):

$$U_{\text{min}} = 8 \cdot 2.7 \text{ V}_{\text{DC}} = 21.6 \text{ V}_{\text{DC}}$$
$$U_{\text{max}} = 8 \cdot 3.9 \text{ V}_{\text{DC}} = 31.2 \text{ V}_{\text{DC}}.$$

For the maximum module current, we use eqn (8.21):

$$I_{\text{max}} = 2 \cdot 1.2 \text{ A} = 2.4 \text{ A}.$$

The capacity of the module is the sum of the capacities of all battery cells:

$$\kappa = 2 \cdot 8 \cdot 4 \text{ Ws} = 64 \text{ Ws}.$$

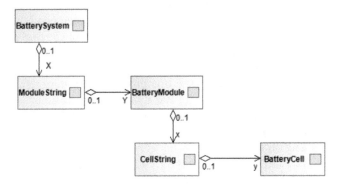

Fig. 8.5 Structural design of battery storage systems: cells form modules, modules form the system. The cells form a module through a $xpys$ connection. The modules form a battery system through a $XpYs$ connection.

To comply with ECS 4, cells are interconnected to form modules. If larger battery systems are to be realized, the modules are connected together to form battery systems. Similarly, the modules are connected to form strings that are linked in parallel. Here we have $XpYs$. In Fig. 8.5, the structure of a `BatterySystem` is shown. A battery system is divided into two system levels. It is composed of one or more `BatteryModules`. These battery modules are composed of one or more `BatteryCells`. When designing storage systems, the battery modules and the interconnection of the cells within the modules are fixed. The system is designed at module level.

The design of a battery system or a module can be described as a combinatorial optimization task with constraints. The largest cost driver in a battery system is the number of cells, which should therefore be minimized. N is the product of series-connected cells x/X and parallel-connected cells y/Y and represents the total number of cells/modules:

$$\min Y = N = x \cdot y. \tag{8.22}$$

There are additional constraints for the particular requirement: the minimum required capacity $\hat{\kappa}_{\min}$, the maximum power $P_{\max}$, and the allowed voltage window $\hat{U} \in [U_{\min}, U_{\max}]$, as well as the maximum allowed current $\hat{I}_{\max}$.

The cell chemistry and the cell or the selected module determine the quantities $U^i_{\min}$ and $U^i_{\max}$, the maximum charge and discharge current $I^i_{\max}$, and the capacity κ^i [Ah].

For any given configuration, then:

$$\hat{I}_{\max} = x \cdot I^i_{\max} \tag{8.23}$$

$$\hat{U}_{\min} = y \cdot U^i_{\min} \tag{8.24}$$

$$\hat{U}_{\max} = y \cdot U^i_{\max} \tag{8.25}$$

$$\hat{\kappa} = y \cdot x \cdot \kappa^i \tag{8.26}$$

$$P_{\max} = U_{\max} \cdot I_{\max}. \tag{8.27}$$

The optimization problem to be solved is thus:

$$\min N = x \cdot y \quad \text{rtb:} \quad
\begin{cases}
x, y \in \mathbf{N} \\
y \cdot U^i_{\min} \geq U_{\min} \\
y \cdot U^i_{\max} \leq U_{\max} \\
x \cdot I^i_{\max} \geq I_{\max} \\
y \cdot U^i_{\max} \cdot x \cdot I_{\max} \geq P_{\max} \\
y \cdot x \cdot \kappa^i \geq \hat{\kappa}.
\end{cases} \tag{8.28}$$

Exercise 8.3 The following requirement has been identified for the electrification of an auxiliary drive of a vehicle: at least a maximum power of $P_{\max} \geq 1$ kW is required. The storage capacity is at least $\hat{\kappa} \geq 10$ kWh. The battery voltage should cover a voltage range of $U_{\min} \geq 300$ V$_{\text{DC}}$ and $U_{\max} \leq 900$ V$_{\text{DC}}$. The maximum current shall be $I_{\max} = 4$ A.

A 2p8s module with $U \in [21.6 - 31.2$ V$_{\text{DC}}]$, $I_{\max} = 2.4$ A, and $\kappa = 382$ Wh with the cells used in the Exercise 8.2 is used.

What could be a suitable configuration?

Solution: We need a capacity of 10 kWh. The single module has a capacity of 0.382 kWh. So the number of modules needed is

$$N = \frac{10 \text{ kWh}}{0.382 \text{ kWh}} = 26.178 \approx 27.$$

Here we have to round up, because the energy content of 26 modules is too small.

We need a maximum current of 4 A. However, a single module can only transmit 2.4 A of current. Therefore, we have to work with two parallel strings. This increases the number of modules to 54, as the number of modules per string must be the same.

We therefore have 27 modules per string. The minimum string voltage is therefore:

$$U_{\min} = 27 \cdot 21.6 \text{ V}_{\text{DC}} = 583.2 \text{ V}_{\text{DC}}.$$

The maximum string voltage—that is, when all cells are fully charged—is:

$$U_{\max} = 27 \cdot 31.2 \text{ V}_{\text{DC}} = 842.4 \text{ V}_{\text{DC}}.$$

The voltage range of the string would be within the required range.

Finally, we check whether the power can still be provided when the battery system is discharged:

$$P = 583.2 \text{ V}_{\text{DC}} \cdot 4.8 \text{ A} = 2{,}799.36 \text{ W}.$$

The battery system with a 2p27s circuit would therefore also meet the power requirements.

Exercise 8.3 shows a design contradiction that is often observed in the design of systems with electrochemical storage: the capacity is determined by the number of cells and by the amount of active material. The voltage level is also determined on the battery side by the cell chemistry and can be raised solely by a series connection. On the connection side, the electronics often require a different, higher voltage level than can be provided by a series circuit. Cell voltage and target capacity simply do not match. To find a solution here, a DC/DC controller is often used to raise or lower the voltage level of the battery. This generates additional costs in the form of further power electronics, but on the other hand these are often compensated by the saved battery costs and the saved battery weight or volume.

8.2.3   The woes of ageing: Battery life and capacity management

An electrochemical cell is a physical system which is not in thermal equilibrium. In thermal equilibrium we would have a voltage that is close to zero—that is, the cell is short-circuited and the reaction in eqn (8.3) is driven only by thermal fluctuations. The moment we charge the cell and interrupt the current flow between anode and cathode, the system is no longer in thermal equilibrium. This is different from previous storage systems. When they were charged, they were still in thermal equilibrium. We have merely changed the physical system to a different state. This fact has consequences for the nature of the cell. The structure of the cell is stressed and loses its ability to store energy over time. The cell ages.

Let us try to approach the phenomenon of cell ageing with a simple experiment. Through magazine articles, internet videos, and rumours among friends, we have heard about the launch of a new mobile phone. Since we are no longer satisfied with our current one—it would be really great to be able to use apps—we decide to buy one. We want to be particularly smart here, and buy not only the mobile, but also a replacement battery. We just want to make sure that we can continue to use the mobile in two to three years by replacing the built-in battery with the spare. We test the energy content of both batteries and find that they have the same energy content. Then we put the replacement battery in a safe, well-tempered place. And we get busy with the new mobile and many great apps.

Over time, we find that we need to recharge the battery more and more often. In the first year, we only needed to charge once a week. Then it was two or three times a week, and now we have reached a point where we have to charge the mobile every evening. We also notice that the mobile gets warmer when charging. We decide to replace the batteries, but before we do, we measure the energy content of both batteries again.

We find that the battery we have been using all this time has lost most of its capacity. This is no surprise. But we also notice that the spare battery, which has been in a safe place all this time, no longer has its initial capacity! Apparently, this battery has also aged. We are furious, because the loss of capacity is measurable, but the remaining capacity is much greater than that of the used battery. So we exchange the batteries.

We realize that there are two different ageing processes that work independently and, in our experiment, on different time scales. One form occurs when the cell is charged and discharging. This ageing process is called 'cycle life', because the life span depends on the number of charging and discharging cycles. The second form depends only on calendar ageing and is called calendaric lifetime (Keil, 2017).

The causes of cycle ageing are, on the one hand, irreversible secondary reactions of the active material that take place in parallel with the main reaction (eqn (8.3)), but also the mechanical stress on the active material, which is in motion during charging and discharging.

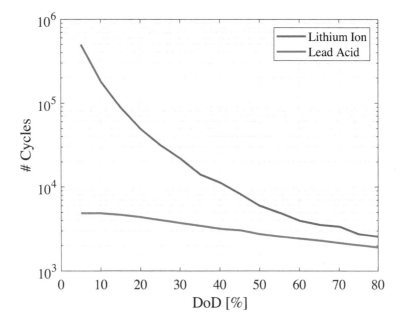

Fig. 8.6 Number of charging cycles of a lithium ion battery and a lead acid battery at different depths of discharge (DoD).

Let us consider here the experiment from Fig. 8.6 (Thiaux *et al.*, 2009). Different depths of discharge (DoD) are plotted on the horizontal axis. The number of cycles is entered on the vertical axis. For the different DoDs, an electrochemical cell was repeatedly charged and discharged. Its capacity was then measured. If this was less than 80% of the initial capacity, the number of cycles was entered in the diagram. This experiment was carried out for both a lithium ion battery and a lead acid battery. We see that the number of cycles depends on the electrochemistry. The lead acid battery can provide a much lower number of cycles. In Sections 8.3 and 8.4, we will take a closer look at the cell chemistries and learn about the different processes that are responsible for the ageing of the cell.

In addition to the absolute number, we will also see a different shape of the number of cycles depending on the DoD. With lead acid we have a linear shape. Since the y-axis has a logarithmic scale, this corresponds to an exponential shape. The lithium ion battery shows a non-linear shape in the logarithmic representation. However, if we also plot the x-axis logarithmically, we see that this shape looks almost linear. This means that the number of cycles in the lithium ion battery follows an algebraic law.

Exercise 8.4 The data from Fig. 8.6 suggest that the cycle life of the lithium ion battery is an algebraic shape—that is,

$$N_{\max} = \gamma \cdot \mathrm{DoD}^{\alpha}.$$

What would α and γ be if we determined these two parameters from the points $(\mathrm{DoD}_1 = 5\%$, $N_{\max} = 497.950)$ and $(\mathrm{DoD}_1 = 60\%, N_{\max} = 3.974)$?

Solution: The logarithm of the above function is as follows:

$$\log N_{\max} = \alpha \log \mathrm{DoD} + \log \gamma.$$

This is a linear equation with α as the slope and $\log \gamma$ as the intercept. The slope is calculated as:

$$
\begin{aligned}
\alpha &= \frac{\log N_{\max,2} - \log N_{\max,1}}{\log \mathrm{DoD}_2 - \log \mathrm{DoD}_1} \\
&= \frac{\log 3.974 - \log 497.950}{\log 60 - \log 5} \\
&= -1.944.
\end{aligned}
$$

We substitute the value α for a given point and determine the value for γ:

$$
\begin{aligned}
\gamma &= \frac{N_{\max,1}}{\mathrm{DoD}_1^{\alpha}} \\
&= \frac{497,950}{5^{-1.944}} = 11,376,300.
\end{aligned}
$$

We can use the formula used in Exercise 8.4 to determine the cycle age of a cell. Assume that a cell is always discharging with the same DoD. This way, we would know how many cycles we have left at any time, because we would only have to subtract the number of cycles already made from $N_{\max}$. It is somewhat more difficult if we work with different cycle depths. But even in this case we can use the approach followed in Exercise 8.4. We just need to weigh and sum up each cycle according to its DoD D_i:

$$a_{\mathrm{Cycle}}(N) = \alpha \sum_{i=1}^{N} D_i^{\beta}. \tag{8.29}$$

a_{Cycle} represents a value for the cycle life age after N cycles. $a_{\mathrm{cycle}} = 0$ means that the cell has not aged yet and a value of 1 means that the end of life criterion is met. This is a relative value. The coefficients β and α must be determined experimentally. See Exercise 8.4 for an analogous approach.

Let us now look at calendaric ageing. This is caused by irreversible decay processes in the active material or electrolyte. Fig. 8.7 shows an experiment on calendar ageing (Schmiegel, 2014). In this experiment, three lithium ion battery cells were charged to

50%. Then they were stored at different temperatures for one week. After this week, their capacity was determined and the relative capacity related to the initial value κ_{BOL} was determined (BOL = Beginning of Life).

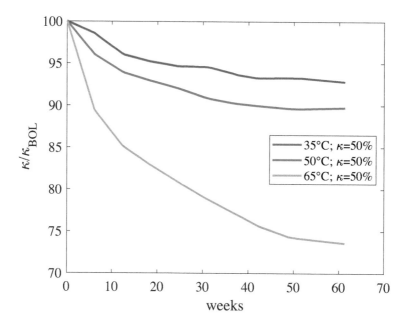

Fig. 8.7 The shape of the capacity of a lithium ion battery at different ambient temperatures over several weeks.

We can clearly see a temperature dependence of the capacity curve. The cell that was tempered to 65°C loses its capacity very quickly and is already below the 80% mark after just under 30 weeks, which is typically regarded as the end-of-life criterion for batteries. In the case of the 35°C cell, we also see a decrease in capacity, but this is significantly lower, and even after 60 weeks there is still a capacity of over 90% of the initial capacity.

How quickly a battery ages calendrically depends on two factors: the ambient temperature and the state of charge. A high temperature means that the likelihood of chemical decay increases, as chemical reactions occur faster at high temperatures than at low temperatures. Similarly, the state of charge has an accelerating effect on ageing. This is because reactions are additionally accelerated when voltage is applied. A simple approach to describe the calendar ageing is to assume the Arrhenius law. Again, we use a_{cal} as a relative numerical value that gives us information about the calendrical age of the cell. Using Arrhenius' law, the approach is:

$$a_{cal}(T) = \int_{t=0}^{T} c_{cal} e^{d_T (T_0 - T(t))} e^{d_U (U_0 - U(t))} \, dt. \qquad (8.30)$$

T_0 corresponds to the upper operating temperature of the cell and U_0 to its maximum open-circuit voltage. The values for the parameters c_{cal}, d_T, and d_U must be determined experimentally. If these are known, the ageing fraction can be determined from the evaluation of the integral.

We now have two simple mathematical descriptions for the ageing of a battery cell. Since we know that both ageing processes occur independently of each other, we can add both effects and thus obtain a statement about the state of health (SOH):

$$\text{SOH} = 1 - (a_{\text{cal}}(t) + a_{\text{Cycle}}(t)). \tag{8.31}$$

The idea in eqn (8.31) is analogous to the individual values of ageing. For a SOH of zero, the cell capacity is equal to the initial capacity. With a value of one, the end of life criteria is fulfilled.

Exercise 8.5 Two identical cells are purchased. One is stored for one year at 25°C with a voltage of $U_{\text{OCV}} = 3.1$ V$_{\text{DC}}$. The maximum voltage of the cell is $U_{\text{OCM}}^{\text{max}} = 3.6$ V$_{\text{DC}}$. The other cell is cycled daily with a DoD of 80%. For cycle ageing, the following holds true:

$$\alpha = 1; \beta = 26.43.$$

For calendar ageing applies:

$$d_{\text{T}} = -0.5; d_{\text{U}} = -3; c = 0.001; T_0 = 20°\text{C}.$$

How much capacity do both cells have at the end of the year? (Note: The end of life criteria here is $\kappa = 0.8\kappa_{\text{BOL}}$)
   Solution: We first determine the cycle aging of the cycled battery after one year:

$$a_{\text{Cycle}} = 365 \cdot 0.8^{26.43} = 1.0022.$$

What is the calendar ageing of the stored battery?

$$a_{\text{cal}} = 365 \cdot 0.001 \cdot e^{-0.5 \cdot (-5) - 3 \cdot 0.5} = 0.9921.$$

Both ageing values are one—that is, the SOH would have a value of zero in both cases, which means that the end of life criteria is fulfilled. Thus, both cells have 80 % of their original capacity left at the end of the year.

Both ageing processes run in parallel. Therefore, the ageing processes in eqn (8.31) add up. This must be taken into account in the system design. Assume that a battery system is planned to have a lifetime of 10 years. A battery technology is used that can run approximately 6,000 full cycles and that has a calendrical lifetime of 10 years at 25°C ambient temperature. The application requires 600 cycles per year and has a temperature management system, so that the battery's ambient temperature remains at 25°C. Both ageing mechanisms by themselves are sufficient to meet the 10-year life requirement. However, according to eqn (8.31), the sum is to be evaluated. This means that after five years, approximately 3,000 full cycles have been run. The SOH would therefore already amount to 0.5. At the same time, according to the calendar ageing, this value is also already at 0.5, and so the system would de facto only run satisfactorily for five instead of 10 years.

8.2.4 Balancing systems

When we looked at the interconnection of double-layer capacitors in Section 7.3, we saw that differences in ageing or production variations lead to changes in capacity. The same is true for electrochemical cells. Things may become even more complex, because cells can also age at different rates. For example, if the temperature in a battery system is not the same for all cells, the result is that some cells age faster than others.

Assume that two cells connected in series have a different SOH. Cell C_1 still has an SOH of one, while cell C_2 already has an SOH of zero. It therefore only has 80% of its original capacity.

Fig. 8.8 shows the voltage curve when charging the cells in relation to the SOC of cell C_1. We recall that the voltage curve of the cell depends mainly on the cell chemistry. So when we charge C_1, we start at a SOC of zero and reach the final charge voltage at a SOC of one. The same applies to C_2, of course. However, it reaches the final charge voltage already at 80%.

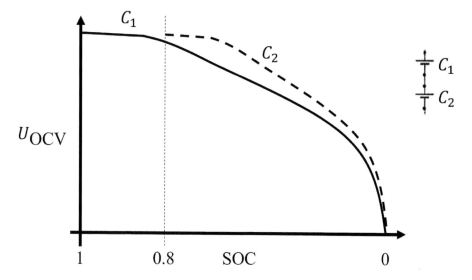

Fig. 8.8 Two battery cells connected in series with different SOH: the old cell C_2 reaches its maximum voltage while the younger cell still has a state of charge of 80%.

It can be seen that the voltage of the cell C_2 rises faster. The open-circuit voltage of C_2 reaches its maximum when the cell C_1 is only 80 % charged.

Unlike a capacitor, the cell is usually still conducting. We can continue to let current flow through the cells and thus charge C_1. There are electrochemical cells that allow current to flow through the cell C_2, although no electrochemical charging reaction takes place. However, this is not without harm to the already old cell C_2. It heats up, and unwanted reactions take place.

There are even cell chemistries that do not allow such overcharging at all. In this case, the charging process must be interrupted and only 80% of the total capacity is available to the user, because the cell Z_1 is not charged any further. This contradicts another requirement for electrochemical storage:

ECS 5: AS A User, I WANT ageing effects and production dispersion that prevent the full use of the active material to be compensated SO THAT I can use as much of the available capacity as possible.

In order to fulfil ECS 5, one uses balancing procedures analogous to the supercap. Passive balancing was already introduced in Section 7.3.

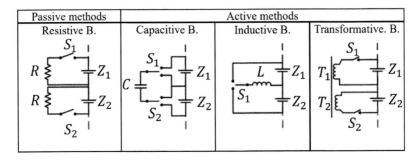

Passive methods	Active methods		
Resistive B.	Capacitive B.	Inductive B.	Transformative. B.

Fig. 8.9 Different types of cell balancing: a distinction is made between passive methods and active methods.

Fig. 8.9 shows different methods of cell balancing. A distinction is made between passive methods and active methods. The active methods include capacitive, inductive, and transformer-supported balancing. The resistive method belongs to the passive balancing. Here, energy is dissipated. With active methods, energy is redistributed.

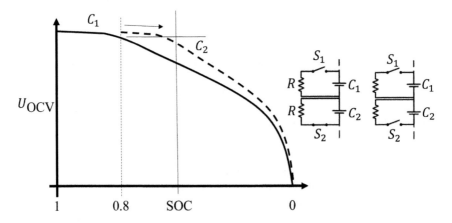

Fig. 8.10 Passive balancing of two cells connected in series with different ageing states: C_1 represents the voltage–capacitance curve of a new cell and C_2 the curve of an aged cell.

Let's look again at passive balancing applied to our example in Fig. 8.8. Fig. 8.10 shows the passive balancing process. In the first step of the charging process, the cell C_2

is charged to 100%, the cell C_1 to 80%. As we did when balancing the supercaps, we now stop charging the two cells and start with the first balancing step. The resistor applied to the cell C_2 is short-circuited with the cell by closing the switch S_2. The cell C_2 is now discharging. We discharge the cell until it has the same cell voltage as C_1. When this is achieved, the switch S_2 on cell C_2 is opened. We now continue charging the two cells until C_1 is charged to 100% again. Then we repeat the balancing process.

This procedure can be realized by a simple setup. For each individual cell, a voltage detection as well as a switch and a resistor are needed. The disadvantage is that we dissipate energy and further cycle the cell, which is already old. This increases the ageing of the cell even more.

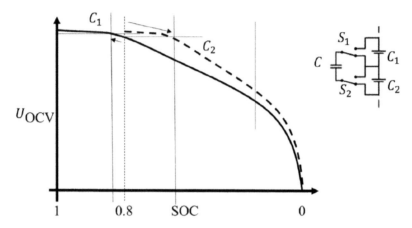

Fig. 8.11 Capacitive active balancing of two cells connected in series with different ageing states: C_1 represents the voltage"=capacitance"=curve of a new cell and C_2 the curve of an aged cell.

With active balancing, the energy is no longer destroyed but actively redistributed. In Fig. 8.11, we have illustrated the procedure for capacitive balancing. In the first step of the charging process, cell C_2 is charged to 100%. The cell C_1 is again at 80%. We switch off the charging current or reduce it. Now the switches S_1 and S_2 are closed and connect the capacitor to the cell. The cell C_2 charges the capacitor and thus reduces its state of charge. When the capacitor is charged, its voltage is between that of C_1 and the end-of-charge voltage of C_2. Now the switches S_1 and S_2 are switched so that the capacitor is connected to C_1. The cell C_1 is charged by the capacitor. If the charging processes are cleverly chosen, one step is sufficient to realize a balance. No energy is lost, and fewer iterations have to be run.

In addition to capacitive balancing, there is also the inductive method. In principle, it works in the same way as capacitive balancing, but here the energy is temporarily stored in a coil. The third variant of active balancing uses a transformer. Here, the cells are connected to a transformer and are charged individually. At the same time, the cells have the option of releasing their energy to another cell via the transformer.

Active balancing methods distribute energy. When implementing an active balancing method, we are therefore faced with the problem of which cells to distribute the energy between. The presented method, or the methods shown in Fig. 8.9 distributed the energy to the nearest neighbour. This means that energy can in principle be transferred indirectly between any cells, but many cells are involved and are cyclized, even if only partially. This is where passive balancing has an advantage. We only have to look at which cell has the highest voltage and how far it deviates from the lowest voltage. Then we reduce the state of charge of the cell with the highest voltage so that it comes close to the lowest voltage (Daowd *et al.*, 2011; Frost and Howey, 2017).

In short, balancing is done at the module level and at the string level. Balancing between the cells is limited to the individual module. Whether controlled balancing takes place at the string level depends on the cell chemistry used. On the one hand, strings are connected and disconnected, i.e. if one string has a higher voltage, it is disconnected and the other string continues to be charged. Alternatively, the equalizing currents that occur are used. Of course, the safety of the battery must be taken into account. At the string level, it is checked whether the voltages between two strings are identical. If this is not the case, equalizing currents occur, which can damage the battery.

8.2.5   Soft turn-off and ageing reserves: Capacity management of battery systems

Both the superconducting current storage choke and the supercap went into saturation during charging and discharging. We could still apply a charging voltage or current to them, but the electrical storage no longer accepted this additional energy. Similarly, we could completely discharge both electric storage components. Overcharge and deep discharge are not allowed for electrochemical storage systems. The effects on the different cell chemistries are different: some chemistries become unstable, others wear out and lose their capacity; what all chemistries have in common is that they do not tolerate overcharging or deep discharging well.

In order not to shorten the life of the cell or, in the worst case scenario, make the cell unstable, we need to avoid deep discharging or overcharging the cell. The most obvious method is to monitor the cell voltage and interrupt the power flow as soon as the cell voltage of a cell within a string meets the termination criterion. This form of hard shutdown introduces significant power management difficulties at the system level. Consider, for example, a solar power storage system connected to a photovoltaic system that provides a household with energy. The household is not connected to a grid. Solar power and battery are the only power supplies. We have a high consumption that cannot be fully covered by the photovoltaic system, so the battery is also discharging. Now we come to the point where the battery is empty. Since not enough power can be provided by the photovoltaic system and the battery together, all consumers switch off. However, the photovoltaic system now supplies electricity that can be stored, because the consumers are switched off. So the battery is charged. After a short period of time, the stored energy is sufficient to cover the consumers. Now, we switch on the consumers again and discharge the battery with the needed power until we have discharged the battery and need to switch off the consumers again.

We can see that a hard disconnection is not reasonable here. Instead, we should work with a hysteresis and a soft control. Here, a safety reserve is defined in the upper as well as in the lower charging range, which the system can use for a short time so that other system components can react. In our example, a message would be sent to the energy management system (EMS) that the lower safety range has been reached and the EMS could then switch off consumers or reduce their power consumption or simply issue a warning.

Over time, electrochemical cells are getting older. There are two effects that then come into play. Firstly, the internal resistance increases, and secondly, the capacity of the cell is reduced. The question for the manufacturer is how to present this to the user. In some products, this fact is not particularly well addressed. In the case of a mobile, we know that a full battery at the beginning means that the mobile will work for a week without recharging, and we have come to terms with the fact that after a while, a full battery will only last for one day. Nevertheless, the battery symbol on the display is full in both cases.

There are products where the user wants to have a transparent view of the battery capacity. For electric vehicles or home storage systems, the user relies on the fact that a full battery always contains a guaranteed amount of energy. To ensure this, an ageing reserve is provided for such battery systems. The user is always provided with the same amount of energy. The loss of capacity, however, is at the expense of the ageing reserve.

We therefore have two perspectives on the total capacity of a battery: the customer perspective and the technical perspective. The relationship between these two perspectives is shown in Fig. 8.12. In the technical perspective, the battery capacity is divided into two safety reserves, which are located at the upper and lower end of the state of charge. In between, there is an ageing reserve. Whether this is in the upper or lower range depends on the technology. Only the remaining capacity is presented to the customer as usable capacity. In the customer perspective, only the usable capacity is shown.

The ratios between the different reserves depend on the technology and the application. As a manufacturer, one strives to keep the safety and ageing reserves as small as possible. The user also wants the usable capacity to be as large as possible. This is because the active material that remains unused in the safety and ageing reserve means additional costs.

8.2.6 System components of electrochemical storage systems

In this section, we consider those system components that are present in all electrochemical storage systems. Furthermore, we consider the responsibilities of these system components in the requirement traceability matrix (RTM). As with all storage systems, the basic requirements G 1 to G 8 are to be considered. Since electrochemical storage systems are always also electrical systems, requirements ES 1 to ES 5 apply. In addition, the requirements for electrochemical storage systems ECS 1 to ECS 5, which we have already described in this chapter, apply. Fig. 8.13 shows the system

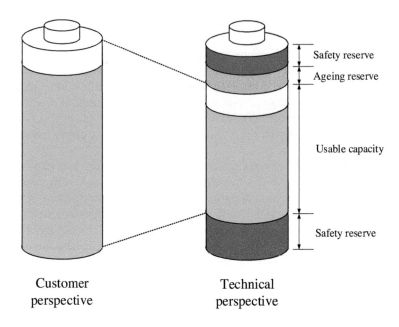

Customer
perspective

Technical
perspective

Fig. 8.12 Customer perspective and technical perspective on the capacity of a electrochemical cell.

components of an electrochemical storage system. The RTM in Tab. 8.2 is required for this. In Fig. 8.13, we have marked in grey those components that generally always have to be present in a storage system and belong to the basic components. Here we have two types of relationships. On the one hand, we have a generalization relationship—that is, a system component derives from the base component. This is the case with the `BatterySystem` component. The other relationship is a «trace» relationship—that is, the system component has the responsibility to fulfil (partial) tasks of the basic component. This is the case with the `EnergyManagementSystem`, which has a relationship with the `PowerDistributionControl`.

Fig. 8.13 is divided into two areas. The upper section shows the electrochemical storage system: the battery. The `BatteryCell` is connected to a `BatteryModule` via an $x pys$ circuit. The modules, in turn, are interconnected by $XpYs$ to form a system. The `BatterySystem` thus represents the `Storage` and has a `TemperatureControl` as well as a `MechanicalConstruction`. We can think of the `BatterySystem` as an independent unit of connected `BatteryCell`.

The area below shows the supporting electronics. Here we have two main components: the `BatteryManagementSystem` and the `EnergyManagementSystem`. These two components can be integrated on the same unit, but should be functionally and architecturally separate. The `BatteryManagementSystem` has three subcomponents: the `BalancingSystem`, the `HealthEstimator`, and the `SOCEstimator`. The `BalancingSystem` ensures good utilization of the available capacity and avoids overcharging of cells with different capacities. `BatteryManagementSystem` and `BalancingSystem`

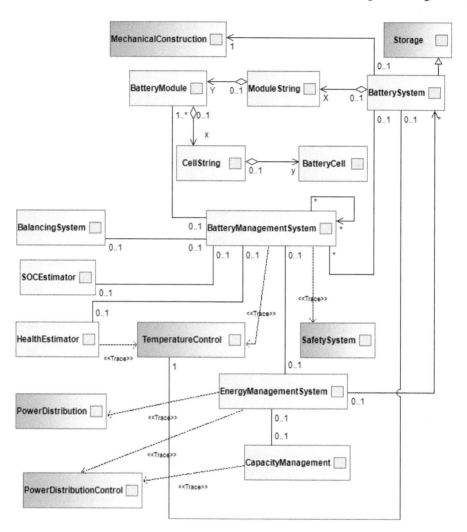

Fig. 8.13 System components of electrochemical storage systems.

are part of the safety concept (G 5). The `BatteryManagementSystem` is therefore derived from the `SafetySystem`. We see that the `BatteryManagementSystem` is associated with the `BatteryModule` and the `BatterySystem`, and can be connected to other `BatteryManagementSystem`. As we have already described in the presentation of the balancing methods, the balancing is done on a single module or string at a time. To avoid high balancing currents flowing between two module strings, another `BatteryManagementSystem` component can be used to switch strings on or off depending on the voltage.

The `HealthEstimator` is a system component that makes statements about the operating parameters of the battery and determines the ageing; thus, it is responsible for G 7.

Table 8.2 Requirement traceability matrix.

	Basic requirements								Electrical systems					Electrochemical storage systems				
	G1	G2	G3	G4	G5	G6	G7	G8	ES 1	ES 2	ES 3	ES 4	ES 5	ECS 1	ECS 2	ECS 3	ECS 4	ECS 5
BatteryCell	x																	
CellString														x			x	
BatteryModule										x	x						x	
ModuleString														x			x	
BatterySystem						x			x	x	x				x		x	
BalancingSystem					x				x		x	(x)		x				
BatteryManagementSystem		x	x		x				x									
HealthEstimator							x											
SOCEstimator				x												x		
EnergyManagementSystem		x					x	x										x
CapacityManagement								x										x

The SOCEstimator implements a method for fulfilling ECS 3 or G 4 and provides a statement about how much energy can still be retrieved from the cell.

The EnergyManagementSystem has only one subcomponent: CapacityManagement. The EnergyManagementSystem is assigned to the BatterySystem and takes over PowerDistributionControl tasks. In order to realize the ECS 5 requirement, the EnergyManagementSystem uses the CapacityManagement. It is responsible for the capacity of the entire storage component. It represents this and ensures that the control behaviour conforms to the application.

The division of tasks between EnergyManagementSystem and BatteryManagementSystem is a separation between application-related and general tasks. The general tasks, which are independent of the specific application, take place in the BatteryManagementSystem. Regardless of whether the battery system is used for a large-scale stationary storage system or for the drivetrain of an electric vehicle, the tasks of the BatteryManagementSystem remain the same. What is different in each case is the way in which the power flows and the capacity of the battery are regulated. This is where different, application-relevant requirements come together, compliance with which is assigned to the EnergyManagementSystem.

Since the BatteryManagementSystem is also responsible for the safety-relevant requirements, by separating the two components, we also have a separation between the part that must be examined and certified by external laboratories within the framework of product liability regulations and safety standards, and any part that can be easily changed.

If we look at Tab. 8.2, we see that requirement ES 5 has no mapping and ES 4 has a mapping that is in brackets. ES 4 deals with the requirement that switching losses should be minimized. In a BatterySystem, there are usually no components that perform switching operations at high frequency. Active balancing systems that work inductively or with a transformer are an exception. For this reason, we have assigned ES 4 to the balancing system, but bracketed it.

ES 5 describes the requirement that voltage must be transformed over different voltage levels. This requirement cannot be realized by a battery system. A power electronic component is necessary for this. This can, but does not have to, be part of the battery system. However, since the choice of the voltage level depends on other system components of the entire storage system, having the power electronic component as part of the battery system is avoided.

8.2.7 Conclusion

In this section, we have described the general characteristics of electrochemical storage. Their basic mechanism is based on the conversion of substances involving electron transport. A distinction is made between electrochemical storage devices whose conversion is not reversible (these are called primary batteries) and those that use reversible conversion as the storage mechanism. These are called secondary batteries and are the focus of this chapter.

We have seen that some effects can be observed that were also seen with supercaps. However, there is a very clear difference. All electrochemical storage systems operate in a voltage range defined by the cell chemistry. The voltage of the cell depends on the state of charge. The relationship between voltage and state of charge is chemistry-dependent and in general not linear.

Similar to supercaps, there may be a need for balancing between different cells. One reason is the ageing of electrochemical cells. Here, a distinction is made between calendar ageing and cycle ageing. Both effects can be separated from each other, but they always occur together. Ageing effects can be minimized by various measures that depend on the cell chemistry.

In the next sections, we will take a closer look at four cell chemistries. We will start with the lead acid battery, the oldest and most widespread electrochemical storage technology. Then we will look at the family of lithium ion batteries, which has enjoyed growing popularity since its introduction and is very common in the consumer sector or in electromobility. In addition to lithium ion batteries, high-temperature batteries are also used for grid storage systems. We will deal here with the sodium sulphur battery and conclude the chapter with a consideration of redox flow technology.

8.3 Lead acid batteries

In this section, we will look at the lead acid battery. This technology was originally developed at the end of the nineteenth century. It is by far the most widely used technology. In vehicles with combustion engines, it is used as a starter battery and drives the starter until the combustion engine is started. In uninterruptible power supplies, it is used as a buffer battery and forklifts use the lead acid battery as a mobile power supply. These applications take advantage of the fact that the battery can deliver high charging and discharge currents for a short time and can be fully charged over a longer period without any problems. Its major disadvantage, its weight, is exploited in forklifts, where the lead acid battery also serves as a counterweight to the load to be carried.

The lead battery has a simple construction and works with available materials: water, sulphuric acid (H_2SO_4), lead, and lead oxide. It can therefore be produced and recycled by relatively simple means, which supports its worldwide distribution.

The basic construction corresponds to the illustration in Fig. 8.14. We use a lead (Pb) plate as the anode. The cathode is also a lead plate, but it is coated with a layer of lead oxide (PbO_2). This layer is part of the active material and is needed for the reaction. Both plates are immersed in a sulphuric acid bath (H_2SO_4). The concentration of the sulphuric acid is reduced because water is added to it. A separator is not needed for a lead acid battery.

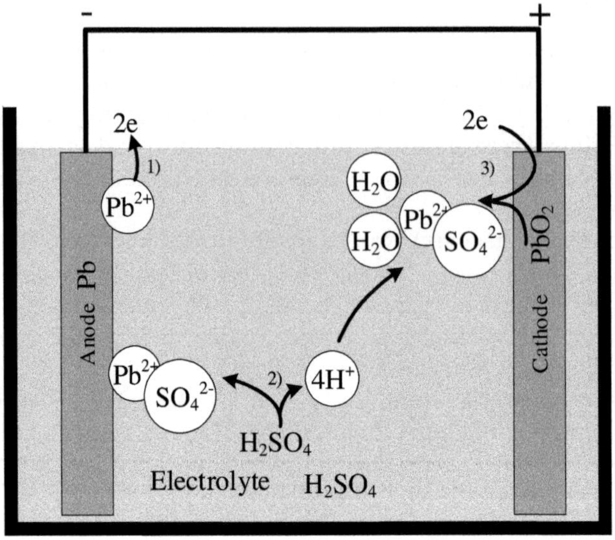

Fig. 8.14 Construction of a lead acid battery and illustration of the electrochemical reactions of a lead acid cell. The anode consists of lead, while the cathode is coated with lead oxide. The electrolyte consists of water and sulphuric acid.

8.3.1 Primary and secondary reactions

We first consider the primary reactions that can be observed during the charging and discharging process in a lead acid battery cell. The individual steps are also shown and numbered in Fig. 8.14. During the discharge process, an electron migrates from the anode side to the cathode side. For this purpose, a lead atom Pb is ionized and gives up two electrons $2\,e$ and becomes a double positively charged lead ion Pb^{2+} (Fig. 8.14 1)):

$$Pb \rightleftharpoons 2\,e + Pb^{2+}. \tag{8.32}$$

A splitting of the sulphuric acid H_2SO_4 now takes place in the electrolyte. The positive lead ion combines with the sulphur oxide SO_4^{2-}, leaving four positive hydrogen ions $4\,H^+$ (Fig. 8.14 2)):

$$Pb^{2+} + H_2SO_4 \rightleftharpoons PbSO_4 + 4\,H^+. \tag{8.33}$$

We see that in this case the electrolyte is not just a medium for transporting charge carriers, but part of the electrochemical reaction. After these two reactions have taken place, we are left with two electrons $(2\,e)$ and four ionized hydrogen atoms $(4\,H^+)$. To get current from the cell, we need a reaction, which binds two electrons on the cathode side. The cathode is coated with a layer of PbO_2. The lead oxide reacts with the four ionized hydrogen atoms $4\,H^+$ and uses the two electrons:

$$PbO_2 + 4\,H^+ + 2\,e \rightleftharpoons 2\,H_2O + Pb^{2+}. \tag{8.34}$$

The double positive lead ion still 'bothers' us because it is still reactive. But from the anode reaction (eqn (8.33)), we know that the ion can react with the sulphuric acid. If we combine eqns (8.34) and (8.33), we get the complete reaction on the cathode side (Fig. 8.14):

$$PbO_2 + 4H^+ + 2e \rightleftharpoons 2H_2O + PbSO_4. \tag{8.35}$$

We can combine the various individual reactions into one overall electrochemical reaction:

$$Pb + PbO_2 + 2H_2SO_4 \rightleftharpoons 2PbSO_4 + 2H_2O. \tag{8.36}$$

If the reaction takes place from left to right, we discharge the lead battery. If the reaction takes place from right to left, we charge the battery. We realize that with this cell chemistry, we have the possibility of directly determining the state of charge. Depending on the state of charge, the concentration of sulphuric acid H_2SO_4 changes. When discharging, the proportion of water increases. When charging, the proportion of sulphuric acid increases. If we are able to measure the concentration of sulphuric acid in the battery, we can make a statement about the state of charge if other effects are not responsible for the change in concentration.

Exercise 8.6 In Exercise 8.7, we had derived a relation between the free enthalpy of reaction ΔG and the voltage of a cell. In Tab. 8.3, the values for the Gibbs energy of the reaction elements of the lead battery reaction are shown (Halka and Nordstrom, 2011). What equilibrium voltage results from these values? Note that according to eqn (8.35), two electrons are involved in one reaction step and use $F = 96{,}485$ As as the value for the Faraday constant.

Table 8.3 Values for the Gibbs energy of the lead battery reaction.

	Pb	PbO_2	H_2SO_4	$PbSO_4$	H_2O
G [kJ/mol]	0	215	744	813	237

Solution: We use the relationship between the Gibbs energy required in the reaction and the electric potential:

$$\Delta G = E_{el} = Q\Delta U = nFz\Delta U.$$

ΔG results from the sum of the Gibbs energy of the individual reactants:

$$\Delta G = 0\frac{kJ}{mol} + 215\frac{kJ}{mol} + 2 \cdot 744\frac{kJ}{mol} - 2 \cdot \left(813\frac{kJ}{mol} + 237\frac{kJ}{mol}\right) = -397\frac{kJ}{mol}.$$

Using the value for ΔG, we can now find the voltage:

$$\Delta U = \frac{\Delta G}{nF} = \frac{-397\frac{kJ}{mol}}{2 \cdot 96{,}485\,As} = 2.05V.$$

Eqn (8.36) shows the primary reaction: it describes the storage of electrical energy. The voltage that we can expect from such a lead acid battery is, following the estimation from Exercise 8.6, about 2 V. In addition to this primary reaction, secondary reactions occur during the operation of the lead acid battery, which reduce the calendrical lifetime and the number of cycles. If a lead acid battery is almost fully charged, a large part of the lead sulphate has already been converted into lead or lead oxide.

If the battery continues to be charged at a high current, the water at the cathode is split by electrolysis:

$$H_2O - 2\,e \longrightarrow \frac{1}{2}\,O_2 + 2\,H^+. \tag{8.37}$$

At the anode, the hydrogen ions again recombine with the electrons to form hydrogen:

$$2\,H^+ + 2\,e \longrightarrow H_2. \tag{8.38}$$

The resulting overall reaction is:

$$H_2O \longrightarrow H_2 + \frac{1}{2}\,O_2. \tag{8.39}$$

If we overcharge a lead acid battery, hydrogen is formed and the amount of water in the lead acid battery is reduced, while the hydrogen is released into the environment. This leads us to the first requirement we have to consider when operating a lead acid battery:

ECS-LA 1: AS A User, I WANT to make sure that I do not overcharge the battery SO THAT the formation of hydrogen and thus the reduction of water in the electrolyte are prevented.

The formation of hydrogen is not the only issue here. The oxygen released in the process reacts with the lead of the anode and forms lead oxide:

$$Pb + \frac{1}{2}\,O_2 \longrightarrow PbO. \tag{8.40}$$

This lead oxide reacts with the sulphuric acid to form lead sulphate and water:

$$PbO + H_2SO_4 \longrightarrow PbSO_4 + H_2O. \tag{8.41}$$

These reactions can be combined to give the following reaction:

$$Pb + H_2SO_4 + \frac{1}{2}\,O_2 \longrightarrow PbSO_4 + H_2O. \tag{8.42}$$

Although this compensates for the reduction in water concentration that occurs in eqn (8.39), this formation of $PbSO_4$ is not a reversible reaction. With an almost fully charged battery and high charging currents, two effects occur: an excess of hydrogen is formed, while oxygen is reduced.

While the reactions in eqns (8.39) and (8.42) mainly occur at high charging currents or in case of an almost fully charged battery, we can reduce this effect by reducing the charging and discharging current to comply with ECS-LA 1. When operating a lead acid battery, corrosion occurs at the electrodes. The lead of the anode reacts with the water and forms lead oxide:

$$Pb + 2\,H_2O \longrightarrow PbO_2 + 4\,H^+ + 4\,e. \tag{8.43}$$

This reaction is also not reversible and leads to permanent damage to the battery. Since lead is needed on the anode for the primary reaction, this reaction reduces the active material. Eqn (8.43) is the main cause of calendar ageing. It also occurs more frequently when high charging currents are present.

When we introduced the power flow diagrams in Chapter 3, we described that storage can lose some of the stored energy over time. The amount of self-discharge depends on the technology. In flywheel storage, the cause of self-discharge is mechanical friction and air resistance. In a current storage device, it is due to the internal resistance of the non-ideal electrical components. In an electrochemical cell, chemical reactions can occur that are either a reversal of the original reaction or a modified form that is more energetically stable than the current state. Such a reaction can be observed in the lead acid battery.

However, they are at a lower energy level and generate heat. They are therefore not dependent on energy supply from the outside. There are two self-discharge reactions that can be observed at the anode and at the cathode. At the cathode, lead oxide (PbO_2) reacts with sulphuric acid (H_2SO_4) to form lead sulphate ($PbSO_4$), water, and oxygen:

$$PbO_2 + H_2SO_4 \longrightarrow PbSO_4 + H_2O + \frac{1}{2} O_2. \tag{8.44}$$

At the anode, the lead reacts with the sulphuric acid to form lead sulphate and oxygen:

$$Pb + H_2SO_4 \longrightarrow PbSO_4 + H_2. \tag{8.45}$$

In both eqn (8.45) and eqn (8.44), no electron flow is required. This reaction takes place solely between the active material in the electrodes and the electrolyte. However, the effect is a decrease in the concentration of H_2SO_4 as well as a contamination of the active material. Moreover, the reactions are not reversible. Fortunately, the formation of $PbSO_4$ is a slow reaction. One observes a self-discharge rate of 0.5–5% per day in lead acid battery. However, the rate of self-discharge increases with the age of the battery.

Exercise 8.7 A lead acid battery is fully charged. It has a storage capacity of $\kappa(0\ \text{d}) = 25$ Ah and a nominal voltage of $U = 24$ V. This battery is fully charged. But already on the second day, a capacity of $\kappa(1\ \text{d}) = 582$ Wh is measured. After how many days has the battery lost half of its original stored energy due to self-discharge?

Solution: First, we determine the capacity in Wh on day zero:

$$\kappa(0\ \text{d}) = 25\ \text{Ah} \cdot 24\ \text{V} = 600\ \text{Wh}.$$

If we know the self-discharge rate α, we are able to calculate the capacity at day x:

$$\kappa(x\ \text{d}) = \kappa(0\ \text{d}) \cdot \alpha^x.$$

With this approach, the self-discharge rate α can be determined from the measurements:

$$\alpha = \frac{\kappa(1\ \text{d})}{\kappa(0\ \text{d})} = \frac{582\ \text{Wh}}{600\ \text{Wh}} = 0.97.$$

With a value for α, we can calculate after how many days the capacity equals 300 Wh:

$$600 \text{ Wh} \cdot \alpha_0^x = 300 \text{ Wh}$$

$$\alpha_0^x = \frac{1}{2}$$

$$\ln(\alpha)x_0 = \ln(\frac{1}{2})$$

$$x_0 = \frac{\ln(\frac{1}{2})}{\ln(\alpha)}$$

$$x_0 = \frac{-0.6931}{-0.0304} = 22.79 \text{ d.}$$

After about 22.79 days, the battery has lost half of its original energy content due to self-discharge.

As we have seen in Exercise 8.7, lead acid batteries are not necessarily suitable for storing energy over a long period of time (days or weeks). For applications where energy is to be stored for a relatively short period of time, for example for a self-sufficient energy supply where the battery is charged during the day with the help of solar or wind power, and the battery is only discharged during the night, a lead acid battery is well applicable. The cost of a lead acid battery and the widespread use of this technology make it especially easy to implement.

In the next section, we will look at the behaviour of the lead acid battery and the requirements for its use, since every electrochemical storage technology has characteristics that must be taken into account when it is used.

8.3.2   Behaviour and requirements: How to handle lead acid batteries

Due to the simple construction and availability of the materials used, lead acid batteries are widely used. However, there are some requirements to be considered for the operation of a lead acid battery which have their origin in the electrochemistry. The main and secondary reactions have consequences for the operation of the lead acid battery. ECS-LA 1 is a first requirement. We will learn about other requirements in the next section, which can be derived from the behaviour and reactions. Let us first look at the observations that led to the requirement ECS-LA 1. If the lead acid battery has reached a high state of charge and is still being charged, eqn (8.39) leads to hydrogen being consumed on the cathode side, while on the anode side the reaction in eqn (8.43) leads to oxygen being produced. If we find a way to transport the oxygen and hydrogen from the cathode to the anode side, then the ageing of the battery could be reduced, since we will only generate water. In sealed lead acid batteries, this approach is realized. In sealed lead acid batteries, the electrolyte is solved in a fleece or gel that conducts the oxygen or hydrogen. This creates an oxygen hydrogen cycle that reduces ageing and allows lead acid batteries to be used with significantly higher charging and discharge currents.

We have herewith a first realization that allows us to comply with ECS-LA 1. Alternatively, we can simply do everything to avoid the reactions. This leads to further requirements that also apply to sealed batteries:

ECA-LA 2: AS A User, I WANT to avoid a permanent high charging and discharging current SO THAT hydrogen generation is avoided.

ECA-LA 3: AS A User, I WANT to avoid a high charging and discharging current when the battery is full SO THAT a reduction in oxygen concentration is avoided. However, at ECS-LA 2 and ECS-LA 3 we are faced with a small difficulty. If a discharge current I_B is present, we can no longer measure the open-circuit voltage of the battery directly. The voltage we measure at the battery U_B is the sum of the open-circuit voltage and the voltage resulting from the internal resistance R_i:

$$U_{\text{Batt}} = U_{\text{OCV}}(\text{SOC}) + R_i I_B. \tag{8.46}$$

When charging the battery we measure a higher voltage, and when discharging we measure a lower voltage. Compliance with ECS-LA 2 and ECS-LA 3 is difficult, because we cannot know exactly what state of charge we have in the charging and discharging process. ECS-LA 2 and ECS-LA 3 are implemented by a charge controller. In order to better understand the method with which a charge controller can work here, we return to our example of the doorman. Until now, his only task was to make sure that there were not too many people in the bar. He had two possibilities: he could count how many were going in or out, or he could close the doors and count them. Now he has been given a new task. He has to make sure that not too many people enter or leave the bar per time. However, the number of people he can let through is supposed to depend on the number of people in the bar. The number of people in the bar corresponds to the battery's state of charge. The number of people going through the door corresponds to the charge or discharge current.

One approach for the doorman is to limit the number of people going in and out of the bar to a constant number. He does this regardless of how many people are in the bar. For the charge controller, this would mean that the charge current is limited, regardless of the state of charge of the battery. Of course, this procedure has the disadvantage that the charging and discharging power is limited. For our doorman, this may mean long queues in front of the bar and a frustrated boss who fears a loss of turnover if there are not enough people in the bar.

After the boss complains to the doorman, he comes up with a better idea: he already k.pdf a record of the number of people in the bar, so he can adjust the number of people entering and leaving through the door depending on the number of people in the bar. This would be equivalent to a charge controller that adjusts the maximum charge and discharge current depending on the estimated or measured state of charge. This procedure only works well for both charge controller and bouncer if the estimate is good. But if the estimate of the number of people in the bar is bad, the doorman runs the risk of letting too many people in and out through the door.

There is a third approach: he looks at how many people are standing near the cloakroom and adjusts the number of people he lets into the bar according to this measurement. While in the first two approaches he regulated the flow of people, he now regulates the difference between those people waiting at the cloakroom and those who want to go to the cloakroom. Of course, this also has an influence on the inflow or outflow of people at the entrance.

To better understand the implementation in a charge controller, let's look at the circuit in Fig. 8.15. We have a voltage source connected to a battery via a DC/DC converter. A voltage sensor is also connected to the battery and its signals are transmitted to a charge controller. The charge controller controls the DC/DC converter. It can specify both a current and a voltage.

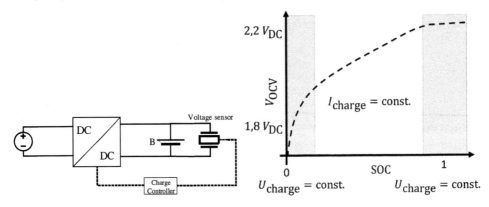

Fig. 8.15 Charging method for lead batteries: at low and high states of charge, a constant voltage charging method is used; in between these, a constant current charging is used.

With the first two methods, the charge controller determines the current. This is called an I charge. The DC/DC converter sets the output voltage on the battery side so that a certain current flows from the voltage source to the battery. In the third method, the voltage is fixed. This is called a U charge. Here, the charge controller proceeds in such a way that it gives the DC/DC converter the target voltage U that is to be set at the battery. The current that is automatically set during charging is proportional to the voltage difference between the DC/DC converter and the battery voltage. The smaller this voltage difference, the lower the current. The battery voltage approaches the target voltage asymptotically. In Fig. 8.15, we have another representation. Here we see the state of charge on the horizontal axis and the cell voltage on the vertical axis. At the two outer areas, i.e. a high or a low state of charge, a U charge takes place when using a lead acid battery. This has the advantage that the battery capacity can be well utilized without violating the requirements for the charging current.

In the medium range, on the other hand, an I charge is used. Constant voltage charging has the disadvantage that the charging power is not fixed but determined by the electrochemical processes. However, in order to be able to control a storage system at the system level, it is necessary to control the power flows. These can be mapped much better by constant current control.

The previous requirements were aimed at reducing the influence of secondary reactions, which mainly occur when the battery has reached a high state of charge and large currents flow. We now want to address the question of what influence the discharge current has on the amount of energy available. For this purpose, we introduce the concept of the C-Rate. The C-Rate is the equivalent of the E-Rate, which we already know from Chapter 3. The E-Rate represents the ratio of charging power and capacity related to a charge within one hour. For a storage with a capacity of 25 Wh and discharged with a power of 5 W, the E-Rate is 0.5. For electrochemical storage, it is common to describe the energy content not only in Wh, but also in Ah. The motivation is that the battery is seen as a voltage source that can provide a current of x A over a certain period of time. The C-Rate analogously indicates the ratio of charging current and capacity related to a charge within one hour.

Exercise 8.8 A lead acid battery has a capacity of 360 Wh, or 30 Ah, and a nominal voltage of $U = 12$ V. It is to be discharged at a final charge rate of 0.75 E. What discharging power in W does this correspond to? What C-Rate does this process correspond to?

Solution: To arrive at the charging power in W, we just need to multiply the capacity by the E-Rate:

$$350 \text{ Wh} \cdot 0.75\frac{1}{\text{h}} = 270 \text{ W}.$$

Now we determine the current we need for this final charging power:

$$I = \frac{270 \text{ W}}{12 \text{ V}} = 22.5 \text{ A}.$$

To determine the C-Rate, we divide the charging current by the capacity and get:

$$\frac{22.5 \text{ A}}{30 \text{ Ah}} = 0.75 \text{ C}.$$

We can refer to the relationship between charging power and capacity as either the C-Rate or the E-Rate. These rates do not always have to be identical. E-Rates are usually used when the relationship between charging power and capacity is to be described at the system level. The C-Rate, on the other hand, is used if you also want to use information about the interconnection of the batteries. It works in the current–voltage diagram.

Let us look at Fig. 8.16. Here a lead acid battery has been discharged at different C-Rates (Reddy, 2011). As we can see, the battery voltage varies as a function of the C-Rate. At a high discharge current, a high C-Rate, the battery voltage U is smaller than at a low C-Rate, a low charge current. We can see here the effects of the internal resistor R_i, which increases or decreases the open-circuit voltage depending on the current and current flow (eqn (8.46)).

The horizontal axis is scaled logarithmically. At a C-Rate of 35 C, the battery is discharged within one minute. At a C-Rate of 0.25 C, the discharge process takes almost 20 hours.

Exercise 8.9 The battery cell used in Fig. 8.16 has a storage capacity of 5 Ah. What is the discharge current at a C-Rate of 35 and a C-Rate of 0.25?

Solution: A C-Rate of 1 would mean that the cell is discharging with a final discharge current of 5 A. With a C-Rate of 35, the current is $35x$ as high—that is, the final discharge current is:

$$I_{35} = I_1 \cdot 35 = 5 \text{ A} \cdot 35 = 174 \text{ A}.$$

Analogously, the current is calculated at a C-Rate of 0.25:

$$I_{0.25} = I_1 \cdot \frac{1}{4} = 5 \text{ A} \cdot \frac{1}{4} = 1.25 \text{ A}.$$

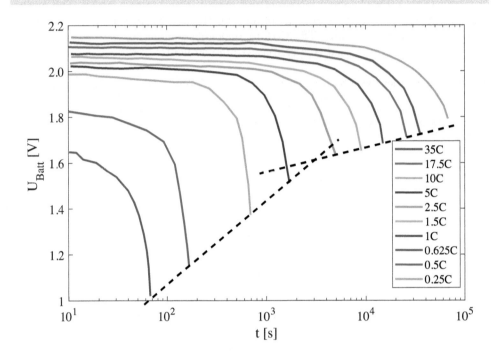

Fig. 8.16 Voltage curve of a discharging process of a lead acid battery at different C-Rates.

The short discharge times at high C-Rates are due to the electrochemistry of the lead acid battery. In a lead acid battery, an electron is released on the anode, turning lead into lead sulphate. The electron moves to the cathode side and causes the formation of lead dioxide and hydrogen. Unlike other cell chemistries, the discharging process does not involve additional direct ion transport. Instead, both electrodes use the substances contained in the electrolyte and located in the immediate vicinity of the electrode. With a high charging current, more active material is needed for a given period, and so it can happen that not enough active material can be added. We experience this phenomenon when we start a car with an old or almost empty starter battery. After the starting attempt has not worked and the starter has made several unsuccessful starting attempts and no longer turns, you switch off the starter and the ignition and wait a moment. After that, the battery supplies a little power again to make another, hopefully successful, attempt to start the car. The starting process in a car requires a lot of power. By waiting, we allow the battery to better distribute the active material in the electrolyte.

At low charging currents, this effect is no longer visible. Here, the active material in the electrolyte has enough time to equalize its concentration. Nevertheless, we observe that the discharge time increases with lower final charge rate. If the explanation for the different behaviour at high charge rates is the concentration difference, then we expect that the discharge time at low C-Rates can be explained by the lower current alone. Indeed, we can confirm this conjecture in Fig. 8.16. At $1/4$ C, we have a discharge time of 20 h. If we double the discharge rate to 0.5 C, we expect to halve the discharge time. The active material has enough time to compensate a difference in concentration. But if we discharge the battery with a higher power, we expect that the energy content will also be used up faster. In fact, we observe a discharge time of about 10 h. And if we double the C-Rate again, we expect a final discharge time of about 5 h, which is also confirmed experimentally.

We have two different physical processes, which take over from each other. At what C-Rate does a change between the two processes occur? If we connect the discharge times at high C-Rates, we see that the discharge times behave linearly in the logarithmic representation. In Fig. 8.16, a straight line is plotted for illustration. We find the same at low C-Rates, but the slope of the straight lines is different. The point of intersection of these two straight lines indicates the range at which the active material can no longer compensate for a difference in concentration quickly enough. In the experiment shown, this point is at about 3 C.

The exact value at which the two effects separate depends on the construction of the battery. In sealed batteries, which use a fleece impregnated with electrolyte to reduce the effects of side reactions, the C-Rate is higher than in an open lead acid battery. Therefore, when designing a storage system with a lead acid battery, one should be a little more conservative and set the limit at about 1 C. This leads us to another requirement:

ECS-LA 4: AS A User, I WANT to make sure that the lead acid battery is preferably operated at a low C-Rate ($\approx$ C $\leq$ 1) SO THAT I have access to all the stored energy.

ECS-LA 4 restricts the scope of lead acid battery to energy applications. To meet power applications, the capacity of the battery has to be increased, or it is accepted that power requests are only possible for a short period of time. This is the case with starter batteries, for example. If the starter motor and engine are well matched, a starting process usually takes a few seconds. During this time, the starter motor drives the engine's flywheel with the help of the battery, which means high power consumption for a short time. However, as soon as the ignition ignites the fuel–air mixture at the right time, the combustion drives the engine and simultaneously charges the starter battery again.

We had seen in 8.2.3 that the number of cycles a battery can provide depends on the DoD. This is also the case with a lead acid battery. In Fig. 8.17, the number of cycles as a function of the DoD is shown for two different lead batteries. As expected, the life of the lead batteries decreases with the size of the DoD. For both batteries, it can be seen that at a DoD below 60%, the lifetime increases more and represents a clear advantage over a higher DoD. When the DoD is reduced from 80% to 60%, the number of cycles for battery 1 increases from 1,000 cycles to 2,000 cycles, which corresponds to a doubling. For battery 2, the number of cycles only increases from 450 to 600. Here, a clearer jump can be seen, with a reduction to 40%, where we observe

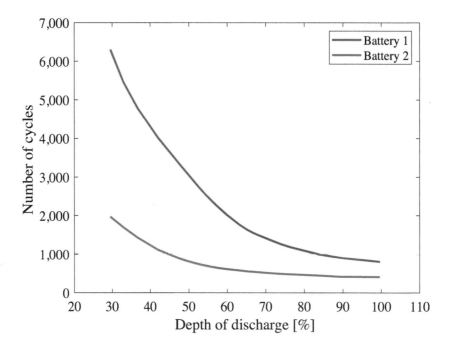

Fig. 8.17 Cycle life for two lead acid batteries as a function of depth of discharge.

almost a doubling of the number of cycles with 1,200 cycles and even a quadrupling for battery 1.

As we have already seen from our previous observations, high charging currents in a full lead acid battery lead to unwanted side reactions. When designing storage systems equipped with lead acid batteries, a lower DoD is used. As a rule, this is 40–50%.

ECS-LA 5: AS A Design Engineer, I WANT to design the system so that the DoD is at its maximum at 40–50% of the installed battery capacity SO THAT side reactions and ageing effects are reduced and the system has a long lifetime.

With ECA-LA 5, we conclude the collection of requirements for the use of lead acid batteries. We have seen that these result from the main and secondary reactions and the construction of the lead acid battery. Although the lead acid battery can only deliver high power for a short time and should be designed for a low DoD, the lead acid battery is one of the most widespread battery technologies. This is due to its simple construction, the availability of raw materials, and the globally established recycling cycles.

In the next section, we turn to an application of lead acid batteries: the power supply of a mobile phone transmitter mast.

8.3.3 Application example: Supply of a mobile radio transmission tower

In recent decades, an interesting technical development has been observed. Regions that previously had a relatively thin communication network built up a dense network within a short period of time. This was achieved through a technological leap. Instead of following the same development s.pdf as other regions, these regions immediately built a mobile phone network. At the heart of these mobile phone networks are self-sufficient mobile radio stations that can be installed in places that have no (or an unstable) electricity supply. These stations can be installed anywhere, allowing a dense mobile network to be established with a few local installations. The needed energy is supplied by diesel generators with fossil fuel. Supplying the generators is not only harmful to the climate but also expensive, because these stations are not always located in good accessible places. Therefore, it makes sense to cover the energy consumption of these stations with renewable energies, such as wind or solar power (Kusakana and Vermaak, 2013; Kumar and Manoharan, 2014).

In this section, we will look at the design of such a mobile station. The station is to be equipped with a solar power system, a diesel genset and a lead acid battery. The solar power system takes care of the energy supply during the day. Surplus power is to be stored in the lead acid battery. If the lead acid battery does not have enough energy stored, the genset is available to ensure the energy supply.

Before we get into the technical implementation and system design, we first start with the requirements analysis.

MB-LA 1: AS A government organization, I WANT a dense mobile network SO THAT government and civilian users can access communications anywhere in my country.

MB-LA 1 makes it clear who the client is. In this example, it is not a private initiative, but a government organization. Hence, not only consumers but also governmental organizations such as security, disaster control, and medical aid services also use the network. This results in higher requirements on the availability of the network in comparison to a purely commercial used network. The mobile network operator who signs a contract with the governmental organization in our example must therefore not only ensure that he solves the technical requirements in a way that makes business sense; he must also ensure that he can guarantee the requirements for the availability of the system.

MB-LA 2: AS A mobile operator, I WANT to build hybrid mobile stations that use solar power but also have a genset to ensure security SO THAT the energy costs are cheaper compared to pure diesel-powered stations, but availability is assured.

MB-LA 2 does not place any requirements on the sizing of the solar power system or the genset. However, the requirement for high availability may lead us to select a larger genset. Therefore, we still need the requirement MB-LA 3:

MB-LA 3: AS A manufacturer, I WANT to keep the performance of the genset as low as possible SO THAT the operating costs and the manufacturing costs are low.

The idea behind MB-LA 3 is to use the lead acid battery, which is not required so far, to store a surplus of solar power on the one hand and to let the genset charge the battery at a lower power on the other hand.

MB-LA 4: AS A Manufacturer, I WANT to integrate a lead acid battery into the system SO THAT I can temporarily store excess solar power and accumulate a stock of energy via a smaller genset that I can consume at a later date.

With this, the main requirements for our system have been compiled. We have here an application where the battery works as a buffer storage. It is charged over a longer period of time with a lower power in order to be discharged at a later time with a higher power. This aspect is similar to the requirements from the opportunity charger in Chapter 3. Of course, there are also a number of other requirements that arise from the perspectives of the different stakeholders.

Exercise 8.10 In this task, we want to derive additional requirements to requirements MB-LA 1 to MB-LA 4. In this case, two requirements each are to be determined for the Service Engineer, the Installer, the Production, and the Development Engineer. (First hint: It makes sense to look at the role description of the actors from Chapter 4 again. Second hint: This task only requires imagination. Of course, if you have not taken one of these roles, you may not know what requirements to have in that role. But you can imagine what one might need. There is no right or wrong here. Don't be afraid of the white paper.)

Solution: A wide variety of requirements can be created for all rollers. This exercise is mainly to train us to think in terms of the different rollers and to think in terms of appropriate requirements. These requirements should then be written down on paper. Let's look at some example requirements:

We start with two requirements for the service engineer:

MB-LA 5 AS A service engineer, I WANT the nominal voltage of the battery module to be at maximum 48 V SO THAT I can safely replace and repair the battery.

The voltage limit of 48 V is a regulatory limit that divides between low voltage and high voltage. 48 V systems are popular for forklifts and other electrified low-floor vehicles. Up to this voltage, people who handle these batteries do not need special training.

MB-LA 6 AS A service engineer, I WANT a DC switch that will release all components from the DC bus SO THAT I can safely perform repairs on the DC bus.

Most standards require disconnecting switches and emergency stop switches; MB-LA 6 represents a refinement. While most standards require disconnection of the individual components and complete shutdown, the service engineer requires that the DC bus can be de-energized by a switch in any case. This is a requirement that is especially important when you consider that we have three voltage sources in our system: the genset, the solar power system, and the lead acid battery. All these sources are located at different locations. The solar power system is somewhere close to the site, but outside of the station. The genset is located near the fuel tanks and the battery system is inside the station. Hence, it is difficult to gain an overview of all sources at the same time.

We now turn our attention to the installer. The installer brings the transmission tower to a location with his team and installs equipment there. Usually, the installation sites are not easily accessible and do not have a general electricity connection.

MB-LA 7: AS AN installer, I WANT the system to consist of two modules—the broadcast system, to which the diesel, the solar power system, and the transmission mast are connected, and the battery—SO THAT I can assemble the system in a few simple s.pdf.

Of course, it would also be conceivable for MB-LA 7 to require the overall system to already have the batteries integrated. In this way, the installer would only have to connect the external components. Since lead acid batteries are heavy, the risk of the battery damaging itself or the system during transport is too great. Therefore, separation makes sense, as it also allows for easier transport and service.

Due to the fact that installation is usually not carried out in easily accessible locations, it makes sense to formulate a requirement that takes into account the fact that the installer and their team still carry out commissioning at the end of a working day and also test it. To save them work here, we come to requirement MB-LA 8:

`MB-LA 8: AS AN installer, I WANT an automated commissioning test that alerts me to errors SO THAT after the system is installed, I can check with little effort that the installation was done without errors.`

We now want to identify requirements for production. In the previous exercises, we looked at general requirements. In this exercise, we will look at requirements that arise from the use of lead batteries. One requirement is that we will probably have to connect different lead acid batteries in parallel during assembly. These should have approximately the same state of charge, otherwise a high balancing current could flow.

`MB-LA 9: AS A production, I WANT to ensure that the lead acid batteries have approximately the same state of charge when assembled SO THAT no high balancing currents flow when the batteries are connected.`

Furthermore, care will be taken in production to ensure that the lead batteries are easy to transport.

`MB-LA 10: AS A production, I WANT the assemblies connecting several lead acid batteries to be equipped with handling support SO THAT my staff can easily stow and transport the assemblies.`

We conclude this exercise with requirements from the design engineer's point of view. As we know, we can use the concentration of sulphuric acid in the lead acid battery to determine the state of charge of the lead acid battery. If we detect the lead acid batteries with sensors for the concentration of sulphuric acid, which send their data to a server, we have the possibility of determining data about the operation of the battery.

`MB-LA 11: AS A Design Engineer, I WANT the system to be equipped with sensors for the concentration of sulphuric acid in the lead acid batteries SO THAT I am able to remotely determine the state of charge of the battery.`

However, the state of charge alone is not enough to be able to completely determine the power flows of the system. Therefore, as a design engineer, we also want data on the production of solar power, the power of the genset, and the required power of the transmission tower:

`MB-LA 12: AS A Design Engineer, I WANT the solar power system to be fitted with an irradiance sensor, transmission tower, and genset with current sensors and the diesel tank with a level sensor SO THAT I have all information about the current power demand, power produced, and the tank remotely available.`

After we have collected the requirements, let's look at possible system configurations. Fig. 8.18 shows a possible realization. Each source or sink is connected to an AC bus via a power electronic component. This constellation represents the most modular realization from a system point of view and the most expensive from a cost point of view, because four power electronic components are needed. We therefore want to look at the characteristics of the sources and sinks in order to identify possibilities for system optimization.

We start with the mobile transmission station. The power demand is made up of two components: the transmitting power to be applied and the cooling requirements of the station. If the mobile radio station is to be installed in regions with a Mediterranean, tropical, or subtropical climate, both the power electronics and the battery must be cooled (Sen, Lal, and Charhate, 2015). In our example, the required transmitting power is 2,260 W. The cooling power required is 1,200 W. The cooling system will

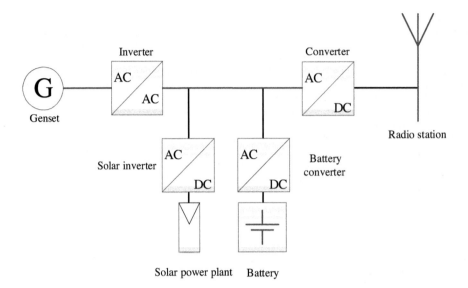

Fig. 8.18 System representation of a mobile radio transmitter mast that is supplied with energy from solar power and a genset, as well as a lead acid battery. In this realization, all sources and sinks are connected via an AC bus. This provides a high degree of modularity.

not run day and night, but is more likely to run in the second half of the day, after the system has heated up due to the ambient temperature plus its own heat losses. At start-up, the cooling system briefly consumes a significantly higher power, as the compressors and pumps have to start up first. The power required for this is 2,400 W. For the transmission unit, on the other hand, we assume a constant load profile—that is, that this power demand is independent of the time of day. The alternating current (AC) bus provides a single-phase AC with a frequency of 50 Hz. The rectifier converts this AC into a direct current (DC) that has a voltage of 48 V. This voltage is used to operate the components of the transmission tower and the cooling system.

Without the support of solar power and lead acid battery, the genset alone provides the power needed by the station. A genset consists of a combustion engine to which an electric motor is connected. If the output current is a three-phase current, the electric motor, which works as a generator, and the speed are set so that the frequency of the AC is 50 Hz or 60 Hz. Alternatively, an inverter is used to provide frequency stability of the AC current regardless of the speed of the motor and also to stabilize the voltage. Usually, a genset is dimensioned so that the operating point is at 60% of the nominal power. Since the total load is 3,460 W, the genset will have a nominal power of 5,766 W. We can see that the 4,660 W power needed for a short time when starting the cooling system could also be supplied by the genset.

The solar power system is realized by interconnected solar modules. A solar module is a very flat diode—that is, a $p-n$ junction. If light shines onto this plate, a voltage is measured. Since a solar module should convert as much light as possible into electrical energy, solar modules are realized as large-area semiconductors. These flat discs, the wafers, are criss-crossed with wires whose job it is to collect the charge carriers. Solar modules have a size of about 1 m × 1.7 m. The power of a module depends on the spectral composition of the light, the angle of incidence, and the irradiation. Standard test conditions (STC) have been agreed upon so that the modules can be compared. With STC, the irradiation is set to a power per area of 1,000 $\frac{W}{m^2}$. The module temperature is 25 °C. To describe the spectrum, the solar spectrum is first normalized and then related to a defined path through the atmosphere. This results in the so-called air mass (AM). An AM of 1 corresponds to the vertical irradiation of the sun at sea level. For STC, an AM of 1.5 is used, which corresponds to an irradiation angle of about 48.2 °C at sea level.

The output of a solar power system depends on various factors (Vanek *et al.*, 2008; Messenger and Abtahi, 2018): the interconnection of the solar modules, the orientation of the modules with respect to the path of the sun, any shading from objects in the vicinity, the temperature of the modules, and the quality of the maximum power point tracking (MPPT). This means that for the design and realization of a solar power system, one must evaluate the specific boundary conditions at the installation site. In our example, we want to simplify the design by assuming that the system is designed so that the solar modules deliver the nominal values for nominal voltage U_{mp} and nominal current I_{mp} determined at STC. We can then connect the modules like battery cells in a first approximation and choose a suitable $xpys$ connection.

Exercise 8.11 A suitable interconnection of modules is to be determined for the mobile radio station. The power of the solar power system should also be able to provide the maximum required power at STC. It must be ensured that the total voltage of the solar power system is between 550 V_{DC} and 850 V_{DC}. The module used has a rated voltage of $U_{mp} = 35$ V_{DC} and a rated current of $I_{mp} = 11$ A.

Solution: We first determine the power that a module can provide under STC:

$$P_{mp} = U_{mp} \cdot I_{mp} = 35 \text{ V}_{DC} \cdot 11 \text{ A} = 385 \text{ W}.$$

In order to also be able to supply the peak power when the cooling system is switched on, we need a power of 4,660 W. We can thus find the number of modules, but we have to round up, because when we round down, the resulting power is below the target:

$$N = \frac{4,660 \text{ W}}{385 \text{ W}} = 12.1 \approx 13.$$

If we connect all 13 modules in series, we get a voltage of $13 \cdot 35$ $V_{DC} = 455$ V_{DC}, which is below the required minimum voltage. Since we also want to be able to generate energy at lower irradiation levels, it makes sense to work with a larger string. If we orient ourselves to the maximum voltage of 850 V_{DC}, we can determine the number of modules. Attention: in this case we have to round down, because rounding up would lead to working above the allowed voltage limit:

$$y = \frac{850 \text{ V}_{DC}}{35 \text{ V}_{DC}} = 24{,}28 \approx 24.$$

Twenty-four modules represents almost a doubling of the number of modules. However, in this case we would still generate the minimum voltage required, even with an irradiance that is only 64% of the STC conditions.

The area exemplified in Exercise 8.11 is 4×6 modules, which corresponds to an area of 6.8 m$\times 6$ m $= 40$ m^2. In Fig. 8.18, the solar power system is connected to the AC bus via a solar inverter. The solar inverter has two tasks. Firstly, it converts the DC into AC. Furthermore, it determines how much current is drawn from the solar modules and adjusts this depending on the solar irradiation so that the maximum power is drawn in each case. This is called maximum power point tracking. By doubling the modules, the installed power of the solar power system is 9,240 W. However, this power only occurs under very good and high irradiation conditions. As a rule, this is the case from midday onwards, when the sun is at its zenith. During the day, the power drops in proportion to the irradiation power. Since the system in this example was chosen to be large, the solar power system will be able to supply the transmitter unit and the cooling system over a long period.

In Fig. 8.18, the lead acid battery is connected to the AC bus via a battery converter. This converts the DC into AC and selects between a current-led and voltage-led charging or discharging mode. In this design, the operational management works in such a way that the various components transfer their power to the AC bus or draw the required power from the AC bus. Energy management in this design only has to ensure the power balance. The components are decoupled, and we have a system that is very close to the implementation of a power flow diagram.

However, this design has several disadvantages. It is noticeable that there are many power electronic components. For each source, sink, or storage there is a converter. Another disadvantage is the large solar power system. The reason that this consists of 24 modules was not that we needed this installed power, but that we needed a voltage greater than 550 V$_{\text{DC}}$. We therefore want to ask ourselves whether we can choose a simpler design with fewer power electronic components and a smaller solar power system. In Fig. 8.19, we have shown an alternative design. The genset is not connected to a DC bus with an inverter, but with a rectifier. Only the transmission tower has a converter. The solar power system and the battery are connected directly to the DC bus. Let us first look at the genset operation. For outputs below 15 kW, gensets

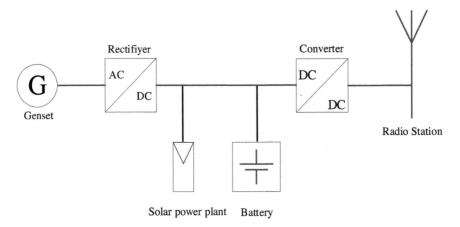

Fig. 8.19 Illustration of a simplified design of a hybrid mobile radio station equipped with a storage system.

can be either single-phase or three-phase. Their output is an AC. Since we need a nominal power of about 5 kW here, we can use a genset with a single-phase designed generator. Depending on the country for which the generator was originally equipped, its amplitude would be between 110 V_{AC} and 240 V_{AC}. Similarly, the frequency varies between 50 Hz and 60 Hz. The output voltage of a passive rectifier is $\sqrt{2} \cdot A_0$, where A_0 is the amplitude of the AC current. Since we are free to obtain the genset, we choose a generator with an AC voltage of 110 V_{AC}. The passive rectifier thus generates a DC voltage of 155 V_{DC}.

The converter of the transmission tower converts the 155 V_{DC} to the voltage level that its components need. We do not need to consider this final conversion, but consider the lead acid battery. A lead acid battery connected to the DC link and charged by the genset undergoes U-charging—that is, the voltage is kept constant and the charging or final discharge current reduces as the battery voltage approaches the DC link voltage. So if we have connected a lead acid battery to the DC link, and if it is suitably wired, a voltage of 155 V_{DC} will be set when the genset is in operation.

Exercise 8.12 A single lead acid battery cell has a voltage window of approximately 1.8 V_{DC} to 2.2 V_{DC}. The genset delivers a voltage of 155 V_{DC}. How many cells must be connected in series for the cells to have a cell voltage of 1.85 V_{DC}, or 2.1 V_{DC} after charging?

Solution: Since the voltages of the individual cells add up, all we have to do here is determine how the voltage is divided among the cells. We have to round up for the upper limit, because this is the only way to ensure that we do not overcharge the cells. For the lower limit, on the other hand, we have to round up, because we do not want to deep discharge the cells.

For the lower limit of 1.85 V_{DC}, we get the following number:

$$N = \frac{155 \ V_{DC}}{1,85 \ V_{DC}} = 83.78 \approx 84.$$

For the upper bound of 2.1 V_{DC}, we need the following number:

$$N = \frac{155 \ V_{DC}}{2,1 \ V_{DC}} = 73.8 \approx 73.$$

From the calculation from Exercise 8.12, we already know that the number of lead acid battery cells must be between 73 and 84 cells if we want to avoid a deep discharge with the help of the genset or prevent the genset from overcharging the lead acid battery.

Let us now turn to the solar power system. We first consider the case where it is connected to the DC bus with the genset alone. The lead acid battery is not connected to the DC bus and should not be considered for the time being. If the sun is not shining, the solar power system does not supply any power and the voltage is zero. If the genset is running, a current could theoretically flow into the solar power system, but since solar power modules are essentially diodes, the modules block, and no current flows.

As soon as there is solar irradiation, the solar power system delivers an output voltage that is proportional to the irradiation. If this voltage is above the DC voltage generated by the genset, a current should flow from the solar power system into the genset. But here the diodes of the rectifier block. The DC link voltage rises unless the transmission tower reduces the power. But what happens if the power required by the transmission tower is higher than the power that can be provided by the solar power system? In this case, the voltage from the solar power generator is reduced, and as soon as the voltage drops below the genset voltage, the power is drawn from the genset again. For the design of the solar power system, we must primarily deal with the question of how high the maximum voltage of the solar power system may become.

Let us now also consider a situation where the lead acid battery is also connected to the DC bus. As soon as the voltage of the solar power system is higher than the battery voltage, the lead acid battery is charged. If the voltage of the solar power system falls below the lead acid battery voltage, a current should theoretically flow from the battery into the solar power system, but this does not happen due to the diode properties of the system.

We see that the design shown in Fig. 8.19 is a reasonable solution. So we have at least two ways to realize a storage system. Let us therefore look at the power flow diagram and ask ourselves how we can map the operating mode of the design of Fig. 8.19.

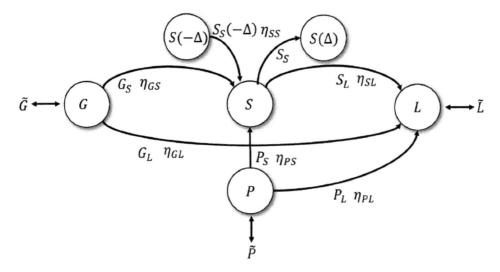

Fig. 8.20 Power flow diagram of the battery-powered hybrid transmission tower. $\tilde{L}$ represents the load profile of the transmission unit including the cooling system. $\tilde{P}$ corresponds to the generation of the solar power system. $\tilde{G}$ describes the consumption of the genset.

The power flow diagram is shown in Fig. 8.20. In this system, the generator serves as the source. Power can only flow from the generator, battery, or solar power system to the load. Power cannot flow back from the storage or load to the generator or from the load to the storage.

We first set up the equation for the consumption node:

$$\tilde{L}\alpha \leq \eta_{\mathrm{GL}}G_{\mathrm{L}} + \eta_{\mathrm{SL}}S_{\mathrm{L}} + \eta_{\mathrm{PL}}P_{\mathrm{L}} \leq \tilde{L}. \tag{8.47}$$

We allow in eqn (8.47) that the power sources and storage do not always have to meet the entire power demand. Instead, we allow a certain percentage of undersupply ($\alpha \in [0,1]$). In principle, we have the option of reducing the transmission power of the transmission tower. If the transmission tower is not installed at a neuralgic point of the mobile network, the result will be a reduced range or fewer parallel calls. However, we also have the option of simply reducing the power of the cooling system.

The power flow for the generator is obtained by summing the outflows to the battery and to the load:

$$\tilde{G} = G_{\mathrm{L}} + G_{\mathrm{S}}. \tag{8.48}$$

Similarly, we can formulate the equations for the solar power system:

$$\tilde{P} \geq P_{\mathrm{L}} + P_{\mathrm{S}}. \tag{8.49}$$

Since it is quite possible that more solar power is available than can be used by the battery and transmission tower, a $\geq$ relationship must be used here.

The storage equation is derived by considering the inflows and outflows:

$$0 = \eta_{\mathrm{GS}}G_{\mathrm{S}} + \eta_{\mathrm{SS}}S_{\mathrm{S}}(-\Delta) + \eta_{\mathrm{PS}}P_{\mathrm{S}} - (S_{\mathrm{S}} + S_{\mathrm{L}}). \tag{8.50}$$

The objective function for a business valuation is given by the fuel cost c_{fuel}:

$$\min Y = c_{\mathrm{fuel}} \int_{t=0}^{T} (G_{\mathrm{L}} + G_{\mathrm{S}})\,\mathrm{d}t. \tag{8.51}$$

Table 8.4 Summary of the power flow equations for the battery-supported, hybrid-powered transmission tower.

Power node	Power flow equation
Load	$\tilde{L}\alpha \leq \eta_{\mathrm{GL}}G_{\mathrm{L}} + \eta_{\mathrm{SL}}S_{\mathrm{L}} + \eta_{\mathrm{PL}}P_{\mathrm{L}} \leq \tilde{L}$
Genset	$\tilde{G} = G_{\mathrm{L}} + G_{\mathrm{S}}$
Solar plant	$\tilde{P} \geq P_{\mathrm{L}} + P_{\mathrm{S}}$
Battery	$0 = \eta_{\mathrm{GS}}G_{\mathrm{S}} + \eta_{\mathrm{SS}}S_{\mathrm{S}}(-\Delta) + \eta_{\mathrm{PS}}P_{\mathrm{S}} - (S_{\mathrm{S}} + S_{\mathrm{L}})$
Yield function	$\min Y = c_{\mathrm{fuel}} \int_{t=0}^{T} (G_{\mathrm{L}} + G_{\mathrm{S}})\,\mathrm{d}t$

In Tab. 8.4, we have summarized the equations again. For the realization of the power flow control of the system in Fig. 8.18, we have a relatively simple algorithm at our disposal:

The algorithm shown in Algorithm 6 can still be extended to include various aspects. For example, it can actively control the state of charge of the battery to ensure that the capacity of the battery does not fall below a certain threshold. We can also require $G_{\mathrm{L}} + G_{\mathrm{S}}$ to have a certain value so that the genset always operates at an efficient operating point.

Algorithm 6 Example of operational management for the system shown in Fig. 8.18. For simplicity, the influence of the efficiencies has been neglected.

while System is running **do**
$\quad L \leftarrow \tilde{L}$
$\quad P \leftarrow \tilde{P}$
$\quad$**if** $P \geq \alpha L$ **then**
$\quad\quad P_L \leftarrow \alpha L$
$\quad\quad P \leftarrow P - \alpha L$
$\quad\quad$**if** $\kappa \leq \kappa_{\max}$ **then**
$\quad\quad\quad P_S \leftarrow P$
$\quad\quad$**end if**
$\quad$**end if**
$\quad$**if** $P \leq \alpha L$ **then**
$\quad\quad P_L \leftarrow P$
$\quad\quad L \leftarrow L - P$
$\quad\quad$**if** $\kappa \geq L \cdot \Delta$ **then**
$\quad\quad\quad S_L \leftarrow L$
$\quad\quad$**else**
$\quad\quad\quad S_L \leftarrow S_S(-\Delta)$
$\quad\quad\quad L \leftarrow S_L$
$\quad\quad\quad G_L \leftarrow L$
$\quad\quad$**end if**
$\quad$**end if**
end while

Exercise 8.13 The genset has a nominal power of 5.8 kW. The constant power losses are $a = 100$ W, the linear losses are $b = 0.01$—that is, 1 % of the power—and the quadratic losses are $c = 8.16 \cdot 10^{-6} \frac{1}{W}$. At what power does the genset have its best efficiency and what is the efficiency?

Solution: If x is the power, then the efficiency η is:

$$\eta = 1 - \left(\frac{a}{x} + b + c \cdot x\right).$$

We determine the maximum via the zero of the first derivative:

$$\frac{d}{dx}\eta(x) = -\frac{a}{x_0^2} + c = 0.$$

The power at which the efficiency becomes maximum is given by:

$$x_0 = \sqrt{\frac{a}{c}} = \sqrt{\frac{100 \text{ W}}{8.16 \cdot 10^{-6} \frac{1}{W}}} = 3{,}500 \text{ W}.$$

The efficiency at x_0 is then:

$$\eta(x_0) = 1 - \left(\frac{100 \text{ W}}{3{,}500 \text{ W}} + b + 8,16 \cdot 10^{-6} \frac{1}{W} \cdot 3{,}500 \text{ W}\right) = 0.96\%.$$

At a power of 3,500 W, the genset thus operates at the optimal working point. This also almost corresponds to the required power when the transmitter and cooling system are running in parallel. When the cooling system starts up, however, a higher power is required for a short time, namely 4,660 W. At this operating point, the efficiency is:

$$\eta(4{,}660 \text{ W}) = 1 - \left(\frac{100 \text{ W}}{4{,}660 \text{ W}} + b + 8.16 \cdot 10^{-6} \, \frac{1}{\text{W}} \cdot 4{,}660 \text{ W} \right) = 95\%.$$

The described control method is described using the power flow diagram. This is possible because all sources, sinks, and storage have power electronics which we can control. With the system from Fig. 8.19 this is not so simple. Since here the behaviour is automatically set via the voltage levels, we have to consider the voltage of the sources and sinks and the storage for the description of the power flows. The only quantities we can control in this case are $\tilde{G}$ and $\tilde{L}$. Algorithm 6 shows a possible version for an operational control.

The operation control in Algorithm 6 is based on checking if the voltage from DC bus drops below $U_{\min}$. If this is the case, the solar power system is not working and the lead acid battery is not discharging. In this case, it is checked whether the load can be reduced. If this is the case, for example the cooling system is switched off, the DC bus voltage increases. If this is not the case, the generator must be switched on. It then remains active until the solar power system raises the voltage above the value of $U_{\max}$.

Algorithm 7 Example of operation control for the system shown in Fig. 8.19. $U_{\min}$ and $U_{\max}$ represent voltage limits of the DC bus, which are used to detect whether the generator is switched on or off.

---

while System is running **do**
  $L \leftarrow \tilde{L}$
  $U \leftarrow U_{\text{DC}}$
  if $U \leq U_{\min}$ **then**
    if L can be reduced by ΔL **then**
      $L \leftarrow L - \Delta L$
    else
      Switch Generator ON
    end if
  end if
  if $U \geq U_{\max}$ **then**
    Switch Generator OFF
  end if
end while

We want to design the lead acid battery. We need the genset to avoid a deep discharge of the lead batteries. It makes no sense energetically to use the genset to fully charge the lead acid battery and then use the battery for power supply. The lead batteries are charged via the solar power system. In Exercise 8.12, we determined that we cannot fully charge more than 73 cells connected in series with the genset at an AC voltage of 120 V_{AC}, and we cannot avoid deep discharge if we have more than 84 cells. For the station we want to use lead acid battery modules. These consist of six interconnected cells that have a nominal voltage of 12 V_{DC}.

Table 8.5 Extractable energy of two lead acid batteries at different discharge rates

Label	$U_{nom}[V_{DC}]$	$\kappa(4C)[Ah]$	$\kappa(2C)[Ah]$	$\kappa(1C)[Ah]$	$\kappa(0,5C)[Ah]$	$\kappa(0,25C)[Ah]$
Module 1	12	50	75	100	105	120
Module 2	6	40	55	70	80	85

Sealed lead batteries are to be used. These are low maintenance, as hydrogen formation and oxygen reduction are reduced, and therefore have a longer service life. They provide different amounts of energy for different discharge rates. Tab. 8.5 shows data for two modules of different nominal voltage at different C-Rates.

Exercise 8.14 The modules shown in Tab. 8.5 are each to be interconnected to form a string that is preserved from deep discharge by the genset. What does this 1pys connection look like? What is the capacity in Wh at nominal voltage?

Solution: The genset provides a voltage of 155 V_{DC}. To find the minimum voltage of the module, we need to find the minimum voltage of a cell with the number of cells of the module. The information presented in Tab. 8.5 does not make any statements about the number of battery cells connected in parallel. However, since we know that a lead acid battery cell has a nominal voltage of 2 V_{DC}, we can derive the number of cells from the nominal voltage of the module. This is six for module 1 and three for module 2. The minimum voltage of module 1 is calculated as $6 \cdot 1.8$ $V_{DC} = 10.8$ V_{DC}. For y we obtain a value of:

$$y = \frac{155 \text{ } V_{DC}}{10,8 \text{ } V_{DC}} = 14.3 \approx 14.$$

We have to round off here because if we connected 15 modules in series, the voltage would be below 1.8 V_{DC}. The nominal voltage of module 1 is:

$$U_{nom} = 14 \cdot 12 \text{ } V_{DC} = 168 \text{ } V_{DC}.$$

The storage capacity is given in Ah in Tab. 8.5. Let's find out from the product of nominal voltage and storage capacity at a C-Rate of 1:

$$\kappa_{M1} = U_{nom} \cdot \kappa(1C) = 168 \text{ } V_{DC} \cdot 100 \text{ Ah} = 16,800 \text{ Wh}.$$

For module 2, the minimum module voltage is set at $3 \cdot 1.8$ $V_{DC} = 5.4$ V_{DC}. Similarly, we determine y:

$$y = \frac{155 \text{ } V_{DC}}{5,4 \text{ } V_{DC}} = 28.7 \approx 28.$$

The nominal voltage is correspondingly smaller: $U_{nom} = 168$ V_{DC}. The capacitance is then

$$\kappa_{M2} = U_{nom} \cdot \kappa(1C) = 168 \text{ } V_{DC} \cdot 70 \text{ Ah} = 11,760 \text{ Wh}.$$

A pure serial string has a storage capacity of 16 kWh or 12 kWh. To evaluate how much storage capacity we need, we first have to look at the consumption of the transmission station. Here we assume that the base load of the transmitting unit is always needed. The cooling unit will only be used from midday until early evening—that is, for about six hours. With these assumptions, the energy consumption E_L is:

$$E_L = 24 \text{ h} \cdot 2{,}260 \text{ W} + 6 \text{ h} \cdot 1{,}200 \text{ W} = 61{,}440 \text{ Wh.}$$

E_L represents the consumption for a full day. For the lead acid battery, however, the task is not to supply the station with energy for a full day, but only in the time between sunset and sunrise. The genset is always available as a backup. Assuming that the cooling system continues to run for two hours after sunset and that the dark phase is 12 hours, this results in a required amount of energy E_B of:

$$E_B = 12 \text{ h} \cdot 2{,}260 \text{ W} + 2 \text{ h} \cdot 1{,}200 \text{ W} = 29{,}520 \text{ Wh.}$$

Exercise 8.15 The energy demand is $E_B = 30$ kWh. The DoD should be 60%. How many strings must be connected for modules 1 and modules 2?
  Solution: We first need to determine the required capacity as the DoD should be 60 %:

$$\kappa = \frac{30 \text{ kWh}}{60 \text{ \%}} = 50 \text{ kWh}$$

From Exercise 8.14 we know the capacity of a string. We need to determine the number of strings. In this case it will be necessary to round up.
  We now determine the number of strings for modules 1:

$$N_{M1} = \frac{50 \text{ kWh}}{16.8 \text{ kWh}} = 2.9 \approx 3$$

The total capacity here is $\kappa = 3 \cdot 16.8 \text{ kWh} = 50.4 \text{ kWh}$, which is very close to the target capacity.
  For module 2, the number of strings is as follows:

$$N_{M2} = \frac{50 \text{ kWh}}{11.76 \text{ kWh}} = 4.25 \approx 5$$

The total capacity here is $\kappa = 5 \cdot 11.76 \text{ kWh} = 58.8 \text{ kWh}$. Relative to a target capacity of 50 kWh, we have here an excess capacity of about 8.8 kWh.

So far, cycle ageing has not been considered. Fig. 8.21 shows the maximum number of cycles at a given DoD. As we already know, the number of cycles reduces significantly as the DoD increases. Since we have designed the battery for a DoD of 60%, we can expect approximately 2,000 cycles.
  We conclude this section with a consideration of the system components and a review of the RTM. Fig. 8.22 shows the system components of the system seen in Fig. 8.19. The main system is the `TransmissionStation`. This consists of five main components: the `GenSet`, the `Battery System`, the `Solarpowerplant`, the `Transmission System`, and the `CoolingSystem`. In addition, the `TransmissionStation` also

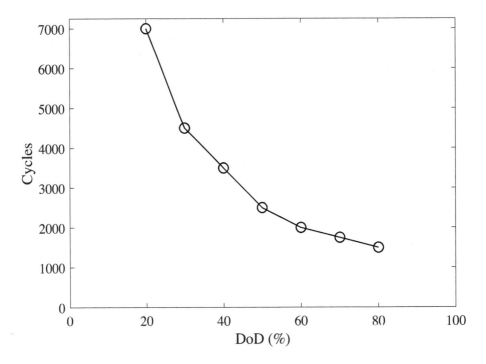

Fig. 8.21 Discharge cycles of the lead acid batteries for different depths of discharge.

includes the SafetySystem. All these five components are related to the general components of the storage system: PowerDistribution, PowerDistributionControl, and MechanicalConstruction. To improve clarity, we do not show aggregations in Fig. 8.22. In addition, the TransmissionStation is linked to an EnergyManagementSystem, which collects and aggregates the data of the various components and can adjust the operating state.

The GenSet consists of three components. The CombustionEngine generates rotational power by burning fossil fuels. This rotational power is converted into electrical power by the Generator. The generator is a special realization of an electric motor. The GenSet also has at least one FuelTank, which is in essence also a Storage. We have not shown this relationship for the sake of clarity.

The Battery System is constructed in a similar way to the representation described in Fig. 8.13 from BatteryCell, where a set of battery cells are connected to a CellThread. The actual Battery Module aggregates these CellThreads. Modules are aggregated analogously to Module Thread and then form the Battery System.

The same logic is used to construct the Solarpowerplant. Here, the individual solar modules are connected to form ModuleThread, which are then connected in parallel to form the Solarpowerplant. The structure here is similar to that of electrochemical or electrical storage.

We will now look at the responsibilities of the various system components for compliance with the requirements. The relationships are summarized in Tab. 8.6.

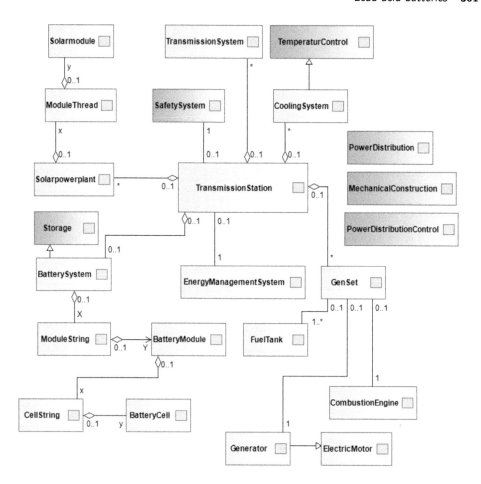

Fig. 8.22 System components of the hybrid-powered, battery-supported transmission tower. The basic components PowerDistribution, MechanicalConstruction, and PowerDistributionControl are shown without connections to the system components, but are linked to the components Battery System, GenSet, TransmissionSystem, and Solarpowerplant.

Table 8.6 Requirement traceability matrix for the transmission tower system.

	Basic requirements								Electrical systems					Electrochemical storage systems				
	G1	G2	G3	G4	G5	G6	G7	G8	ES 1	ES 2	ES 3	ES 4	ES 5	ECS 1	ECS 2	ECS 3	ECS 4	ECS 5
BatteryCell	x																	
CellString														x			x	
BatteryModule										x	x						x	
ModuleString														x			x	
BatterySystem					x				x	x	x					x	x	
BalancingSystem				x					x		x	(x)		x				
BatteryManagementSystem		x	x		x				x									
HealthEstimator						x												
SOCEstimator				x													x	
EnergyManagementSystem		x				x	x											x
CapacityManagement							x											x

The Battery system and its subcomponents are responsible for the requirements G 1 to G 8, ES 1 to ES 5, and ECS 1 to ECS 5. This has already been described in Tab. 8.2. Therefore, the analysis is simplified to the requirements for lead acid batteries, ECS-LA 1 to ECS-LA 5, as well as those requirements resulting from the application MB-LA 1 to MB-LA 12.

ECS-LA 1, ECS-LA 2, and ECS-LA 3 aim to avoid side reactions. These occur mainly at high charging currents and during high or deep discharge of the battery. Since we have not integrated any active power electronics in the simplified system that actively avoid this, we have to coordinate the components Solarpowerplant, GenSet, TransmissionSystem, and CoolingSystem in such a way that these unwanted system states do not occur or their effects are kept low. Therefore, all four system components share this responsibility.

The ECS-LA 4 and ECS-LA 5 requirements are focused on operating the lead acid battery at a low C-Rate and not choosing a DoD that is too large. We have considered both of these in the design of the Battery System, and therefore we assign these two requirements to the Battery System.

We now want to look at the requirements that arise from the application. In doing so, we use, among others, the requirements elaborated in the solution of Exercise 8.10. MB-LA 1 must be fulfilled by the TransmissionSystem. After all, we are building the station so that we can realize a dense mobile network. In MB-LA 2 the mobile operator requires that the station is supplied with a solar power plant and a genset; logically, this is a requirement that is fulfilled by Solarpowerplant and GenSet. Similarly, MB-LA 3 is assigned to GenSet and MB-LA 4 to Solarpowerplant and Battery System.

In MB-LA 5, a requirement is placed on the Battery System. However, our battery system does not have a nominal voltage of 48 V_{DC}; the individual modules are even significantly lower than this. Therefore, there is now a requirement to ensure that the installer cannot come into contact with the full DC voltage. Here, compliance with MB-LA 6 helps, where all DC sources are switched off with the help of a switch. This is a requirement of the Power distribution.

Requirements for the construction of the system, such as MB-LA 7, must be met by the Mechanical construction. In this case, we must ensure that this component occurs with every system component.

We assign the requirement for a commissioning operation, MBA-LA 8, to the Energy Management System. This has the overall control and can manage and read out all the individual components.

MBA-LA 9 to MBA-LA 11 are requirements for the construction of the Battery System and thus clearly assigned. Requirement MBA-LA 12, which requires a set of sensors to be read and utilized by the Energy Management System, must be met by the various subsystems.

We see that all requirements are assigned to one or more system components. In this case, it is noticeable that the structure of the RTM is much simpler than what we have already considered in other systems. This suggests that the simplified system shown in Fig. 7 has significant advantages.

8.3.4 Conclusion

In this section, we, have dealt with the lead acid battery in detail. We have learned about the main and secondary reactions and looked at which charging and discharging strategy we must use to operate the lead acid battery in order to ensure a long service life. This has shown that it makes sense to oversize lead acid batteries in relation to their capacity, as this significantly increases the number of cycles. In addition, unwanted side reactions are reduced, as the relative charging and discharging currents are also lower with an oversized lead acid battery.

As an application example, we looked at a transmission tower powered by a genset, as well as a solar power system. The solar overproduction should be stored in a lead acid battery.

In the next section, we will look at another very common cell chemistry: the lithium ion battery. This is not one but a family of cell chemistries, but they all use the same principle.

8.4 Lithium ion batteries

Lithium ion batteries are one of the most commonly used electrochemical storage technologies. A number of the applications presented in Section 2.2 now work with electrochemical storage based on lithium ion cell chemistry. This includes in particular mobile applications. Be it in consumer electronics or mobile tools, lithium ion batteries have been increasingly used in electromobility for several years. This is due on the one hand to the higher energy density compared to other technologies and on the other hand to a sharp drop in costs caused by a growing demand and a number of technology and production improvements.

In this section, we will look at lithium ion batteries. These are not based on a single cell chemistry, but a whole family of cell chemistries, although they are very similar in their basic functioning. After we have described the cell chemistries, we will look at the requirements that must be met when operating systems with lithium ion batteries. Due to their cell chemistry, lithium ion batteries are not as easy to handle as lead acid batteries. To ensure safe operation, lithium ion batteries have additional requirements.

We conclude this section with an example application for lithium ion batteries: the residential solar storage system for home applications. These systems have become widespread in recent years. While the first home storage systems used lead acid batteries, these have now been almost completely replaced by systems with lithium ion batteries.

8.4.1 The chemistry of lithium ion batteries

Lithium ion batteries are a family of batteries that differ in their electrochemical reactions. More specifically, they differ in the anode and cathode materials they use. However, the primary reactions are so similar that these different batteries are all collectively called lithium ion batteries.

Fig. 8.23 shows the basic structure of a lithium ion cell. It consists of two electrodes surrounded by an electrolyte and separated by a separator. The separator is

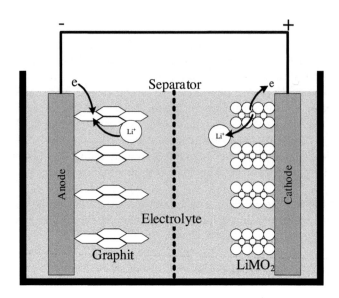

Fig. 8.23 Design of a lithium ion battery cell. The cathode is coated with a lithium metal oxide that releases lithium ions during the charging process. The anode is coated with a material that absorbs these ions again. In this example, graphite is used as the anode material. The anode and cathode are divided from each other by a separator.

not permeable for electrons, but allows ions to pass through. The two electrodes are coated with a different material: the anode with graphite and the cathode with a lithium metal oxide ($LiMO_2$). When the battery is charged, x lithium ions are released from the $LiMO_2$, and at the same time x electrons are released:

$$LiMO_2 \rightleftharpoons Li_{1-x}MO_2 + xLi^+ + x\,e^-. \tag{8.52}$$

The lithium ion diffuses through the separator to the anode, while the electron flows across the conductor to the anode. At the anode, the lithium ion then combines with the electron and remains in the anode material:

$$C + xLi^+ + x\,e^- \rightleftharpoons Li_xC. \tag{8.53}$$

The total reaction is:

$$LiMO_2 + C \rightleftharpoons Li_xC + Li_{1-x}MO_2. \tag{8.54}$$

Unlike a lead acid battery, the electrolyte only serves as a transport medium. In a lithium ion battery, all reactions take place in the anode and cathode material. In short, two charge carriers, the $x\,Li^+$ ions and the $x\,e^-$ electrons, are only separated from each other to recombine on the other side of the battery. Various metal oxides and cathode materials can be used for this reaction. The different types of lithium ion cells differ in the combination of metal oxide and anode material. In Fig. 8.24, some

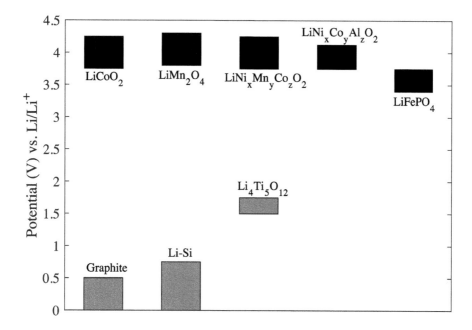

Fig. 8.24 Cathode and anode materials for lithium ion batteries.

materials and their electrical potentials are shown. We see that the cathode material has an electric potential that lies between 3 V and 4.25 V—in other words, to release a lithium ion from this material, we need a voltage that lies in this range. We have to compare this potential value with the potential values of the anode material. By far the most common material is graphite. It has a potential that lies between 0 V and 0.6 V. If we combine, for example, lithium iron phosphate ($LiFePO_4$) with graphite (C), we expect a cell voltage that lies between 2.8 V and 3.5 V, although we cannot see the exact shape of the voltage from this diagram.

The discharging process of a lithium ion cell is based on ion transport. Unlike a lead acid battery, where the reaction takes place on the surface of the electrodes, here the ions are absorbed into the interior of the crystal. Therefore, the crystal structure of the active material is important for the characteristics of the cell chemistry. In the case of cathode material, a distinction is made between three basic types of materials that have different lattice structures. With olivine lattices, the lithium ions have only one degree of freedom for their movement. Batteries with $LiFePO_4$ have such a crystal lattice. A two-dimensional movement is possible with the layered construction of the crystal. Batteries with a $LiMO_2$ compound belong to these cathode materials. Cobalt ($LiCoO_2$), nickel, manganese ($LiNixMnyCozO_2$), or aluminium ($LiNixCoyAlzO_2$) can each be used as the metal oxide. So-called spinel lattices (LiM_2) allow a three-dimensional movement of the lithium ions; manganese and nickel are used here.

The function of the cathode material is to embed the lithium in its crystal lattice. During charging and discharging, the lithium ion detaches from the lattice and moves through the separator to the anode. Ideally, we would assume that we could in principle dissolve out all the lithium that is in the $LiMO_2$. But this is not the case. If we did this, we would dissolve the crystal structure of the cathode material and a stable $LiMO_2$ would theoretically become a lithium oxide (LiO_2), which has a different crystal structure and is chemically more stable. For a lithium ion cell to work, the crystal structure of the cathode material must be preserved. Therefore, not all of the lithium can be removed from the cathode material. Excessive charging can cause permanent damage to the cell, as the degradation can be irreversible. For example, in $LiCoO_2$ cells only 50% of the existing lithium ions are used, and in $Li_{1-x}Ni_{0.33}Mn_{0.33}Co_{0.33}O_2$ only 66%.

The anode material is also integrated into the crystal lattice. The absorption capacity of the anode material partly limits the energy density. Carbon dominates the anode material. We have already seen in Chapter 7 that carbon has an extremely large surface per volume, which is advantageous for the absorption of lithium ions. Lithium ion cells use either natural carbon or artificially produced carbon. Artificially produced carbon has a more homogeneous crystal structure. The structure of natural carbon is amorphous—that is, there are zones or areas in which a homogeneous crystal structure can be recognized, but this is repeatedly broken through. Since the absorption capacity of carbon is limited, other substances are also used as anode material. Mixtures of silicon and carbon or pure silicon are used. Cells with silicon can hold considerably more lithium ions in their grid than carbon can. However, the volume of the anode material changes, which puts a mechanical stress on the active material.

Besides carbon and silicon, lithium titanium oxide ($Li_4Ti_5O_{12}$) is also used as anode material. This material has the advantage that there is no change in volume when lithium ions are stored. Therefore, the cycle stability of batteries with this anode material is significantly higher compared to anodes made of carbon or silicon. Cells with $Li_4Ti_5O_{12}$ as anode material are therefore suitable for applications with high charging power. However, the energy density of these cells is not as high because the potential difference between the anode and cathode material is lower.

8.4.2 Behaviour and requirements: How to handle lithium ion batteries

In this section, we will look at the behaviour of lithium ion batteries and derive requirements for their use. Lithium ion batteries are a family of cell chemistries that differ in the combination of anode and cathode materials. The cell chemistry defines the shape of the $U_{OCV}(\kappa)$ curve. This is characteristic for the respective combination of cathode and anode material. Fig. 8.25 shows curves for different cathode materials; the anode material is always graphite. The curve is shown as a function of the DoD. Information about the energy content is missing in this representation. Two things are remarkable in this diagram. The voltage range that can be realized with lithium ion battery cells is between 2.6 V and 4.6 V. This means that lithium ion battery cells have an energy content of 1.5 V. This means that lithium ion battery cells have a higher voltage range than lead acid batteries, whose nominal voltage is 2 V. Furthermore, it is noticeable that the shape of the $U_{OCV}(\kappa)$ curves depends on the chosen combination of anode and cathode material. The shape of a cell with $LiFePO_4$ differs considerably from the shape of $LiCoO_2$. The combination of lithium iron phosphate ($LiFePO_2$) with

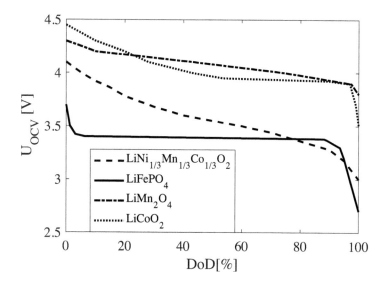

Fig. 8.25 Open-circuit voltage U_{OCV} of different lithium ion batteries depending on the depth of discharge. The anode material is always graphite, while the cathode material varies.

graphite has a wide range in which the open-circuit voltage is constant and does not change, although the battery is being charged or discharging. In the combination of lithium manganese oxide ($LiMn_2O_4$) with graphite, on the other hand, we have the situation that each state of charge can be assigned a unique voltage. But even here, there are areas where the change of the open-circuit voltage depending on the state of charge is sometimes stronger and sometimes very small. This leads us to the first requirement in the realization of a storage system based on lithium ion cells. While with lead acid batteries we always have clear information about the battery's state of charge based on the voltage and also the concentration of sulphuric acid, some cell chemistries are not directly accessible.

ECS-LIB 1: AS A developer, I WANT to implement a state-of-charge estimator that estimates the state of charge sufficiently well SO THAT I am able to determine the state of charge even for cell chemistries with a flat U_{OCV} curve.

In Section 8.2.2 we talked about the different methods of state-of-charge estimation. With lithium ion batteries, reliable state-of-charge estimation is even more important, because lithium ion cells must neither be deep discharged nor overcharged. We have already mentioned that we must not dissolve too many lithium ions from the metal oxide. This can either lead to parts of the cathode material no longer being able to reabsorb lithium ions or to the cathode material becoming chemically unstable and decomposing. Therefore, we need to avoid overcharging. This is different from a lead acid battery. Here, a cell could certainly be overcharged. Violation of the ECS-LA 1 to ECS-LA 3 requirements did occur, but this could be accepted if appropriate measures were taken to avoid the effects of side reactions. This is no longer possible with lithium ion battery cells. Above a voltage of 4.3 V, there is a risk of the electrolyte beginning to oxidize. Such a reaction leads to a chain reaction called thermal runaway.

In case of a thermal runaway, heat is generated by the oxidation of the electrolytes. If the temperature increases to above 100 °C, lithium is released at the anode. This produces heat and hydrocarbon compounds that lead to further decomposition of the anode. The additional heat that occurs in the process ensures that various reactions also take place at the cathode. Oxygen is produced, which increases the oxidation at the cathode. A number of gases are formed and further heat is released. These self-reinforcing reactions lead to a cell fire.

Requirement G 6 demands that a storage system is safe. Therefore, new requirements are derived from G 6 for lithium ion battery cells:

ECS-LIB 2: AS A developer, I WANT to ensure that no thermal runaway occurs SO THAT there is no danger for man and machine from my energy storage system.

ECS-LIB 3: AS A developer, I WANT to ensure that if a thermal runaway does occur, people are protected from the immediate hazards SO THAT people can remove themselves from the hazardous area and are safe.

ECS-LIB 2 and ECS-LIB 3 can also be summarized verbally as follows: no thermal runaway may occur, and if this unlikely event does occur, no one may be harmed. Compliance with ECS-LIB 3 depends on the application of the energy storage system and enables different approaches. For mobile applications, such as mobile phones, laptops, or electrically operated tools, it is important to ensure that the thermal runaway does not lead to an explosion of the device but occurs slowly and visibly. This ensures that people recognize the danger and move away from the hazard area. With electric vehicles it is similar: you have to make sure that the driver and passengers have enough time to leave the vehicle. In the case of stationary small and large storage facilities, the realization of ECS-LIB 3 is somewhat more difficult; on the one hand, the energy quantities are considerably larger, and on the other hand, these storage facilities are stationary. Here, the measures range from elaborate fire protection technology to encapsulated areas that are simply supposed to burn down.

But ECS-LIB 3 addresses the question of 'what if'. Of course, we have to address the question of how a thermal runaway can be prevented in the first place. This is the requirement of ECS-LIB 2. ECS-LIB 2 is a requirement that the system actively try to avoid this dangerous condition.

In order to be able to answer the question of how to avoid entering this state, let us look at the question of what can cause a thermal runaway in a lithium ion battery cell. The answer is simple: we must avoid an internal or external short circuit. This is because this short circuit causes a high current to flow and drive the chemical chain reaction of a thermal runaway.

The fact that an internal or external short circuit must be avoided also applies to all other electrochemical cells. However, the hazard potential of a lithium ion battery cell is considerably higher than that of other cells. The electrochemical energy stored in a lithium ion battery cell is three times higher than the amount of energy that can be extracted electrically.

The risk of an external short circuit can be reduced with a number of technical measures. At a system level, batteries are equipped with switches on both electrical poles so that separation from the overall system is redundant. At a cell level, further measures are implemented in the design. A PTC, a thermistor, is inserted into the cell. A PTC resistor is a thermistor whose resistance value increases at high temperatures. If the cell heats up, the resistance increases and thus reduces the current. In this way, additional heating due to excessive current is to be prevented. The second element is a fuse. If the PTC is not sufficient and the current flow nevertheless becomes too high, this fuse is intended to prevent an additional current flow. Manufacturers have various solutions here. For example, the fuse function can be transferred to the connection wires or to the connection strip (busbar). The latter is a measure that then already acts at the module or system level.

All these measures intervene in the event of an external short circuit. However, it is more difficult to avoid an internal short circuit. This occurs when the separator is cut or loses its insulating properties. This can happen through mechanical faults, for example if a battery cell is damaged by external force. But it can also happen chemically. If, for example, a cell is deep discharged or overcharged, a degradation process can occur in both cases, which then leads to an internal short circuit. For this reason, care must always be taken when operating lithium ion battery cells that the specified voltage window is not exceeded.

Another cause or effect that can lead to degradation of the active material or the separator is excessive heating of the cell. In this case, reactions can occur which lead to a thermal runaway.

If a thermal runaway happens, a cascade of chemical reactions begins that unfolds over a period of minutes. The process consists of three phases. In the first phase, the cell begins to drive the combustion process itself. However, the voltage slowly drops, while at the same time the cell slowly heats up. At this point, the separator is still partially intact. In the second phase, the voltage drops very quickly and the cell begins to heat up noticeably and rapidly. In this phase, the separator collapses. Once the separator has dissolved, the third phase begins. The electrolyte begins to dissolve and an unstoppable fire develops.

To reduce the effects of the third phase, lithium ion battery cells are equipped with a pressure-relief valve. This opens as soon as the internal pressure that builds up in the first two phases exceeds a critical value. It allows electrolyte and hot gases to escape and is intended to prevent an explosion of the cell, which can lead to a chain reaction in a battery system consisting of many cells.

In addition to the design measures described above, there is another measure that must be taken into account when operating lithium ion battery cells:

ECS-LIB 4: AS A developer, I WANT to ensure that during operation all cells are always within the permitted voltage and temperature range SO THAT the likelihood of an internal short circuit is reduced.

It is important to understand that ECS-LIB 4 is both a hard and a soft requirement. It is hard when the limits are exceeded too far. In this case, an internal short circuit will usually occur. But these limits are often so clear and far from normal operating points that you never reach this hard limit. More subtle are the soft limits. For a short time, it is possible to cross these limits. But each boundary crossing increases the likelihood and there may well be no temporal correlation between the boundary crossing and the actual event.

To comply with ECS-LIB 4, battery systems using lithium ion battery cells are equipped with a monitoring system, which is usually integrated into the battery management system (BMS). Voltage and temperature are then monitored at cell, module, and system levels.

In Section 8.2.2, it was mentioned that a safety margin is kept at both the upper and lower capacity limits to avoid hard control and to ensure safety. Whereas with lead this was only necessary to reduce ageing effects, it is absolutely required for system safety and the useability of lithium ion battery systems. The same applies to the 'balancing' functionality. An unbalanced battery system can exhibit highly undesirable behaviour. Assuming that the cells of a battery system of an electric vehicle are not completely 'balanced', for example all cells are half full, but only one cell is almost empty, the state of charge display shows the user the sum of all states of charge of all cells. Since each individual cell has a small storage capacity, the single almost empty cell does not have a significant influence on the value of the state of charge. The driver therefore sees that he has a half-full 'tank', and drives off. However, the one almost empty cell is now discharged in the same way as all other cells. Since this cell would now fall out of the permitted voltage range if it were to be further discharged, the BMS must react. The simplest option would be to stop discharging the entire battery. In fact, this is a common reaction of such systems. However, in the case described here, this would lead to braking or at least coasting of the vehicle, which might put the driver at additional risk. The requirement is therefore derived:

ECS-LIB 5: AS AN engineer, I WANT to ensure that when the voltage and temperature limits of individual cells are reached, the overall system is still controllable SO THAT the overall system does not enter an uncontrolled and unsafe state.

ECS-LIB 5 is a requirement that must be met at the level of the entire system. This is because knowledge about the entire system is needed to assess what is an uncontrolled or unsafe state for an entire system. Since knowledge about the specific application is needed here, ECS-LIB 5 is most appropriately assigned to the EMS.

Let us now look at the behaviour of lithium ion battery cells at different C-Rates. From ECS-LIB 4, we can already conclude that lithium ion battery cells are not suitable for very high C-Rates. This is because a high current also leads to an increasing temperature, which can cause damage within a cell. In fact, lithium ion battery cells are suitable for charge rates of up to 2C and discharge rates of up to 4C, where the higher C-Rates have a strong influence on ageing. In Fig. 8.26, we have performed a similar experiment to that seen in Fig. 8.16. However, the C-Rates range from 0.2C to 2.7C. In this experiment, we measured the energy content rather than the discharge time. That is, we first fully charged a lithium ion battery cell and then discharged it with a constant current. By measuring the open-circuit voltage and current, we were able to draw a curve $U_{OVC}(\kappa)$. The plot corresponds to the one in Fig. 8.25, but the discharge current was varied and the cell chemistry remained the same. In this experiment, $LiCoO_2$ was the cathode material and C the anode material (Reddy, 2011).

We also see here that the amount of energy we can extract from a lithium ion battery cell depends on the charge rate. The higher the charge rate, the lower the

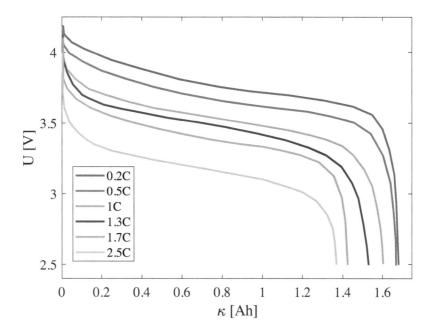

Fig. 8.26 Discharge Capacity of an Lithium Ion Battery cell with $LiCoO_2$ as an cathode and C as an anode material at different C-Rates.

amount of energy. Above a charge rate of 1C, the effect is strong. As can be seen in Fig. 8.26, the difference in capacity between 0.2C and 0.5C is no longer noticeable. The difference in capacity between these maximum values and a charge rate of 1C is 5%. The difference in capacity between 0.2C and 2.6C is 20%.

This observation is different from that regarding the lead acid battery. There we could distinguish between two areas. At low charging rates, the amount of energy was approximately the same. At high charging rates, the charging rates differed greatly. It must be clear to us that the charging rates used in the experiment in Fig. 8.16 were considerably higher than the charging rates used here.

With a lead acid battery, the limiting element was the supply of active material in the electrolyte. Specifically, it had to be ensured that sufficient sulphuric acid could come into contact with the electrodes. Lithium ion battery cells do not have such an element. As can be seen in the primary reaction in eqn (8.54), a lithium ion separates from the active material on one electrode and migrates to the other electrode to combine with the other material on the electrode. Limiting factors can therefore only be transport or absorption. However, we have not mentioned any side reaction that binds a lithium ion during transport. Therefore, it can only be an effect during storage. And this is exactly the reason for the different energy contents. At higher charge rates, the lithium ions do not find a place to store themselves, or do not find a way to release themselves from the material. If we then add more energy to the system, there will be unwanted heating, which we want to avoid because of ECS-LIB 4.

In summary, it can be said that lithium ion battery cells are ideally only used up to a maximum of 1C. It is possible to use higher C-Rates. However, you should check the manufacturer's specifications. 1C does not sound like much at first. But since the voltage is significantly higher than with lead acid battery cells, the extractable power is also higher. At the same time, the current we have to carry to transport this power is lower.

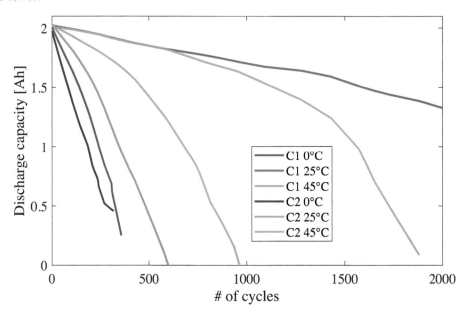

Fig. 8.27 Number of cycles taken from a lithium ion battery cell at different operating temperatures. The discharge rates were 1C and 2C respectively.

In Fig. 8.6, we had already seen that the number of cycles that can be discharged from lithium ion battery cells depends on the DoD and can be significantly higher than that of lead acid battery cells. In addition to the dependence on the DoD, the number of cycles also depends on the temperature of the cells. In Fig. 8.27, we observe the following experiment. We take a lithium ion battery cell and cycle it at different temperatures. We count the number of cycles (Wang *et al.*, 2016). The highest number of cycles is obtained at a temperature of 25 °C and a discharge rate of 1C. If we do not change the temperature, but double the number of cycles, we see that for this cell the number of cycles is significantly reduced. This is also seen with all other temperature pairs. So the cycle life depends not only on the DoD, but also on the charge rate. However, we recognize another effect in this measurement. We have the longest cycle life at 25 °C. If we increase the temperature to 45 °C, the life span is reduced even if we leave the charge rate at 1C. But if we cool down the battery cell, and in this case we go down with the temperature to 0°C, then we see that even then the life span reduces considerably. Now, 0 °C and 45 °C are perhaps two extreme temperature differences and we can ask ourselves whether the ratio is not a little better at 13 °C or 37 °C. In fact, analyses show that for different temperature ranges, the life expectancy of the battery cell is slightly better. They show that there is a temperature optimum for different lithium ion battery cells and that this value depends on the construction of the cell and its chemistry. However, the optimal temperature window for most lithium ion cells is in the range of 20 °C to 25 °C.

With the calendar ageing of lithium ion cells, we had already seen in Fig. 8.7 that a lower temperature always correlates with a longer lifetime. This is different for the temperature dependence of the cycle ageing. But since both ageing effects overlap, we have to set a requirement for the operating temperature that takes into account both observations:

ECS-LIB 6: AS A User, I WANT to ensure that the cell temperature in a temperature range of 20°C to 25°C SO THAT the calendar ageing does not accelerate too much and the cycle life remains high.

The calendar life depends not only on the temperature, but also on the voltage. In lithium ion battery cells, we observe that a high cell voltage leads to a faster decomposition of the active material. We can explain this by the fact that at higher temperatures or higher voltages there is more thermal energy in the active material, which leads to an acceleration of decay reactions. This, taken together with the observation that cycle life depends on the DoD and is greater when the DoD is smaller, leads to another requirement for the operational management of lithium ion battery cells:

ECS-LIB 7: AS A User, I WANT to keep the mean state of charge of the battery as low as possible SO THAT the influence of the state of charge on calendar ageing and cycle ageing is kept low.

This requirement has an impact on the system design and the charging strategy. If the capacity is defined, it is the EMS that tries to keep the average state of charge low with suitable measures. However, there is a contradiction here that must be solved for the respective application. If the average state of charge is to be as low as possible, the capacity of the battery must be greater than would actually be necessary for the application. However, a larger capacity also increases the costs of the storage system. How this conflict is resolved in practice depends on the application.

The handling of lithium ion batteries requires more focus on the safety of the operating parameters than with a lead acid battery. On the other hand, the energy density and voltage level of lithium ion batteries is significantly higher. Unlike lead acid batteries, which can be discharged at very high C-Rates even for a short period of time, the C-Rates are limited. Nevertheless, lithium ion batteries are becoming more widely used. In the next section, we will look at the design of such an application based on lithium ion batteries.

8.5   Application example: Residential solar storage system

In this section, we want to design a residential solar storage system that uses lithium ion cells as storage technology. A residential solar storage system combines a solar power plant with a battery. The basic idea is that a surplus of solar power is temporarily stored in a battery to supply the consumers in the household at times when not enough solar power is produced. This helps to minimize energy costs.

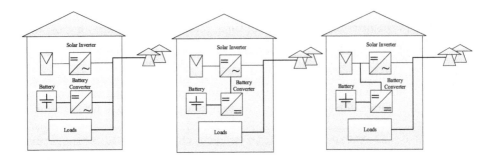

Fig. 8.28 Topologies of residential photovoltaic storage systems. A distinction is made between systems in which the battery is charged via the AC bus and those in which it is charged directly from the solar power system via the DC bus.

Fig. 8.29 An inverter, as shown in Fig. 8.28, consists of a DC/DC converter and an inverter.

Again, We have several degrees of freedom to realize the system. The main system components are the solar power system, the loads in the household, the battery, and the grid. In Fig. 8.28, we have shown three common realizations. The most modular version is the first configuration. Here, all system components are linked via an AC bus. This means that both the solar power system and the battery are equipped with an inverter that is connected to the household power grid, the AC bus. The advantage of this architecture is that the solar power system and the battery system are independent components. The power of the solar inverter is based on the size of the solar power system. The power of the battery inverter is based on the charging power of the battery. Both components can be optimized on their own. The disadvantage is that the losses are higher when storing solar power, since the solar power must first be transformed into AC, then back into DC and then back into AC for use. The other two systems try to compensate for this disadvantage. In both systems, the battery is directly connected to the solar power system via a DC bus. In the first case, the battery is connected to the DC bus of the solar inverter via a DC/DC converter. In the second variant, the DC/DC converter is connected between the solar power system and the solar inverter. In both cases, the loss for storing the energy is reduced.

Exercise 8.16 The inverters shown in Fig. 8.28 are to be understood as a combination of a DC/DC converter and an inverter (Fig. 8.29). In the solar inverter, the DC/DC converter is responsible for the MPPT. It ensures that current and voltage are taken from the solar power system in such a way that the power taken is at a maximum and transfers this power to the DC link. The inverter takes the power from this DC link and feeds it in to the AC grid. In the battery inverter, the DC/DC converter is used to transfer the battery voltage, which depends on the state of charge, to the DC link voltage. The inverter takes the power here and transfers it to the AC grid.

For simplicity, we make the assumption that the loss coefficients for DC/DC converters and inverters are the same, regardless of which system component they are to be used for. In Tab. 8.7, we have shown the coefficients. We consider the use case of feeding $P = 3{,}500$ W solar power into the battery first and then discharging it completely. How much power arrives in the household and what is the resulting efficiency? This is to be calculated for all three systems.

Table 8.7 loss coefficients of inverter and DC/DC converter.

component	$a[\mathrm{W}]$	$b[.]$	$c[\frac{1}{\mathrm{W}}]$
DCDC	100	0.01	10^{-8}
ACDC	50	0.02	$1.5 \cdot 10^{-8}$

Solution: We start with the AC system, the first of the three systems. Transfer from solar plant to DC bus:

$$P_{\mathrm{DCbus}} = P - \left(a_{\mathrm{DCDC}} + b_{\mathrm{DCDC}}P + c_{\mathrm{DCDC}}P^2\right) = 3{,}364 \text{ W}.$$

Transfer to the AC bus:

$$P_{\mathrm{DCAC}} = P_{\mathrm{DCbus}} - \left(a_{\mathrm{ACDC}} + b_{\mathrm{ACDC}}P_{\mathrm{DCbus}} + c_{\mathrm{ACDC}}P_{\mathrm{DCbus}}^2\right) = 3{,}247.5 \text{ W}.$$

Transfer from the AC bus to the DC link of the battery inverter:

$$P_{\mathrm{DCbus}} = P_{\mathrm{DCAC}} - \left(a_{\mathrm{ACDC}} + b_{\mathrm{ACDC}}P_{\mathrm{DCAC}} + c_{\mathrm{ACDC}}P_{\mathrm{DCAC}}^2\right) = 3{,}132.3 \text{ W}.$$

Transfer from DC link to battery:

$$P_{\mathrm{bat}} = P_{\mathrm{DCbus}} - \left(a_{\mathrm{DCDC}} + b_{\mathrm{DCDC}}P_{\mathrm{DCbus}} + c_{\mathrm{DCDC}}P_{\mathrm{DCbus}}^2\right) = 3{,}009 \text{ W}.$$

Transfer from the battery to the DC link:

$$P_{\mathrm{DCbus}} = P_{\mathrm{bat}} - \left(a_{\mathrm{DCDC}} + b_{\mathrm{DCDC}}P_{\mathrm{bat}} + c_{\mathrm{DCDC}}P_{\mathrm{bat}}^2\right) = 2{,}870 \text{ W}.$$

Transfer from DC bus to AC bus—that is, to load:

$$P_{\mathrm{DCAC}} = P_{\mathrm{DCbus}} - \left(a_{\mathrm{ACDC}} + b_{\mathrm{ACDC}}P_{\mathrm{DCbus}} + c_{\mathrm{ACDC}}P_{\mathrm{DCbus}}^2\right) = 2{,}763 \text{ W}.$$

The total efficiency of this storage process is $eta_{\mathrm{conf1}} = \frac{2{,}763 \text{ W}}{3{,}500 \text{ W}} = 78.94\%$.

Now we perform the calculation for the second system. Here, the power is first transferred to the DC link via the DC/DC converter of the solar inverter. This is followed by a transfer via the DC/DC converter of the battery inverter to the battery. This is used again for the feed-in, and then the feed-in takes place via the inverter.

DC bus:

$$P_{\text{DCbus}} = P - \left(a_{\text{DCDC}} + b_{\text{DCDC}}P + c_{\text{DCDC}}P^2\right) = 3,364 \text{ W}.$$

Transfer from DC link to battery:

$$P_{\text{bat}} = P_{\text{DCbus}} - \left(a_{\text{DCDC}} + b_{\text{DCDC}}P_{\text{DCbus}} + c_{\text{DCDC}}P_{\text{DCbus}}^2\right) = 3,231 \text{ W}.$$

Transfer from the battery to the DC link:

$$P_{\text{DCbus}} = P_{\text{bat}} - \left(a_{\text{DCDC}} + b_{\text{DCDC}}P_{\text{bat}} + c_{\text{DCDC}}P_{\text{bat}}^2\right) = 3,098 \text{ W}.$$

Transfer from DC bus to AC bus—that is, to load:

$$P_{\text{DCAC}} = P_{\text{DCbus}} - \left(a_{\text{ACDC}} + b_{\text{ACDC}}P_{\text{DCbus}} + c_{\text{ACDC}}P_{\text{DCbus}}^2\right) = 2,986 \text{ W}.$$

The overall efficiency of this storage process is $eta_{\text{conf2}} = \frac{2,986 \text{ W}}{3,500 \text{ W}} = 85.33\%$.

By leaving out two converter stages, the efficiency increases as expected. More power reaches the loads. In the third system, the number of transfers is reduced again; the solar power is initially stored directly in the battery via the DC/DC converter. When discharging, the DC/DC converter of the solar inverter can also be neglected if the output voltage of the DC/DC converter sets the voltage equal to the DC link voltage.

Transfer from the solar power generator to the battery:

$$P_{\text{bat}} = P - \left(a_{\text{DCDC}} + b_{\text{DCDC}}P + c_{\text{DCDC}}P^2\right) = 3,364 \text{ W}.$$

Transfer from the battery to the DC link:

$$P_{\text{DCbus}} = P_{\text{bat}} - \left(a_{\text{DCDC}} + b_{\text{DCDC}}P_{\text{bat}} + c_{\text{DCDC}}P_{\text{bat}}^2\right) = 3,231 \text{ W}.$$

Transfer from DC bus to AC bus—that is, to load:

$$P_{\text{DCAC}} = P_{\text{DCbus}} - \left(a_{\text{ACDC}} + b_{\text{ACDC}}P_{\text{DCbus}} + c_{\text{ACDC}}P_{\text{DCbus}}^2\right) = 3,116 \text{ W}.$$

The total efficiency of this storage process is $eta_{\text{conf3}} = \frac{3,116 \text{ W}}{3,500 \text{ W}} = 89.04\%$.
We see that in terms of efficiency, system configuration 3 has the best efficiency.

We start with a focus on the applicable requirements. Of course, the general basic requirements G 1 to G 8 apply, as well as the general requirements for electrical storage systems, ES 1 to ES 5. Since we want to realize an electrochemical storage system with lithium ion battery cells, requirements ECS 1 to ECS 5 and ECS-LIB 1 to ECS-LIB 8 apply. The second step is to get a collection of user requirements. We want to focus on a few here. The first requirement is derived from the objective of using the system.

RSS 1: AS A customer, I WANT the energy mix of solar power and electricity from the utility to be optimally mixed SO THAT my household's energy costs are minimized.

At first glance, this requirement seems obvious. But there could also be other goals. For example, it could be a goal that as much solar power as possible shall be consumed by the household. This could be reasonable, if no feed-in tariff is available. On the other hand, if feed-in tariffs and electricity price vary in time, the residential solar storage system could participate actively on the energy market. However, to simplify our example, we assume that the feed-in tariff and electricity price do not vary.

We also add another user requirement, which requests the storage system to perform peak shaving:

RSS 2: AS A grid operator, I WANT the feed-in power at the grid connection point to always be less than 60% of the installed power of the solar power system SO THAT the distribution grid is relieved during the midday period when solar power production reaches its maximum.

Solar power systems are characterized by their maximum feed-in capacity. If a 8 kWp system is installed, this means that the system achieves a maximum feed-in power of 8 kW. In order to be compliant with RSS 2, the maximum feed-in power shall be less than 8 kW$_p$ · 60% = 4.8 kW. The intention of this requirement is to reduce the noon-time feed-in peak by forcing the solar storage system to shift power injection to a later time. Studies have shown that voltages rise in the distribution networks on sunny days, because many solar power plants feed large amounts of solar power into the network at the same time (Stetz, 2014; Von Appen *et al.*, 2014). With RSS 2, this effect shall be reduced. Thus, the purpose of this requirement is to stabilize the distribution network.

Peak shaving is not only a requirement for stabilizing the grid; the user may also have an interest in performing peak shaving. In some countries, households also have to pay for higher power connections; by applying peak shaving, a photovoltaic system can be installed with a higher nominal power than the grid connection may allow.

In some countries like Germany, Austria, and Switzerland, all households have a three-phase domestic grid connection. Since there are only a few households with consumers who need all three phases at the same time, asymmetrical consumption can occur—that is, more electricity is consumed on some phases, less on others. If the feed-in of the solar power system or the storage is also single-phase, the voltage level of the three-phase can become asymmetric (Von Appen *et al.*, 2012). For this reason, we introduce RSS 3 as a requirement for single-phase electricity storage systems.

RSS 3: AS A Grid operator, I WANT the voltage unbalance between two phases to be below 4.7 kVA SO THAT the solar power feed-in of many neighbouring plants does not lead to a voltage unbalance in the distribution grid.

Exercise 8.17 In this task, we want to derive additional requirements to the requirements RSS 1 to RSS 4. In this case, three requirements each are to be determined for the product manager and the installer. (First hint: It makes sense to look at the role description of the actors from Chapter 4 again. Second hint: This task only requires imagination. Of course, if you have not taken on one of these roles, you may not know what requirements to have in that role. But you can imagine what one might need. There is no right or wrong here. Don't be afraid of the white paper.)

Solution: The product manager manages the product portfolio of their division. For them, solar power storage is one product among many, and they have to see that this product fits into their strategy. They are interested in marketing the product well. As they know that each country has different grid codes, they want to keep the technical barrier to entry as low as possible.

RSS 4: AS A product manager, I WANT the feed-in characteristics of the solar power storage system to be adapted with little effort SO THAT we can sell the solar power storage system quickly in different countries.

Since the grid codes are similar in many respects, this task means that the developers have to recognize the similarities, identify variants, and optimize the process of qualification.

Next, the product manager looks at the living conditions in the different countries. These depend on many factors: the culture, the history, the food, and the climate. It is clear that in countries where the temperature is above 30 °C for most of the year, the electricity consumption is different from those in which the temperature is between 15 °C and 25 °C throughout the year. Solar power production is also different. The regulatory framework is different, solar irradiation is different, consumption and grid connection conditions are different. As a result, the product manager does not feel able to define one system size for all countries, and instead requests:

RSS 5: AS A product manager, I WANT the maximum power of the solar power system and battery to be scalable SO THAT I can bring out a product modification for different markets with little effort.

This approach has the charm of being able to quickly adapt to the needs of the market. However, it has the disadvantage, like any modularity, that the product costs increase because the interfaces become more complex.

Since a storage system is not only defined by the maximum power of the various components, but also by the storage capacity, it seems advisable to demand a certain flexibility here as well:

RSS 6: AS A product manager, I WANT the battery size to able to be adjusted at a later time SO THAT customers can also upgrade their storage after purchasing the system.

Compliance with this requirement ensures that the appropriate storage size can be determined with the customer during the sales talk. In addition, they can be given the prospect of being able to install additional storage at a later date.

We now turn to the requirements of the installer. For the installer, it is important to install as many devices as possible at the same time. At the same time, he wants to keep the use of personnel to a minimum. Perhaps his installation company does not have many employees or he wants to be able to serve different customers in parallel. Therefore, he initially has a very simple requirement:

RSS 7: AS AN installer, I WANT the equipment to be delivered in components that have a maximum weight of 20 kg SO THAT a single person is able to carry and install them.

The second requirement of the installer relates to the connection technology. Here, too, he would like to keep the effort and the required technical know-how as low as possible, because then he can use his employees flexibly and does not have so much effort himself when installing a device. He therefore formulates the requirement as follows:

RSS 8: AS A installer, I WANT the connections to be easily accessible and not different from the connection technology of a solar power system SO THAT my employees easily understand how to connect the device.

And since everything is so simple, he has one last requirement to make his job even easier:

RSS 9: AS AN installer, I WANT the system to have a diagnostic function SO THAT I am alerted to errors after the installation is complete.

The diagnostic function also provides benefits in terms of customer acceptance. He can then show the customer directly that the system reports on its own that the installation was successful.

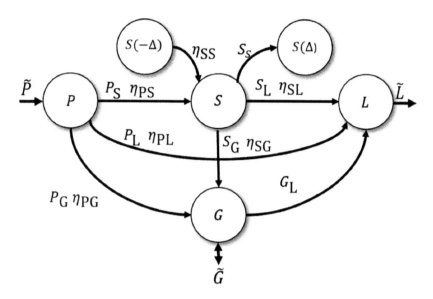

Fig. 8.30 Power flow diagram of a residential solar storage system.

After we have compiled the requirements, we turn to the power flow diagram. In Fig. 8.30, the power flow diagram is shown. It is independent of the technical realization shown in Fig. 8.28.

For the description of the power flows in this system, we start with the production of solar power, P. Since the solar power production is given, we label this as $\tilde{P}$. The equation for the solar power node reads as:

$$\tilde{P} \geq P_{\mathrm{L}} + P_{\mathrm{S}} + P_{\mathrm{G}}. \tag{8.55}$$

The amount of solar power transferred from the generator to the grid P_{G}, the loads in the households P_{L}, and the storage P_{S} need to be equal or less than the available solar power $\tilde{P}$. With a solar power system, we have the advantage that we do not have to use all the power available through irradiation. The light separates the charge carriers in the p–n junction: if they do not recombine due to a current flow, they recombine due to their thermal diffusion. Hence, any solar inverter is capable of reducing the amount of power taken from the generator by forcing the maximum power point tracker away from the maximum, without harming the photovoltaic system.

The requested load profile is labelled as $\tilde{L}$. We assume, that the user always want the loads to be covered:

$$\tilde{L} = \eta_{\mathrm{SL}} S_{\mathrm{L}} + \eta_{\mathrm{GL}} G_{\mathrm{L}} + \eta_{\mathrm{PL}} P_{\mathrm{L}}. \tag{8.56}$$

The load $\tilde{L}$ is covered by the solar generator P_{L}, the storage S_{L}, and the grid G_{L}. The amount of power from solar and the storage might be limited, but in our case the grid is a unlimited power source. Therefore, we can assume that the loads are always covered. This may not be the case if we use this combination of system components for an off-grid or island grid application. In such cases, we need to change the equation into an inequality.

The equation of the storage node is determined by the power inflow and outflow:

$$0 = \eta_{PS}P_S + \eta_{SS}S_S(-\Delta) - (S_S + S_G + S_L).\tag{8.57}$$

The storage can be charged from the solar generator P_S and discharged to cover the loads S_L or feed power into the grid S_G. $S_S(-\Delta)$ and S_S reflect the property of any storage technology to transfer power from the past to the future.

We do not allow the storage from the grid to be charged, since its economically viability depends on the feed-in tariff, the electricity price, and the transfer losses, and in our case this power flow has no economical value.

Exercise 8.18 Assuming that the sale of stored electricity would always be possible—that is, we would be allowed to charge the storage from the grid and later sell this energy again—then the question still remains whether this also makes sense economically. In this exercise, we assume that the feed-in tariff is fixed.

The feed-in tariff is $c_{fi} = 12\frac{\$ct}{kWh}$. The efficiency of the charging process from the grid equals $\eta_{GS} = 98\%$ and the efficiency of the injection equal $\eta_{SG} = 98\%$. Self-discharge of the storage shall be neglected. What is the necessary electricity price c_{gc} to make the resale of solar electricity profitable?

Solution: In order for the charging of the storage via the grid and the subsequent consumption to be profitable, the price difference between c_{fi} and c_{gc} must be large enough to compensate for the charging and final charging losses:

$$c_{fi} \cdot \eta_{SG} \cdot \eta_{GS} - c_{gc} \geq 0$$

$$\Rightarrow c_{gc} \leq c_{fi} \cdot \eta_{SG} \cdot \eta_{GS} = 11.52\frac{\$ct}{kWh}.$$

In this example, we are using an electricity price of $24\frac{\$ct}{kWh}$. Therefore, a power flow from the storage to the grid will violate RSS 1.

Eqns (8.55), (8.56), and (8.57) describe the power flows, which can be observed in this system. In the next step, we formulate the boundary conditions derived from the user requirements.

RSS 2 is represented by the following equation:

$$-AC_{max} < \eta_{PG}P_G + \eta_{SG}S_G - G_L \leq 0.6P_{max}.\tag{8.58}$$

AC_{max} is the maximum power flow provided by the grid connection and P_{max} is the maximum feed-in power of the solar generator.

Requirement RSS 3 depends on the technology of the residential solar storage system. The power flow between the solar generator and the storage is not relevant in the case of a DC system. In this case, the total number of power flowing into the grid needs to be less than 4.6 kW. Hence, the boundary conditions for a DC system equals:

$$\eta_{PG}P_G + \eta_{PL}P_L + \eta_{SL}S_L + \eta_{SG}S_G \leq 4.6 \text{ kW}.\tag{8.59}$$

Table 8.8 Equations describing the power flow diagram of a residential solar storage system.

Balances/requirements	Equations
Power flow of the solar plant	$\tilde{P} \geq P_{\mathrm{L}} + P_{\mathrm{S}} + P_{\mathrm{G}}$
Power flow of the load node	$\tilde{L} = \eta_{\mathrm{SL}} S_{\mathrm{L}} + \eta_{\mathrm{GL}} G_{\mathrm{L}} + \eta_{\mathrm{PL}} P_{\mathrm{L}}$
Power flow of the storage node	$0 = \eta_{\mathrm{PS}} P_{\mathrm{S}} + \eta_{\mathrm{SS}} S_{\mathrm{S}}(-\Delta) - (S_{\mathrm{S}} + S_{\mathrm{G}} + S_{\mathrm{L}})$
Grid connection condition	$-AC_{\max} < \eta_{\mathrm{PG}} P_{\mathrm{G}} + \eta_{\mathrm{SG}} S_{\mathrm{G}} - G_{\mathrm{L}} \leq 0.6 P_{\max}$
Voltage unbalancing in case of a DC system	$\eta_{\mathrm{PG}} P_{\mathrm{G}} + \eta_{\mathrm{PL}} P_{\mathrm{L}} + \eta_{\mathrm{SL}} S_{\mathrm{L}} + \eta_{\mathrm{SG}} S_{\mathrm{G}} \leq 4.6 \text{ kW}$
Voltage unbalancing in case of an AC system	$\eta_{\mathrm{PG}} P_{\mathrm{G}} + \eta_{\mathrm{PL}} P_{\mathrm{L}} + \eta_{\mathrm{PS}} P_{\mathrm{S}} \leq 4.6 \text{ kW}$
	$\eta_{\mathrm{SG}} S_{\mathrm{G}} + \eta_{\mathrm{SL}} S_{\mathrm{L}} \leq 4.6 \text{ kW}$

For AC systems, the power flow between the solar power plant and the storage system is relevant. In the case of AC systems, we have two devices which can feed power into the grid. The solar generator and the storage system act independently and they also may inject power on different phases into the grid. Therefore, two boundary conditions are necessary: one for the photovoltaic generator,

$$\eta_{\mathrm{PG}} P_{\mathrm{G}} + \eta_{\mathrm{PL}} P_{\mathrm{L}} + \eta_{\mathrm{PS}} P_{\mathrm{S}} \leq 4.6 \text{ kW}, \tag{8.60}$$

and one equation for the storage system,

$$\eta_{\mathrm{SG}} S_{\mathrm{G}} + \eta_{\mathrm{SL}} S_{\mathrm{L}} \leq 4.6 \text{ kW}. \tag{8.61}$$

Note, that in eqns (8.59) to (8.61), the power flow is always reduced by the efficiency, since this is the amount of power which is transferred into the grid.

In Tab. 8.8, the equations for the power flow and the boundary conditions for systems with AC bus or DC bus are shown again. The different technical realizations are adjusting the efficiencies and the choice of boundary conditions.

The next step is to determine the size of the storage and the necessary charge and discharge power. Therefore, we need to analyse the load and production profile. In Fig. 8.31, the load profile of a private household is shown. Fluctuations have been removed by averaging over the time and the same days. In order to get a mean value for Saturday at 15:00, we took the values of the two Saturdays before and after and the values from 14:30 to 15:30. This step is necessary, since load profiles reflects the behaviour of the persons in this household, and this behaviour is similar on the same days, but may differ between the different working days and obviously the weekend.

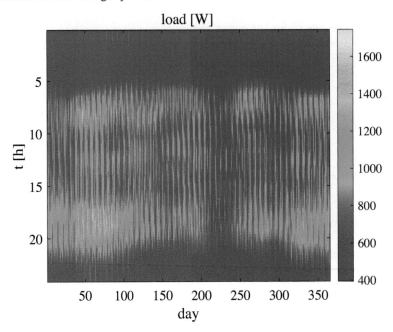

Fig. 8.31 Load profile of a household. Each point in the dataset has been averaged over the same day and the same hour, to remove fluctuations.

From Figure 8.31 one can read that this household has a base load of 400 W. This amount of power is always needed, regardless of the time of day or season. We can also see that the consumption is higher during the morning at 08:00 and the noon time after 12:00.

The averaged data give us some information about the way people in this household lives. A noticeable observation is that a first consumption maximum occurs between 05:00 and 10:00. It seems obvious to link this with breakfast. However, consumption is higher in the winter months than in the summer months. It can be assumed that hot water is also produced by electricity here. There is another maximum at lunchtime, but it is not as pronounced. Only at around 20:00 does consumption increase again. The dependence on the season is somewhat greater, which is probably due to the demand for electric light. Furthermore, one can see that every sixth and seventh day has a higher consumption. This may be due to the differences in behaviour at the weekend.

Due to the high base load and the strong consumption in the evening and morning hours, this household is very well suited for the application of a home storage system. At midday, when the sun is at its strongest, relatively little consumption is observed. The household can only use a little of the solar electricity directly. The high base load of approximately 400 W ensures that there is consumption in the evening and morning hours. Let us now turn to the production profile.

A solar production profile is shown in Fig. 8.32. Again, we have removed fluctuations by averaging over the hours and the days, but this time, we took the two days before and after, since the weather is supposed to be similar up to two days before

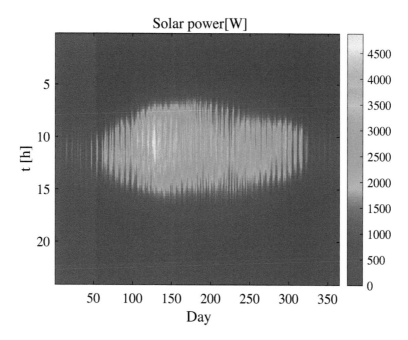

Fig. 8.32 Solar production profile of a household. Each point in the dataset has been averaged over two days and one hour, to remove fluctuations.

and after a particular day. This picture shows that the maximum solar production is at around noon and this is quite high from February to October.

If we compare Figs 8.31 and 8.32, we see that production does not correlate with demand. Consumption is high in the morning and evening, while production peaks at noon. In the colder and darker winter months, consumption is high, but the sun is generally too weak, so production is not high enough here either. Although this relationship between excess production in summer and demand in winter exists, residential photovoltaic storage systems are not used as seasonal storage. Their use is usually limited to compensating for a shift in production and load over one or more days of the week.

We can already estimate the smallest maximum charge rate from the RSS 2 requirement. If the storage system must always be able to ensure that the feed-in power P_S does not exceed 60% of the photovoltaic power, it follows that the 40% excess power must be charged into the storage system. We can therefore already derive a smallest maximum charging rate for the storage from the size of the photovoltaic generator:

$$P_S^{\max} \geq 40 \ \% \cdot P_{\max} \cdot \frac{\eta_{PG}}{\eta_{PS}}. \tag{8.62}$$

This power is necessary to satisfy NA 2 under the assumption that we always want to be able to charge the battery with its maximum power.

Exercise 8.19 The solar generator has a peak power of 8 kW$_p$ with an efficiency of $\eta_{PG} = 85\%$. The battery and the used power electronics have a combined efficiency of $\eta_{PS} = 83\%$. What is the maximum required charging power of the battery?

Solution: We assume that the battery must be able to store the peak power of the solar generator. However, since the solar inverter and battery charge controller already have transmission losses, the required charging power does not correspond to the full peak power of the generator, but can be corrected for the losses:

$$P_S^{max} \geq 40\ \% \cdot \frac{0.85}{0.83} \cdot 8\ kW_p = 3.28\ kW_p.$$

Instead of analysing the entire annual course of load and solar production, we can estimate these quantities by analysing the production and load of a single day. For the determination of the discharge power and the storage capacity, the typical load flow of a residential solar storage system should be considered. As can be seen in Fig. 8.33, the storage is charged during the day (κ in Fig. 8.33). Discharging begins as soon as the sun has set, first covering the loads in the evening and then the base load during the night, which is always there. In order to use the storage system sensibly, it should therefore be at least large enough to cover the base load.

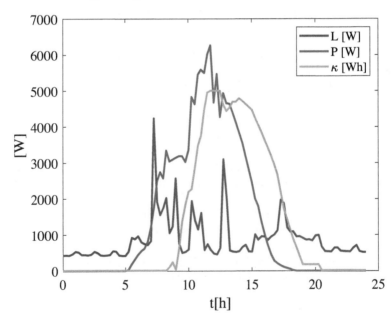

Fig. 8.33 Power flows and storage capacity of a solar power storage system: P corresponds to the power of the solar system, L describes the household consumption. The state of charge is represented as κ.

Exercise 8.20 We wanted to estimate the required capacity of an household. Therefore we took some measurements and generated the table shown in Tab. 8.9. From these data, we want to derive the needed storage capacity.

Table 8.9 Power consumption of a household after sunset.

Time	18:00	20:00	21:00	23:00	06:00
Power	500 W	400 W	300 W	300 W	300 W

Solution: To determine the storage capacity, we assume that the power consumptions are constant between the measurement points and integrate over time:

$$\kappa = \int_{t=18:00}^{t=6:00} L \, dt$$
$$= 2 \text{ h} \cdot 0.5 \text{ kW} + 1 \text{ h} \cdot 0.4 \text{ kW} + 10 \text{ h} \cdot 0.3 \text{ kW}$$
$$= 4.4 \text{ kWh}.$$

In the next exercise, we will look at a battery with two different cathode materials. One cell uses lithium iron phosphate (LiFeP, or LFP for short). The other cell uses nickel manganese cobalt (NiMnCo, or NMC for short). Both cells are available as cylindrical cells in a 26650 format—that is, 26 mm in diameter and 650 mm in height. In Tab. 8.10, some technical data of the cells are shown.

Based on these technical data, the required number of cells can now be determined.

Table 8.10 Technical data of the lithium ion battery cells used in this application example.

	LFP	NMC
Voltage range	$[2.6 \, V_{DC}; 3.65 \, V_{DC}]$	$[3 \, V_{DC}; 4.2 \, V_{DC}]$
Nominal capacity	3.3 Ah	4.4 Ah
Maximum charge power	$[0.5C; 1C]$	$[0.5C; 1C]$

Exercise 8.21 Over night, or better, during the dark phase of the day, the household needs a capacity of $\kappa = 4.4$ kWh. The average DoD of the cells should be 80%. What voltage range can be achieved with the cells shown in Tab. 8.10?

Solution: The energy content of the cells is approximated by the average voltage and the nominal capacity:

$$\kappa_{LFP} = \frac{2.6\,V_{DC} + 3.65\,V_{DC}}{2} \cdot 3.3\,Ah = 10.31\,Wh$$

$$\kappa_{NMC} = \frac{3\,V_{DC} + 4.2\,V_{DC}}{2} \cdot 4.4\,Ah = 15.8\,Wh.$$

Hence, we can calculate the number of cells for each technology to reach this amount of capacity:

$$N_{LFP} = (4{,}400\,Wh)/(10.31\,Wh) = 426.77 \approx 427$$

$$N_{NMC} = \frac{4{,}400\,Wh}{15.8\,Wh} = 278.48 \approx 279.$$

Note that in this case, we have to round up to ensure that we really have enough capacity. The maximum voltage results when these cells are completely connected in series.

$$427s1p : U_{LFP} \in [N_{LFP} \cdot U_{LFP}^{min}; N_{LFP} \cdot U_{LFP}^{max}] = [1{,}110.2\,V_{DC}; 1{,}558.55\,V_{DC}]$$

$$279s1p : U_{NMC} \in [N_{NMC} \cdot U_{NMC}^{min}; N_{NMC} \cdot U_{NMC}^{max}] = [837\,V_{DC}; 1{,}172.8\,V_{DC}].$$

The storage system should be designed so that the storage capacity can be upgraded (RSS 6). For this reason, the battery cells are not aggregated individually, but in modules. This is the usual system design that we already know from lead batteries. We set the voltage of these modules to approximately 48 V.

Exercise 8.22 The battery modules shall consist of two parallel strings with a nominal voltage of 48 V_{DC}. What is the storage capacity of the 48 V modules that use these cells?

Solution: The 48 V size refers to the average voltage level. In order to calculate the number of cells, we just need to divide this voltage by the average cell voltage. We need to round off, to ensure that we are not violating this average target voltage.

The result for the LFP modules is thus:

$$N_{LFP}^{module} = \frac{48\,V_{DC} \cdot 2}{3.65\,V_{DC} + 2.6\,V_{DC}} = 15.36 \approx 15 \Rightarrow 15s2p$$

$$\Rightarrow \kappa_{LFP}^{module} = 15 \cdot 2 \cdot 10.31\,Wh = 309\,Wh$$

$$\Rightarrow U_{LFP}^{module} = 15 \cdot [2.6\,V_{DC}; 3.65\,V_{DC}] = [39\,V_{DC}; 54.75\,V_{DC}].$$

And for the NMC modules, the following module properties result:

$$N_{NMC}^{module} = \frac{48\,V_{DC} \cdot 2}{4.2\,V_{DC} + 3\,V_{DC}} = 13.3 \approx 14 \Rightarrow 14s2p$$

$$\Rightarrow \kappa_{NMC}^{module} = 14 \cdot 2 \cdot 15.8\,Wh = 442.4\,Wh$$

$$\Rightarrow U_{NMC}^{module} = 14 \cdot [3\,V_{DC}; 4.2\,V_{DC}] = [42\,V_{DC}; 58.8\,V_{DC}].$$

The 48 V modules with NMC cells have a higher capacity than the modules using LFP cells. The voltage is determined by the cell chemistry, the capacity by the amount of active material. If the capacity per cell increases, the voltage level remains the same, but the energy content increases.

In this application, the maximum voltage of the battery system shall be below $850\,V_{DC}$. This voltage results from the maximum input voltage range of the inverter used. In the lower range, the voltage is freely selectable, since a DC/DC converter is used. However, the conversion ratio should be kept as low as possible to avoid additional losses.

Exercise 8.23 The storage system shall have a capacity reserve of 20%. The maximum battery voltage shall be less than $850\,V_{DC}$. What will a reasonable battery module configuration look like?

Solution: A capacity of $\kappa = \frac{4.4\,\text{kWh}}{0.8} = 5.5\,\text{kWh}$ is required. To find the number of modules, we divide this capacity by the capacity of the modules. Here, too, we have to round up, as we do not want less than the desired number of modules:

$$N_{\text{LFP}}^{\text{module}} = \frac{5.5\ \text{kWh}}{0.309\ \text{kWh}} = 17.79 \approx 18$$

$$N_{\text{NMC}}^{\text{module}} = \frac{5.5\ \text{kWh}}{0.442\ \text{kWh}} = 12.4 \approx 13.$$

If these battery modules are connected in series, the resulting voltage values equal:

$$18s1p\colon U_{\text{LFP}} = 18 \cdot [39\,V_{DC}; 54.75\,V_{DC}] = [702\,V_{DC}; 985.5\,V_{DC}]$$
$$13s1p\colon U_{\text{NMC}} = 13 \cdot [42\,V_{DC}; 58.5\,V_{DC}] = [546\,V_{DC}; 760\,V_{DC}].$$

The modules with NMC cells can be used in a 13s1p configuration; the modules with LFP cells cannot, because the voltage window exceeds the allowed maximum voltage. Therefore, we split the string into two halves and find that a 9s2p configuration is required here. The resulting battery system voltage is then:

$$9s2p\colon U_{\text{LFP}} = 9 \cdot [39\,V_{DC}; 54.75\,V_{DC}] = [351\,V_{DC}; 492.75\,V_{DC}].$$

The lifetime of the battery cells is of great importance in making reliable business case for a residential solar storage system. We expect to have 230 full cycles per year—that is, the storage is completely charged and discharged on 230 days each year. Of course, this value only applies to a storage system where the PV generator is large enough to produce enough surplus power. This is usually the case if the self-consumption is 30% without a storage system. For self-consumption values above 50%, the number of annual cycles is reduced, as there is not always enough surplus power that can be stored.

Exercise 8.24 How many cycles can the NMC and LFP cells provide under the assumption that the relationship between number of cycles and DoD shown in Fig. 8.34 is correct? (For simplicity, both cell chemistry shall have roughly the same cycle life.)

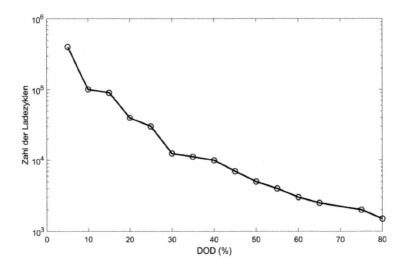

Fig. 8.34 Cyclelife of the two cells used in this example. For simplicity, we assume that both cell chemistries have the same cycle life.

Solution: To determine the cycle life of both cells, we need to calculate the average DoD:

$$\text{DoD}_{\text{LFP}} = \frac{4.4\,\text{kWh}}{5.688\,\text{kWh}} = 77.3\% \Rightarrow 3{,}125 \text{ cycles}$$

$$\text{DoD}_{\text{NMC}} = \frac{4.4\,\text{kWh}}{5.2\,\text{kWh}} = 84.6\% \Rightarrow 2{,}900 \text{ cycles}$$

At 230 full cycles per year, this would correspond to a cycle life of 13.58 years for LFP cells and 12.6 years for NMC cells.

In addition to cycle lifetime, the calendar life must also be considered. Relatively little information about the calendar life of a cell is available from the manufacturer or in the data sheets. A simple orientation of the estimated lifetime are the applications of the cells. Cells which are used for the energy supply of electrical tools or in consumer electronics are designed for a system life of three to five years. Cells used in electromobility have a system lifetime of 10 to 15 years.

Exercise 8.25 Both cells have a calendar life of 20 years. How long is the expected battery system life?
   Solution:

$$\text{SOH}_{\text{LFP}} : 1 - x_{\text{EOL}}\left(\frac{1}{20} + \frac{1}{13.58}\right) = 0 \Rightarrow x_{\text{EOL}} = \frac{1}{\left(\frac{1}{20} + \frac{1}{13.58}\right)} = 8.08 \text{ years}$$

$$\text{SOH}_{\text{NMC}} : 1 - x_{\text{EOL}}\left(\frac{1}{20} + \frac{1}{12.6}\right) = 0 \Rightarrow x_{\text{EOL}} = \frac{1}{\left(\frac{1}{20} + \frac{1}{12.6}\right)} = 7.73 \text{ years.}$$

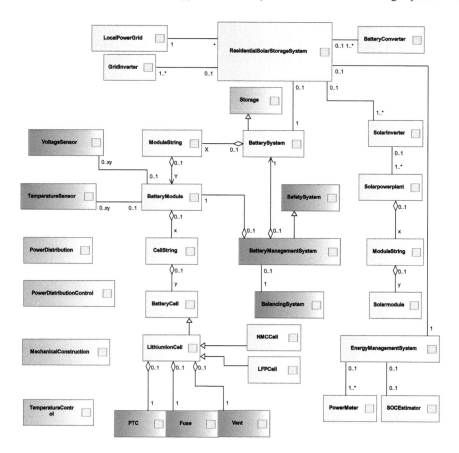

Fig. 8.35 System components of the residential PV storage system. Components that belong to the safety system are marked in orange. In order to maintain a clear overview, the connections to the basic system components are not shown.

We will conclude our investigation of the residential solar storage system by looking at the system components and the RTM. In Fig. 8.35, the system components are shown. The BatterySystem component plus the BatteryConverter are associated with the storage S node of the power flow diagram, the SolarInverter, and the components of the solar power plant of production P. The household is represented by the system component LocalPowerGrid. To keep the diagram a little clearer, we have not shown the connections of the system components to basic system components. We have also marked in orange those components that are responsible for safety. Since lithium ion batteries always carry the risk of a thermal runaway, additional system components have been integrated here. These are responsible for requirements ECS-LIB 2 and ECS-LIB 3. The safety concept works here on all three system levels of the battery system: the cell level, the module level, and the battery system level. At the cell level, it is the PTC, Fuse, and Vent components. At the module level, it is the BatteryManagementSystem that evaluates the data from the VoltageSensor and the `TemperatureSensor` to ensure compliance. The `BalancingSystem` as part of the `BatteryManagementSystem` is also responsible for the safety of the modules.

At the battery system level, the `BalancingSystem` and the `BatteryManagementSystem` reappear. Here it monitors the complete module thread and aggregates the data from the BMS operating at the module level.

Exercise 8.26 In Fig. 8.35, we were not shown connections to the basic system components. Which system components fulfil tasks of these components or are derived from them?
  Solution:

• `Power distribution`: This basic component includes all system components that distribute power. We do not list all cables, fuses, etc., here, but limit ourselves to the top-level system components. The two components that are also represented in the power flow diagram `BatterySystem`, `SolarInverter`, and `LocalPowerGrid` belong to this. One could also add all the subcomponents of the `Solarpowerplant` and the `BatterySystem`.

• `PowerDistributionControl`: The `BatteryManagementSystem` and the `EnergyManagementSystem` take control of the power flows here.

• `MechanicalConstruction`: All components that have a mechanical construction can be listed here. But it makes sense to concentrate on the most important components: `BatterySystem`, `Solarpowerplant`, `SolarInverter`, and `BatteryConverter`.

• `TemperatureControl`: The `Temperaturecontrol` is done via the `BatteryManagementSystem` in the `BatterySystem`. Furthermore, the two power electronic components `SolarInverter` and `BatteryConverter`, also have a `TemperatureControl`.

Since we do not want to restrict ourselves to a specific system topology in Fig. 8.35, we have assigned the components `SolarInverter` and `BatteryConverter` as independent components of the `ResidentialSolarStorageSystem`. Only a system description on a deeper, more detailed system level then makes it necessary to define the topology and adapt the description.

The `EnergyManagementSystem` also has a number of `Power Meters` in addition to the `SOC Estimator`. These are used to determine the power flows in the household. Often, one `PowerMeter` is already sufficient. This is connected to the household's grid connection point and only measures whether power flows into the grid or into the household. However, if billing functions are to be realized, or if the household network has a more complex structure, more than one `PowerMeter` may be needed.

Let us now look at the RTM. Here we only consider the requirements that are relevant for lithium ion batteries and for the application. However, we point out that requirements G 1 to G 8, as well as ES 1 to ES 5 and ECS 1 to ECS 5, also need to be checked.

Tab. 8.11 shows the RTM. It is noticeable that not all components from Fig. 8.35 can be found here. For clarity, we have hidden some subsystem components. The responsibilities of these components are dealt with in Exercise 8.27.

ECS-LIB 1 deals with the task of estimating the state of charge. This task is assigned to a `SOC Estimator`, as in all battery systems. Here we have defined the `SOC Estimator` as a subcomponent of the `EnergyManagementSystem`. The basic idea of

Table 8.11 Requirement traceability matrix for the residential PV storage system

	Lithium Ion Batteries							Application Requirements								
	ECS-LIB 1	ECS-LIB 2	ECS-LIB 3	ECS-LIB 4	ECS-LIB 5	ECS-LIB 6	ECS-LIB 7	RSS 1	RSS 2	RSS 3	RSS 4	RSS 5	RSS 6	RSS 7	RSS 8	RSS 9
GridInverter									X	X	X			X	X	
Solarpowerplant														X	X	
SolarInverter											X			X	X	
BatterySystem		X	X									X	X	X	X	
BatteryManagementSystem		X		X										X		X
EnergyManagementSystem		X			X	X	X	X						X	X	X
SOCEstimator	X															
PowerMeter																
BalancingSystem													X			
TemperatureControl					X											
MechanicalConstruction			X								X		X	X	X	

this assignment is that the SOC Estimator can also use knowledge about the application to estimate the battery condition. Furthermore, since operation management is implemented by the EnergyManagementSystem and this depends on the state of charge, an assignment of the SOC Estimator to the EnergyManagementSystem makes sense. However, there are also systems in which the SOC Estimator is assigned to the BatteryManagementSystem. This is justified by the fact that the BatteryManagementSystem has the knowledge about its subcomponents and this knowledge can be used for a better estimation. At this point, we see that there is not always a unique solution in system architecture, and sometimes the realization has an influence on it.

ECS-LIB 2 and ECS-LIB 3 deal with the important safety function in systems with lithium ion batteries. ECS-LIB 2 deals with the task of preventing the occurrence of thermal runaway. This task is assigned to the BatterySystem and its subcomponents, as well as the BatteryManagementSystem and the EnergyManagementSystem. If a thermal runaway does occur, damage can actually only be reduced by the construction of the BatterySystem and the MechanicalConstruction. Therefore, these two components are responsible for ECS-LIB 3. In order to avoid thermal runaway, ECS-LIB 4 must be complied with. This is also monitored by the BatteryManagementSystem.

We had already talked about the fact that the division of labour between the BatteryManagementSystem and the EnergyManagementSystem is based on the fact that the EnergyManagementSystem focuses on the application, which is always changing. Therefore, it is also the responsibility of the EnergyManagementSystem to comply with ECS-LIB 5 to ECS-LIB 7. Please note that the temperature management is also partly the responsibility of the TemperatureControl.

Exercise 8.27 In Tab. 8.11 some subcomponents of the battery system are not shown. However, these have responsibilities for the basic requirements ECS-LIB 1 to ECS-LIB 7. What does an RTM look like only for the components shown in Fig. 8.36?

Solution: We had already established that a clear assignment of the state of charge estimate is not possible, as there are good reasons for locating it both in the EnergyManagementSystem and in the BatteryManagementSystem. Therefore, we have entered both possibilities in our sample solution, Tab. 8.12. For the ECS-LIB 2 requirement, which relates to the

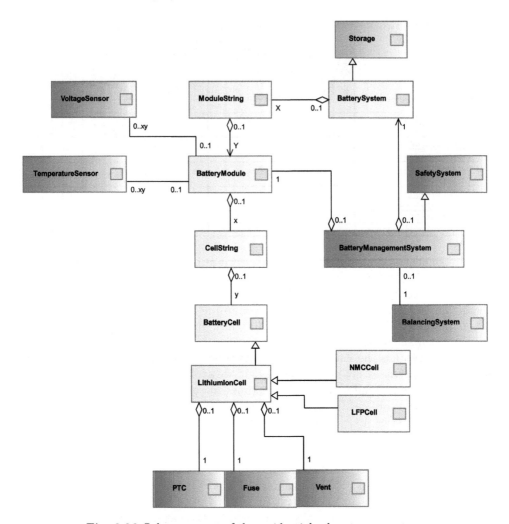

Fig. 8.36 Subcomponent of the residential solar storage system.

prevention of a thermal runaway, a number of components are responsible. The main component is the BatteryManagementSystem, which monitors the battery condition at module and system levels, using measurements from the VoltageSensor and TemperatureSensor.

If the BatteryManagementSystem does not detect a possible thermal runaway, the PTC and Fuse components can intervene at the cell level. In the unlikely event that an event does occur, the Vent and other constructive measures in the Battery Cell try to avoid a chain reaction. This is where the BatteryModule intervenes; this must be constructed in such a way that no chain reaction takes place.

The VoltageSensor and TemperatureSensor support the BatteryManagementSystem in complying with ECS-LIB 4. There are different implementations. In some battery systems, each individual cell is monitored. Other realizations monitor only sub-areas at string or module level.

ECS-LIB 5 and ECS-LIB 7 are assigned to the application and should be the responsibility of the EnergyManagementSystem.

ECS-LIB 6, temperature range compliance, is naturally assigned to the overall system. However, BatterySystem and BatteryModule take this requirement into account by building the internal mechanical construction and internal cooling accordingly.

Table 8.12 Requirement traceability matrix for battery system subcomponents.

	Lithium Ion Batteries						
	ECS-LIB 1	ECS-LIB 2	ECS-LIB 3	ECS-LIB 4	ECS-LIB 5	ECS-LIB 6	ECS-LIB 7
BatterySystem						X	
BatteryModule			X			X	
BatteryCell			X				
PTC		X					
Fuse		X					
Vent			X				
VoltageSensor		X		X			
TemperatureSensor		X		X			
BatteryManagementSystem	(X)	X		X			
BalancingSystem							
EnergyManagementSystem	(X)				X		X

Let us now consider the application requirements. The basic requirement for the residential solar storage system RSS 1 is the responsibility of the `EnergyManagementSystem`. Through the `PowerMeter` and information from the `SolarInverter`, `GridInverter`, and `BatteryManagementSystem`, it has all the information needed to optimize the power flows.

The `GridInverter` is responsible for the interface to the distribution grid and compliance with the grid codes. Therefore, it is also responsible for requirements RSS 2 to RSS 4. For RSS 4, the `MechanicalConstruction` also plays an important role.

The requirements for scalability of battery power, capacity, and solar power system, RSS 5 and RSS 6, are the responsibility of the affected components `SolarInverter` and `BatterySystem`. Whereas RSS 6 additionally involves the `BalancingSystem`, the `EnergyManagementSystem` and the `BatteryManagementSystem`, since a change in the battery configuration must be communicated to these components and they may also have to change their behaviour. It goes without saying that the `MechanicalConstruction` is also adapted here.

RSS 7 was a requirement for the maximum weight of the components. Therefore, the physically tangible components are also responsible here.

RSS 8 adds the `EnergyManagementSystem` it has to be informed about the power flows via the `PowerMeter` and is connected to the main components.

RSS 9, the requirement for a diagnostic programme, is the responsibility of the two components `BatteryManagementSystem` and `EnergyManagementSystem`, as they have the knowledge about the subcomponents and their interaction and can also check them.

8.5.1 Conclusion

In this section, we have learned about lithium ion battery cell chemistry. The basic reaction is that a lithium metal oxide on the cathode releases lithium ions and electrons.

The released electron recombines with the lithium ion in the anode material. Unlike the lead acid battery, the electrolyte is not part of the reaction. It has a transport function.

Since there are a number of different anode and cathode materials, lithium ion batteries are part of a family of cell chemistries, but they all have similar properties.

One of the most important new requirements for the design of a storage system using lithium ion cells is the safety requirement. Since the cathode material becomes thermally unstable under certain boundary conditions, care must be taken to ensure that it is not damaged. Care must also be taken to ensure that the cells are within the safe temperature and voltage range. Additional system components are necessary for this.

The advantage of the lithium ion battery is its higher energy and power density: one can roughly speak of a factor of three compared to lead acid batteries. The voltage of these cells is higher than that of lead batteries, so it is easier to use them with higher voltages. This makes the required current smaller, which reduces costs and temperature development.

In the next section, we will look at high-temperature batteries, specifically the sodium sulphur battery. This battery has properties in addition to those of the lead acid battery or the lithium ion batteries presented here, and has been widely used in stationary applications.

8.6 High-temperature batteries

The previous cell chemistries, lead acid and lithium ion batteries, operated at room temperature. It was even a requirement for long-term use that the temperature of the battery be kept at room temperature if possible. Their basic construction was also comparable: an anode and a cathode are placed in an electrolyte. A separator is fixed between them, which prevents electrons from being exchanged through the electrolyte between the anode and the cathode, but allows ions to be transferred.

In this section, we learn about a new type of battery chemistry, the high-temperature batteries. High-temperature batteries, like the sodium sulphur battery described in this section, differ in their operating temperature range. This is so high that the anode and cathode material becomes liquid. The charge carrier transport can therefore take place directly via the anode and cathode material. An electrolyte is no longer required.

8.6.1 Primary reaction

If we look at a sodium sulphur cell at room temperature, we notice that this cell does not contain any liquids. The classical construction of a cell, which consists of two conductors in an electrolyte, cannot be recognized at first. The cell consists of two areas separated by a wall of aluminium oxide (Al_2O_3). This construction is shown in Fig. 8.37. The inner region is filled with sodium Na. This is the anode material. The outer region is filled with sulphur S. This is the cathode material.

We now heat the cell. At a temperature of 97.79 °C, the sodium first starts to melt. At 115.12 °C, the sulphur also melts. But even at this temperature, little happens when we apply a voltage. Only at a temperature of 300 °C do we find that we can charge and discharge the cell. At this temperature, the Al_2O_3 becomes conductive for ions

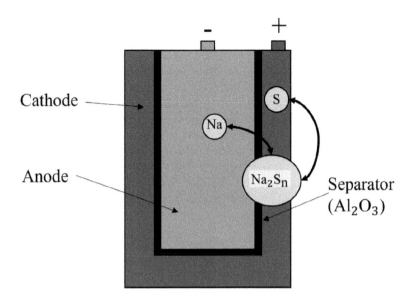

Fig. 8.37 Structure of a sodium sulphur battery cell. Instead of two electrodes, the cell uses an anode and a cathode material, which are both melted. In this case the anode material is sodium and the cathode material is sulphur. A layer of aluminium oxide (Al_2O_3) is used as a separator. It allows only ions to pass; hence, the electron needs to move via the terminals of the cell.

and we can realize ion transport through this separator, while the electrons have to travel via the anode and cathode.

The sodium sulphur cell works similarly to a lithium ion cell. During charging and discharging, two sodium atoms are ionized on each side. On the anode side, two electrons are taken from two sodium atoms, resulting in two sodium ions:

$$2\,Na - 2\,e \rightleftharpoons 2\,Na^+. \tag{8.63}$$

Analogous to the lithium ion cell, the electron travels across the electrical conductors and the sodium ions through the separator to the cathode side. In the process, sulphur and sodium combine to form sodium sulphide:

$$nS + 2\,Na^+ + 2\,e \rightleftharpoons Na_2Sn. \tag{8.64}$$

Here $n = \{3, 4, 5\}$. The total reaction is:

$$2\,Na + nS \rightleftharpoons Na_2Sn. \tag{8.65}$$

This reaction produces a heat of 300 to 350 °C. This heat ensures that sodium and sulphur remain liquid. The cell therefore only needs to be heated when it is neither charged nor discharged—that is, when the reaction is not taking place. So we have to supply additional energy to the cell when there is no charging or discharging. One could interpret this as a kind of self-discharge. However, this interpretation is misleading. In the case of a self-discharge, the amount of stored energy also decreases. This is not the case here. When the cell gets cold, the stored energy is frozen. The energy has therefore not been lost.

A requirement thus arises for the operation of a sodium sulphur cell. It must be ensured that sodium and sulphur are molten and that the Al_2O_3 is ion-conducting.

ECS-NAS 1: AS A operator, I WANT to ensure that the temperature of the cell is greater than 300℃ SO THAT sodium and sulphur are liquefied and the aluminium oxide is ion-conducting.

With this formulation, of course, the use of waste heat and the generation of maintenance heat is involved. No matter how the heat is supplied to the cell, the operator must maintain the temperature at 300 °C to comply with ECS-NAS 1.

We had seen in eqn (8.65) that a different number of sulphur atoms are involved in the reaction $n = \{3, 4, 5\}$. This number depends on the cell voltage. The higher this is, the more sulphur atoms are bound.

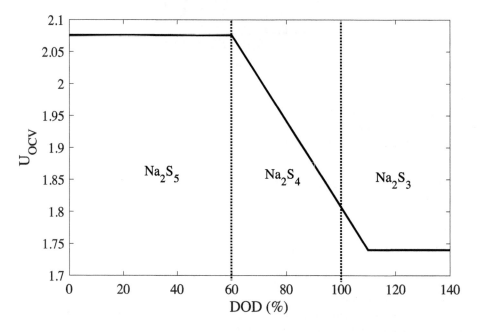

Fig. 8.38 Open-circuit voltage of a sodium sulphur cell as a function of DoD.

In Fig. 8.38, we performed the following experiment: we take a fully charged cell, measure the open-circuit voltage U_{OCV}, and discharge it for a short moment. Then we measure the open-circuit voltage U_{OCV} again.

With a fully charged cell, we measure a voltage of 2.08 V_{DC}. So the voltage is slightly higher than that of a lead acid battery cell. When discharging, we find that Na_2S_5 is degraded first. The open-circuit voltage remains almost constant. At a DoD of about 60%, the voltage starts to drop. We find that we are now primarily generating Na_2S_4. The voltage drops. However, we can still discharge the battery and get more energy than expected, because the depth of charge is larger than 100%. The voltage drop goes up to a voltage of 1.75 V_{DC}. If we reach this voltage, we can still discharge the cell further. However, in doing so, we deplete Na_2S_3 and NaS_2 is formed. We can continue discharging until we reach a DoD of 140%. In doing so, the cell heats up and the internal resistance of the cell increases.

If we nevertheless discharge the cell as far as we can, we unfortunately find afterwards that we cannot charge it again. The reason is that the reaction for the formation of Na_2S_2 is not a reversible reaction. So we have severely damaged the cell by deep discharging it.

ECS-NAS 2: AS AN operator, I WANT to ensure that no deep discharge occurs SO THAT the formation of Na_2S_2 is avoided.

Since we should still have some Na_2S_4 at a DoD of 100%, there is no operational requirement to provide a larger safety capacity here. However, the BMS must ensure that no single cell is deep discharged.

8.6.2 Requirements and system components

Having already introduced two requirements arising directly from cell chemistry, in this section we will go deeper into properties and the system components required to operate a sodium sulphur battery.

We investigate the charging and the discharging of a sodium sulphur battery cell (Reddy, 2011). The discharge rate was $C\frac{1}{3}$, and the charge rate was $C\frac{1}{5}$. Both processes were thus carried out with a relatively low current.

The open-circuit voltage is shown in Fig. 8.39, in addition to the charging and discharging voltage. This is independent of whether we are charging or discharging, and we can see that the open-circuit voltage follows the shape shown in Fig. 8.38.

Let us first look at the charging process. From the formula

$$U_{Bat} = U_{OCV} + R_{Bat}I_{Bat},$$

we see that the charging voltage U_{Bat} must be higher than the open-circuit voltage U_{OCV} during charging. The voltage increase depends on the internal resistance of the battery R_{Bat} and the charging current I_{Bat}. This can also be seen in Fig. 8.39. As can be expected, the voltage initially increases linearly until the formation of NA_2S_4 is completed and the formation of NA_2S_5 begins. The transition appears to start at 80% during charging, although from Fig. 8.39 and looking at the open-circuit voltage, we expect the transition to start at 60%. The reason for this can be seen by looking at the shape of the internal resistance of the battery R_{Bat}. Here we see that it increases slightly, which leads to a correction of the charging voltage.

From a DoD of 60%, the charge voltage is independent of the DoD. We have a similar situation here as that when, for example, the lithium ion cell is used with $LiFePO_4$ as the cathode material.

As soon as we have charged the battery almost completely, which corresponds to a DoD of 5%, the charge voltage increases significantly and reaches its maximum value of 2.3 V at 0%. The cause is a steep increase in the internal resistance. This is due to the fact that when the battery is almost fully charged, there is no longer enough active material available to transport the charge carriers. This is in clear contrast to previous cell chemistries. With both lead acid batteries and lithium ion batteries, there was enough active material to transport charge carriers even when the battery was fully charged, even at the expense of the battery's lifetime. This is not the case

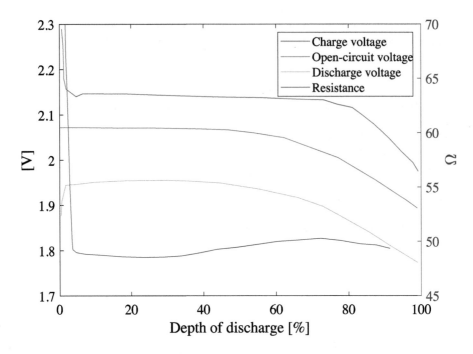

Fig. 8.39 Charging and discharging behaviour of a sodium sulphur cell. The discharge rate was $C\frac{1}{3}$, the charge rate was $C\frac{1}{5}$.

with sodium sulphur batteries. When a single cell is almost fully charged, its internal resistance increases significantly.

Exercise 8.28 Let's investigate the influence of a fast-increasing resistance on two parallel cells in a small experiment. We have connected two cells in parallel (Fig. 8.40). One cell has a state of charge of 95%, the other is just at 50%. We cannot really measure the state of charge of both cells by measuring the open-circuit voltage, because both are already in the state that only Na_2S_5 is formed during charging—that is, $U_1 = U_2 = 2.07$ V_{DC}. However, the internal resistance of the first cell is already $R_1 = 70$ mΩ. For the second cell, it is $R_2 = 35$ mΩ. The voltage source used for charging has a charging voltage of $U_c = 2.08$ V. What is the charging current of the two cells I_1 and I_2? What is the total charging current IC?

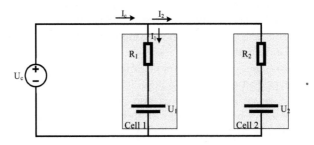

Fig. 8.40 Charging two NAS battery cells connected in parallel.

Solution: The voltage applied to each of the two cells is equal to the voltage of the voltage source:

$$U_c = U_1 + R_1 I_1$$
$$U_c = U_2 + R_2 I_2.$$

We can rearrange both terms respectively to I_1 or I_2 and thus obtain:

$$I_1 = \frac{U_c - U_1}{R_1}$$
$$= \frac{2.08 \text{ V}_{DC} - 2.07 \text{ V}_{DC}}{70 \text{ m}\Omega} = 0.142 \text{ A}$$
$$I_2 = \frac{U_c - U_2}{R_2}$$
$$= \frac{2.08 \text{ V}_{DC} - 2.07 \text{ V}_{DC}}{35 \text{ m}\Omega} = 0.285 \text{ A}.$$

The charging current is the sum of the two individual currents to:

$$I_c = I_1 + I_2 = 0.428 \text{ A}.$$

The results from Exercise 8.28 show that with sodium sulphur cells connected in parallel, the charging current is reduced proportionally when the cells are almost fully charged, so that the cells which are not yet fully charged can continue to be charged. This considerably reduces the effort required by the BMS to monitor the charging process.

Let us now look at the discharging process. Here we observe that the discharge voltage becomes lower than the open-circuit voltage of the cell. We know this behaviour from other cell chemistries. However, up to a DoD of 5%, we observe a slight increase in the discharge voltage. This effect is due to the behaviour of the internal resistance. Since the internal resistance drops sharply at a DoD of 5%, the voltage rises also.

Since there are no significant secondary reactions in the sodium sulphur battery, sodium sulphur batteries have a very long calendar life. As we have seen, these cells protect themselves from overcharging. Only deep discharge permanently damages the cell, and it is likely that the formation of Na_2S_2 is a major reason for the reduction in cycle life.

Let us now look at the behaviour of the cells during repeated charging and discharging. In Fig. 8.41, the cell was repeatedly charging and discharging (Sterner and Stadler, 2017). In this experiment, the capacity of the cell and the efficiency of the cell were measured after several charging and discharging cycles. The efficiency refers to the ratio of the charged energy and the extracted energy. The energy we need for temperature management was not taken into account here.

We see that the capacity decreases approximately linearly. The same applies to the efficiency, which is about 92% at the beginning and just under 91% after 6,000 cycles.

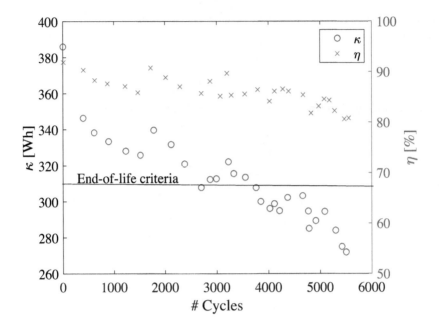

Fig. 8.41 Coulomb efficiency and storage capacity of a sodium sulphur battery as a function of charge and final charge cycles.

Exercise 8.29 Tab. 8.13 shows some data points from Fig. 8.41. What is the capacity and efficiency loss per charge and discharge cycle? (Note: It is perfectly fine if this task is solved using a spreadsheet or a small program.)

Table 8.13 Measurement data for cycle and efficiency test.

Cycles	0	1,200	2,400	3,600	4,800	5,500
$kappa$[Wh]	385	328	320	313	284	271
η[%]	92.50	87.38	87.38	85.81	82.27	81.09

Solution: We are only interested in the deviation of the capacity and the efficiency in therms of the number of cycles. There are a number of techniques for determining this from the data points. We want to use a shortened version of a robust estimator, which is a powerful but seldom-used technique. We take two data points $\{x_1, y_1, x_2, y_2\}$ each and determine the slope

$$a_{1,2} = \frac{y_2 - y_1}{x_2 - x_1}.$$

We do this with all possible combinations of data points and obtain a set $\{a_i\}$. We then take the mean from this set of slope values to obtain an estimate.

Table 8.14 Calculations of the slope through different pairs of points and the resulting mean.

	$a_{1,2}$	$a_{1,3}$	$a_{1,4}$	$a_{1,5}$	$a_{1,6}$	Avg.
a_κ	−0.0475	−0.02708	−0.02	−0.02104	−0.02073	−0.02727
a_η	−0.00426	−0.00213	−0.00186	−0.00213	−0.00208	−0.00249

In Tab. 8.14, the slopes for different pairs of points have been determined. The first data point was always the initial value for the efficiency and the capacity. We see that if we exclude the pair $a_{1,2}$, the slopes are all close to each other.

We see that the capacity loss is $-0.02\frac{\text{Wh}}{\text{Cycle}}$. If we assume that we have about 250 cycles per year and normalize this to the initial capacity of 385 Wh, we get a relative annual capacity loss of:

$$\frac{\Delta \text{SOC}}{a} = -\frac{0.02\frac{\text{Wh}}{\text{Cycle}}}{385 \text{ Wh}} \cdot \frac{250\text{Cycle}}{a} = -\frac{1.3\%}{a}.$$

We lose about 1% of capacity per year.

Sodium sulphur cells have a very high cycle life. If we look at the data from Fig. 8.41 and apply an end-of-life criteria of 80%, the number of possible cycles is 3,000–4,000 cycles. This is otherwise only achievable with higher-quality lithium ion battery cells. However, sodium sulphur cells have the disadvantage that the charging power is limited to just $C\frac{1}{6}$ to $C\frac{1}{5}$. The charging power can be up to $C\frac{1}{3}$, but this is very low compared to the lead acid battery, where we can realize up to C35, or lithium ion cells, where C4 is possible at discharging. Sodium sulphur cells are therefore more suitable for energy applications—that is, applications with low E-Rates.

Let us now ask ourselves what other requirements must be observed for the application of sodium sulphur cells. In these cells, the primary reaction ensures that the operating temperature is maintained—that is, if the cell is regularly charged or discharged, the reaction in eqn (8.65) releases sufficient thermal energy to maintain it. In operation, the task of temperature management is now to prevent heat loss. This can be achieved by thermal insulation measures. In this way, the temperature is maintained for a long time, even when no charging or discharging takes place. Only in phases in which the battery does not have the appropriate temperature, for example in the start-up phase, is it necessary to reach the operating temperature with a heater.

This requirement is realized by means of a suitable operating control system. Similar to the previous constructions, a charge controller and suitable safety limits are used in the lower SOC range. This is not absolutely necessary in the upper SOC range. Sodium sulphur cells have the property that the internal resistance increases considerably at a state of charge of 100%—that is, as soon as only Na_2S_5 has formed in the cell. This simplifies the parallel connection of sodium sulphur cells. Those cells that are already fully charged during the charging process have a greater resistance than those cells that are not yet fully charged, so that the current is only used to charge these cells.

The sodium sulphur cell only works when both sodium and sulphur are melted and the separator has reached a temperature of 300 °C. An operational sodium sulphur battery therefore consists of 300 °C hot liquid material inside. This must, of course, be taken into account by the safety system. For example, if a cell is damaged during operation, the hot, liquid active material may escape and damage other cells within the module. Since the active material is electrically conductive, an internal short circuit occurs, which heats up the cells even more. A chain reaction occurs that damages the battery and can lead to a fire. Therefore, the following requirements apply:

ECS-NAS 3: AS A design engineer, I WANT to design the cell in such a way that no hot, active material can leave the cell SO THAT in case of failure, there is no danger to adjacent cells.

ECS-NAS 4: AS A design engineer, I WANT to implement measures to prevent a short circuit from occurring in the event of leakage of liquid active material SO THAT in the unlikely event of leakage of hot active material, there will be no chain reaction.

Here, again, we have two safety requirements building on each other. The first deals with the task of preventing the event. The second deals with the task of limiting the damage of the event in case that the event still occurs.

The measures taken to fulfil these requirements are similar to those for lithium ion cells: on the one hand, suitable overpressure or overflow valves are implemented in the cells to enable a controlled discharge of the material. Furthermore, suitable fuses are used to prevent short-circuit current.

Sodium sulphur cells are energy cells. They are operated with a charging capacity of $C\frac{1}{6}$ to $C\frac{1}{5}$. Since this means that the ratio between capacity and power is five or six to one, systems with sodium sulphur batteries tend to be oversized. Another disadvantage is that during idle times, when the battery is not being charged or discharged, heating elements have to maintain the temperature. If the energy required for this is taken from the battery, self-discharge will accelerate.

8.6.3   Application example: Integrating a sodium sulphur battery in a wind farm

In the following example, a sodium sulphur storage system will be used to enhance the use of a wind farm on an island (Haessig *et al.*, 2013; Rodrigues *et al.*, 2014).

The concept is shown in Fig. 8.42. The consumers were previously supplied by a power plant and a wind farm. Both are connected to the distribution grid. The task of the power plant is to cover the base load and form the grid—that is, it generates an AC grid to which the generators, the wind farm, and the consumers are connected. The power plant also has the task of ensuring the grid stability so that the grid codes are met: the frequency and amplitude in the AC grid are stabilized by the power plant. Of course, the grid codes also apply to the wind farm and the battery, but they are still too small in terms of their output to build up the entire grid and completely take care of the stabilization. However, both are subject to the grid connection conditions of the grid operator, who will specify in his grid code how the wind farm and battery are to behave in the event of a deviation (Brundlinger, 2019)

Let us first look at the production curve of a wind turbine. In Fig. 8.43, the wind data are taken from a 10 m height during a year. The location is in Hawaii (Power Data access viewer). One can also see in these data a dependence of the wind speed on the time of day. The wind speed tends to be higher at midday or in the second half

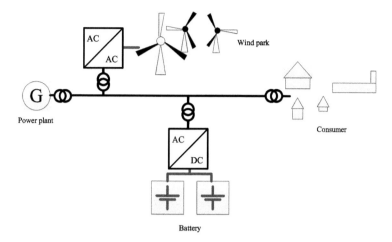

Fig. 8.42 Illustration of the application example for sodium sulphur batteries: consumers are supplied with electricity via a power plant and a wind farm. The battery is interposed and buffers energy.

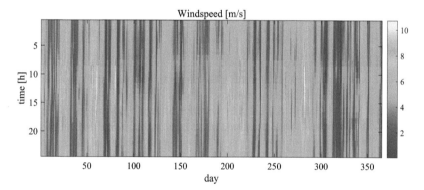

Fig. 8.43 Windspeeds at a height of 10 m during a complete annual cycle.

of the day than in the morning. It can also be seen that the differences between the days are greater at this location. A seasonal dependency, as we were able to recognize in the solar power data—which is, of course, also due to the choice of location—is not recognizable in this dataset. This is good, because we can assume a reliable power supply over the course of the year. The seasonal changes are less pronounced.

Let us consider the load profile of a small town in Hawaii (OEDI Data Lake), which is shown in Fig. 8.44. This is aggregated data that was then normalized to an individual household. The aggregation involved recording the electricity consumption of many different households. Then the total consumption of the households was distributed proportionally to the number of households. The aggregation of the households has the effect that the load curve looks smoother than in Fig. 8.31. Fluctuations caused by behavioural changes of single persons are not significant anymore. Nevertheless, some statements that we know from Fig. 8.31 remain. We see that consumption increases in the evening hours. This is due to the fact that electric light and presumably also more consumer electronics are used in the evening hours. Unlike our solar power storage example, here there is a slight increase in consumption in the midday hours. But the consumption increases at around 08:00 in the morning. We can also see that there is a difference in consumption between the weekdays and the weekend.

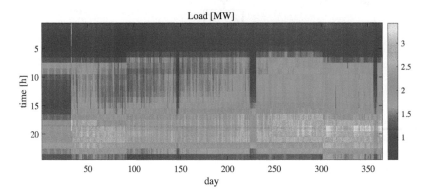

Fig. 8.44 Load profile of a small town in Hawaii.

If we compare the two profiles, we can also see here that the wind still blows when there are no consumers, and, conversely, there is consumption when there is no wind. This already results in the first requirement for this system:

`NAS-WP 1: AS AN operator, I WANT to temporarily store surplus wind power SO THAT at a later time this stored energy can be used for supply.`

Since the battery allows us to call up the power when we need it, we can further optimize the interaction between wind farm, power plant, and battery. For example, we can use the storage to reduce short-term power fluctuations so that the load on the power plant is reduced. This is particularly necessary when the output of the wind farm fluctuates strongly.

`NAS-WP 2: AS AN operator, I WANT to use the storage for compensating power fluctuations SO THAT I do not have to use the power plant for this compensation.`

We can interpret NAS-WP 1 as an energy demand. In the solar power storage example, this was the case. We had an energy application that operated within a time scale of minutes. Sudden fluctuations in solar power production could always be compensated by the grid, especially since the output of the solar power system and the consumption of the household are in the kW range, while the output of the power plant is in the MW range.

Here, in our example, we are dealing with an island grid application. The power plant output, the output of the wind farm, and the consumption of the city are in the same order of magnitude. We are talking about a power in the range of 3 MW for the wind farm, and the consumption is up to 1.5 MW, which means that the power plant must also have a power of about 3 MW.

In such a system, the coordination of power distribution between sources and sinks must be much more precise. Even a short-term overproduction or undersupply of power can lead to a grid failure within a few grid periods (20 ms). This must be avoided at all costs, and leads to NAS-WP 3:

NAS-WP 3: AS A grid operation, I WANT power plant, wind farm, and battery storage to avoid an overproduction or undersupply already at the time scale of a grid period SO THAT there is no grid failure.

NAS-WP 2 and NAS-WP 3 are related to each other. Since wind power plants do not generate continuous power, but rather wind power that varies depending on the weather, this means that the voltage fluctuates for the grid. If the share of wind power is only a small fraction of the total power production, these fluctuations are compensated by the inertial masses of the power plant. Voltage fluctuations are absorbed by the torque of the generator in the power plant. If the share of wind power increases, the load on the power plant's generator also increases, because it turns at a higher frequency. To reduce this load, the storage can be used by temporarily compensating these fluctuations.

Exercise 8.30 After we have gathered application-specific requirements, let's look again at the requirements of other stakeholders. In this task, two requirements each are to be determined for the utility and the service manager. The service manager is the person who has to coordinate the service of all three systems: power plant, battery, and wind farm. (Again, it helps to be aware of the stakeholder's tasks and objectives.)

Solution: The role of the utility, the energy supplier, is to ensure that customers are supplied with electric power. In doing so, its goal is to optimize the Levelized Cost of Energy (LCOE).

NAS-WP 4: AS A utility, I WANT operations management to assess both losses and operating costs in determining the optimal power flow combination SO THAT the LCOE is optimized.

A second requirement arises from the utility's supply mandate.

NAS-WP 5: AS A utility, I WANT to ensure that the availability of electric power is above 99.9%, with the single event not exceeding $\frac{1}{2}$ h SO THAT I meet my legal mandate.

An availability of 99.9% sounds extremely high at first. However, it means that 8.76 h of outages are allowed per year. If you divide this by the maximum length, this is 16 events per year.

The service manager is tasked with restoring supply within $\frac{1}{2}$ h in the event of a failure of any of the three power plant components. We focus in this task on the requirements it places on the battery.

NAS-WP 6: AS A service manager, I WANT to be able to switch off individual parts of the battery SO THAT I can perform maintenance on partial batteries and at the same time other parts of the battery still provide a partial function.

Previous battery storage systems were relatively small in storage capacity. Their capacity was mostly in the kW range. In terms of volume, we could speak of one or more refrigerators. Here the case is different. We are talking about a battery capacity of several MW. Such a battery would not be realized monolithically. Instead, according to the requirements of NAS-WP 6, the battery would be divided into smaller, functional blocks. These small blocks would then be scaled to the desired total power.

The next requirement will try to reduce the workload on the service manager and reduce the likelihood of critical events.

NAS-WP 7: AS A service manager, I WANT to be able to analyse the operating data of individual modules in the different battery blocks SO THAT I can plan preventive maintenance.

Preventive maintenance means that you carry out maintenance before the fault occurs. An example of this is when we change the oil in our car. Usually we don't wait until the engine oil stops lubricating or the oil level is too low, but we measure the oil level regularly and perform an oil change even if no faults occur. We do this because we know that an engine failure will lead to a longer and more expensive breakdown. It is similar with the battery systems. Since there is a high demand on the availability of the grid here, the service manager wants to fix the faults before they occur.

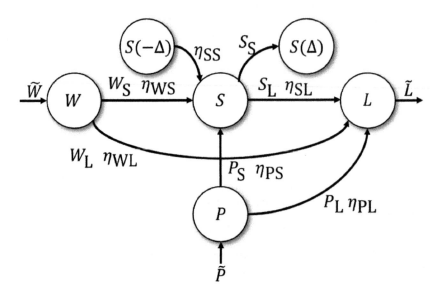

Fig. 8.45 Power flow diagram for the integration of a battery storage system into an island grid with a wind turbine. The node W represents the power of the wind farm. L represents the power demand of the consumers. S corresponds to the storage. P here is the power plant.

Fig. 8.45 shows the power flow diagram for our example. The node W represents the power of the wind farm. Its production profile is represented by $\tilde{W}$. The aggregation of all consumers on the island is described by L with the load profile $\tilde{L}$. We assume that the decentralized energy generation units already present on the island are so small that $\tilde{L}$ is always positive—that is, there is always also consumption and no power needs to be stored. Alternatively, we can also say that negative consumption can be ignored because the power must not be fed into the grid. The power plant is represented by the node P. $\tilde{P}$ represents the required production of the power plant.

At first glance, the power flow diagram resembles that of the solar power storage system. However, the power path from the wind turbine and the storage to the power plant is missing in this flow diagram. In the case of the solar power storage, we considered the grid as a kind of infinite power source and sink. Any surplus was fed into the G node, or in the case of an undersupply from the battery and solar plant, the grid served as a power source. This is not the case in our example. The only component that can store power is the battery. The wind power must either be consumed immediately, which corresponds to the power path $W_{\rm L}$, or stored, which corresponds to the path $W_{\rm S}$.

Let us now look at the individual power flow equations. We start with the node for the wind farm. The production profile $\tilde{W}$ represents the possible production. This power can be distributed to the storage or the load. As with a solar power plant, there is no need to use all the power as well. For wind turbines, there are several ways to reduce the power produced when wind is present. Firstly, a wind turbine can be turned away from the wind. Similar to a sailing ship, which aligns its sails parallel to the wind so that the wind can no longer exert any force on the sails, the rotor blades and the rotor nacelle of a wind turbine can be turned so that the force of the wind is no longer used to generate energy. Furthermore, with a wind turbine, there is also the possibility of reducing the power fed into the grid with the help of power electronics. The power flow equations for the wind turbine are therefore:

$$\tilde{W} \geq W_{\rm S} + W_{\rm L}. \tag{8.66}$$

Let us next consider the power node of the power plant. Unlike the wind plant, all the power $\tilde{P}$ produced must be dissipated. This is because in a power plant, fossil or nuclear fuels generate heat that vaporizes water to drive a turbine. If the generated kinetic energy is not removed, the rotation of the turbine increases until it is mechanically damaged. Before this happens, another effect occurs: if the power produced is not taken off, the frequency of the grid voltage increases. This can damage equipment on the grid. Therefore, operators of power generation units that feed electricity into the distribution or transmission grid are subject to the grid code, which specifies how electricity is to be fed into the grid. Therefore, for the production node P, the following applies:

$$\tilde{P} = P_{\rm S} + P_{\rm L}. \tag{8.67}$$

For consumers, the current load should always be met by wind power, storage, or the power plant:

$$\tilde{L} = \eta_{\rm WL} W_{\rm L} + \eta_{\rm SL} S_{\rm L} + \eta_{\rm PL} P_{\rm L}. \tag{8.68}$$

In eqn (8.68), unlike solar power storage, the power flow from the grid to the load is assigned with an efficiency. From a utility perspective, this view is justified. NAS-WP 4 requires an optimization of the LCOE. This means that the energy supplier also has an interest in ensuring that inefficient transmission paths must be included in the cost calculation. In the case of solar power storage, the homeowner only pays for transmission losses within the house, and these can be neglected.

The power flow equation for the storage is given by the sum of the inflows and outflows from the storage:

$$0 = \eta_{\rm WS} W_{\rm S} + \eta_{\rm SS} S_{\rm S}(-\Delta) + \eta_{\rm PS} P_{\rm S} - (S_{\rm L} + S_{\rm S}(\Delta)). \tag{8.69}$$

NAS-WP 2 calls for the compensation of power fluctuations. This refers to fluctuations over a time scale of a few minutes, during which the power increases or decreases sharply. These are caused by sudden gusts of wind or unanticipated peaks and troughs in consumption. These power fluctuations must be compensated by the power plant via P_L if no storage is available for buffering. The NAS-WP 2 requirement can therefore be interpreted as compensating for the temporal fluctuations of P_L. We have two possibilities to describe this mathematically.

The first variant interprets power fluctuations as deviations from the temporal mean $\bar{P}_L$,

$$\Delta_{\max} P_L \leq \|P_L - \bar{P}_L\|, \tag{8.70}$$

where $\bar{P}_L$ is the moving average over the period T:

$$\bar{P}_L = \frac{1}{T} \int_{\Delta t = -T}^{\Delta t = 0} P_L \mathrm{d}\Delta t.$$

The description in eqn (8.70) requires that these deviations are smaller than ΔP_L. However, in its implementation it means that we have to include the moving average when calculating the power flows, and this includes values of past power flows. Another disadvantage is that these boundary conditions depend on the parameter T. The shorter T is, the more sensitive eqn (8.70) becomes to variations in P_L.

The second approach looks at the requirement from a different perspective. The challenge is not the power variation per se. A power plant, with its large rotating masses, has a buffer that can compensate smaller power fluctuations. Again, let's do a little thought experiment on this. Assume that we have a constant P_L. This means that the generator and the turbine of the power plant rotate at a constant speed, ω. The power-generating moment of the generator T_{Gen} is the same as the driving moment T_{Turb} of the turbine.

Now the power demand in the grid suddenly increases. The generator's power-generating moment also suddenly increases by ΔT_{Gen}. If the driving moment of the turbine is able to readjust, everything is in equilibrium. But if this is not the case, this additional rotational energy must come from somewhere. There is energy in the rotating mass of the generator and the turbine, and this energy can be extracted. As a result, the rotational speed and the electrical frequency are reduced. But this goes hand in hand with an increased mechanical load on the mechanical components. Furthermore, the steam circuit must now be made to increase the driving torque of the turbine. This stresses the thermal components of the power plant.

Let us now do the same experiment. In this case, the power demand increases by $\frac{1}{4}\Delta T_{\text{Gen}}$ in each of four small s.pdf. In this case, the stress on the components of the power plant is no longer as severe. The sudden removal of power produces a smaller reduction in frequency because the change in power is smaller.

In this picture, NAS-WP 2 can be considered a demand on the time change of P_L:

$$\frac{\mathrm{d}}{\mathrm{d}t} P_L \leq \Delta_{\max} P_L. \tag{8.71}$$

We can also describe eqn (8.71) in the time-discretized form. Then the formulation is:

$$\frac{P_{\mathrm{L}}(-\Delta) - P_{\mathrm{L}}}{\Delta} \leq \Delta_{\max} P_{\mathrm{L}}. \tag{8.72}$$

This second formulation only needs information from the previous timestep $(P_{\mathrm{L}}(-\Delta))$, which can be solved with less calculation effort.

Let us use the NAS-WP 4 requirement to define our objective function. NAS-WP 4 requires the optimization of operating costs with the aim of reducing the LCOE. To calculate the operating cost, we assign a tariff to the power plant, the wind farm, and the storage: c_P, c_W, c_S. We can determine the tariff via the LCOE of the individual power plant:

$$c_W = \frac{\sum_{t=0}^{t=T_{\mathrm{EOL}}} W_{\mathrm{S}}(t) + W_{\mathrm{L}}(t)}{\sum_{t=0}^{t=T_{\mathrm{EOL}}} (1+r)^{-t}(c_{grid} - c_{\mathrm{coW}})(W_{\mathrm{S}}(t) + W_{\mathrm{L}}(t))} \tag{8.73}$$

$$c_S = \frac{\sum_{t=0}^{t=T_{\mathrm{EOL}}} S_{\mathrm{L}}(t)}{\sum_{t=0}^{t=T_{\mathrm{EOL}}} (1+r)^{-t}(c_{grid} - c_{\mathrm{coS}})(S_{\mathrm{L}}(t))} \tag{8.74}$$

$$c_P = \frac{\sum_{t=0}^{t=T_{\mathrm{EOL}}} P_{\mathrm{S}}(t) + P_{\mathrm{L}}(t)}{\sum_{t=0}^{t=T_{\mathrm{EOL}}} (1+r)^{-t}(c_{grid} - c_{\mathrm{coP}})(P_{\mathrm{S}}(t) + P_{\mathrm{L}}(t))}. \tag{8.75}$$

r is the market interest rate, which we want to be fixed over the lifetime of the power plant component, T_{EOL}. c_{grid} is the average electricity price realized through the sale of electricity to customers. We have defined the running costs as mean operating costs c_{coW}, c_{coS}, and c_{coP} to make the formulae clearer here.

At first sight, the optimization task of NAS-WP 4 seems simple to implement. Whenever we adjust the power flows, we just have to look to minimize the total cost Y:

$$\min Y = c_W \cdot (W_{\mathrm{S}} + W_{\mathrm{L}}) + c_P \cdot (P_{\mathrm{S}} + P_{\mathrm{L}}) + c_S \cdot (S_{\mathrm{S}} + S_{\mathrm{L}}). \tag{8.76}$$

The consideration is a local one. It always focuses on only the current point in time. If, on the other hand, a production and consumption forecast is available that is sufficiently accurate for a time horizon of $\tilde{T}$, future events can be taken into account:

$$\min Y = \sum_{t=0}^{t=T+\tilde{T}} Y(t). \tag{8.77}$$

It can be seen that this type of optimization is always advantageous when the forecast is good (Schmiegel and Kleine, 2014).

The equations required for the modelling are summarized again in Tab 8.15. Of course, there are also boundary conditions for the value ranges of the power flows and the capacity, as well as values for the efficiencies.

Table 8.15 Power flow equations for a stand-alone grid equipped with a wind farm, a battery, and a power plant.

Description	equation
Wind park	$\tilde{W} \geq W_S + W_L$
Power plant	$\tilde{P} = P_S + P_L$
Storage	$0 = \eta_{WS} W_S + \eta_{SS} S_S(-\Delta) + \eta_{PS} P_S - (S_L + S_S(\Delta))$
Power fluctuations averaged	$\Delta_{\max} P_L \leq \| P_L - \bar{P}_L \|$
Power fluctuations gradient (continuous)	$\frac{d}{dt} P_L \leq \Delta_{\max} P_L$
Power fluctuations gradient (discrete)	$\frac{P_L(-\Delta) - P_L}{\Delta} \leq \Delta_{\max} P_L$
Operating costs of the components	
Wind	$c_W = \dfrac{\sum_{t=0}^{t=T_{EOL}} W_S(t) + W_L(t)}{\sum_{t=0}^{t=T_{EOL}} (1+r)^{-t}(c_{grid} - c_{coW})(W_S(t) + W_L(t))}$
Storage	$c_S = \dfrac{\sum_{t=0}^{t=T_{EOL}} S_L(t)}{\sum_{t=0}^{t=T_{EOL}} (1+r)^{-t}(c_{grid} - c_{coS}(S_L(t)))}$
Power plant	$c_P = \dfrac{\sum_{t=0}^{t=T_{EOL}} P_S(t) + P_L(t)}{\sum_{t=0}^{t=T_{EOL}} (1+r)^{-t}(c_{grid} - c_{coP})(P_S(t) + P_L(t))}$
Total costs without forecast	$\min Y = c_W \cdot (W_S + W_L) + c_P \cdot (P_S + P_L) + c_S \cdot (S_S + S_L)$
Total costs with forecast	$\min Y = \sum_{t=0}^{t=T+\tilde{T}} Y(t)$

Exercise 8.31 In order to get a feeling for the required power and the required storage capacity, a simple operation management is to be defined. NAS-WP 4 is solved in a simplified way by assuming that c_W and c_S can be neglected. The storage capacity is also assumed to be arbitrarily large. The influence of the efficiencies is also to be neglected. NAS-WP 2 should also not be observed. What can a simple operation management look like?

Solution: Here we realize a simple algorithm for the operational management of island grid supply:

Setting Baseload to P_0: $\tilde{P} \leftarrow P_0$
$S_S \leftarrow 0$
Setting the time increment in this calculation to one minute: $\Delta t \leftarrow \frac{1}{60}$
Setting initial capacity: $\kappa = 0$
while $\tilde{P} \neq 0$ OR $\tilde{L} \neq 0$ OR $\tilde{W} \neq 0$ **do**
    $R \leftarrow \tilde{W} - (\tilde{L} + P_0)$
    if $R \geq 0$ **then**
        $W_L \leftarrow \tilde{L} - P_0$
        $W_S \leftarrow \tilde{W} - W_L$
        $P_L \leftarrow P_0$
        $S_L \leftarrow 0$
        $P_S \leftarrow 0$
    else
        $W_L \leftarrow \tilde{W}$
        if $\frac{\kappa}{\Delta t} > -R$ **then**

$$S_L \leftarrow \frac{\kappa}{\Delta t} - W_L$$
$$P_L \leftarrow \tilde{P_0}$$
else
$$S_L \leftarrow \frac{\kappa}{\Delta t}$$
$$P_L \leftarrow \tilde{P_0} - (W_L + S_L)$$
end if
end if
$$\tilde{P} \leftarrow P_S + P_L$$
$$S_S \leftarrow \frac{\kappa}{\Delta t} + W_S + P_S - S_L$$
$$\kappa \leftarrow S_S \cdot \Delta t$$
$$\tilde{P} \leftarrow P_S + P_L$$
end while

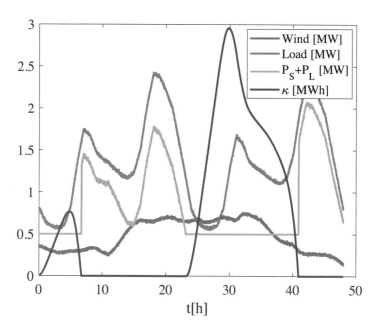

Fig. 8.46 Wind production and demand, and power plant production and capacity trend. (Maximum wind capacity 1.5 MWh, maximum consumption 3.5 MWh, maximum production 4.5 MWh, baseline production 0.5 MWh).

In Fig. 8.46, we have shown the time flow over two days as an example. The maximum wind output was 1.5 MWh, the maximum consumption was 3.5 MWh, the maximum production was 4.5 MWh, and the baseload production was 0.5 MWh.

At the beginning of the day, the baseload supplies the load. The wind power produced is lower than the baseload. The wind energy is stored temporarily. Around 08:00, when the consumption is higher than the baseload, the storage is discharged. However, the power plant must increase its feed-in power at around 10:00, as the storage is discharged here. This increase takes place quickly. To simplify the algorithm, we neglect NAS-WP 2 in this operation. If we wanted to comply with this requirement, we could do so by having the power plant increase its feed-in power a little earlier.

From 22:00, the storage is charged intensively—on the one hand from the power plant, as consumption decreases from 24:00, but then also during the day, as wind production increases.

We see that the battery capacity that would be needed to store any surplus is about 3 MWh. We have not used a capacity limit here. This constraint, made to simplify the solution of Exercise 8.31, allows us to estimate the maximum power and maximum storage capacity for different system configurations.

To get an idea of what storage capacity is reasonable, in Fig. 8.47 we have shown the distribution of charging states. In 97% of cases, the battery has its small state of charge. The frequency of higher states of charge is exponentially distributed. From 5 MWh this is very low. It therefore makes sense to set the storage capacity of the battery at 5 MWh if financial or technical reasons do not play a role and we really want to play it safe.

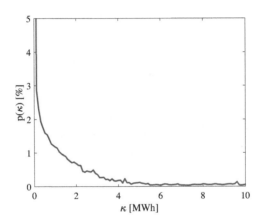

Fig. 8.47 Distribution of capacity throughout the year.

In Fig. 8.48, we show the distribution of the charging and final charging powers. It is first noticeable that the charging powers tend to be low and the frequency decreases to higher powers. Very low charging powers are clearly more likely. This is related to our operational management, in which we prioritize the direct consumption of wind power and therefore always store only the leftovers in the battery. For the discharging power, we observe a broader distribution. The frequency of power above 0.50 MW is considerably higher than for charging. If we wanted to cover the entire required range, we would need a maximum charging power of 1.5 MW.

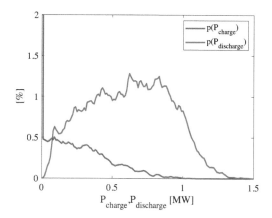

Fig. 8.48 Distribution of battery charging and discharging rates throughout the year.

Table 8.16 Technical data of sodium sulphur cells.

Label	Capacity [Ah]	Max. charge current [A]	Max. discharge current [A]	Voltage [V_{DC}]
Type 1	248	62	100	[1.9; 2.05]
Type 2	280	56	84	[1.9; 2.05]

Exercise 8.32 We assume that the same cells as in Fig. 8.39 are used. These allowed a maximum discharge rate of $\frac{1}{5}C$ and a charge rate of $\frac{1}{3}C$. What capacity is necessary to meet the requirements of Fig. 8.48?

Solution: We have observed a maximum charging capacity of 1 MW. This results in a capacity requirement κ_{charge} of:

$$\kappa_{charge} = 1 \text{ MW} \cdot 5 = 5 \text{ MWh}.$$

The discharge rate is a maximum of 1.5 MW and the capacity requirement here is:

$$\kappa_{discharge} = 1.5 \text{ MW} \cdot 3 = 4.5 \text{ MWh}.$$

So to meet the requirement for charging and discharging, a capacity of 5 MWh is sufficient, which also matches well the required capacity from Fig. 8.47.

We now have all the key data we need to design the storage. We know the capacity we are looking for, and the power we need. Since the cell chemistry determines the voltage window, we know that the open-circuit voltage is between 1.9 V_{DC} and 2.05 V_{DC}.

Tab. 8.16 shows the technical data of the two cells available for selection. For our application, we use modules that are interconnected to form a system. The two modules differ in the voltage range but not in the provided power of 50 kW.

Exercise 8.33 The module of Type 1 cells shall have a nominal open-circuit voltage of around 48 V_{DC}. The module of Type 2 cells shall have a nominal voltage around 24 V_{DC}. The nominal maximum charging power should be 50 kW.

How must the cells of Type 1 and Type 2 be connected? What is the capacity and voltage range of the modules?

Solution: We first determine the number of cells required for the two modules, which must be connected in series:

$$y_{Type1} = \frac{48 \ V_{DC} \cdot 2}{1.9 \ V_{DC} + 2.05 \ V_{DC}} = 24.3 \approx 24.$$

In this case, we round off the number. Knowing y, we can also find the voltage range of the module:

$$U_{Type1} = 24 \cdot [1.9; 2.05] \ V_{DC} = [45.6; 49.2] \ V_{DC}.$$

For the Type 2 modules,

$$y_{Type2} = \frac{24 \ V_{DC} \cdot 2}{1.9 \ V_{DC} + 2.05 \ V_{DC}} = 12.15 \approx 12,$$

The resulting voltage window is then:

$$U_{Type2} = 12 \cdot [1.9; 2.05] \ V_{DC} = [22.8; 24.6] \ V_{DC}.$$

Next, we determine the number of parallel strings. First, we determine the nominal power that one string can provide:

$$P_{Type1} = 62 \ A \cdot 48 \ V_{DC} = 2{,}976 \ W$$

$$P_{Type2} = 56 \ A \cdot 24 \ V_{DC} = 1{,}344 \ W.$$

Thus we can find the number of parallel strings. In this case, we round up, as we do not want to fall short of the nominal power. The result is then:

$$x_{Type1} = \frac{50 \ kW}{2{,}976 \ kW} = 16.8 \approx 17$$

$$x_{Type2} = \frac{50 \ kW}{1{,}344 \ kW} = 37.2 \approx 38.$$

To determine the capacity of the modules, we determine the capacity of the cells and multiply it by the number of cells. For the capacity of the module with cells of Type 1, the following then applies:

$$\kappa_{Type1} = \frac{1.9 + 2.05}{2} \ V_{DC} \cdot 248 \ Ah \cdot 17 \cdot 24 = 199.83 \ kWh.$$

The following applies to the module with cells of Type 2:

$$\kappa_{Type2} = \frac{1.9 + 2.05}{2} \ V_{DC} \cdot 280 \ Ah \cdot 38 \cdot 12 = 252.16 \ kWh.$$

The module with Type 1 cells delivers more power with less installed capacity. Roughly, we have a ratio of four to one. For module 2, the ratio is five to one. This is also reflected in the number of cells. The module with Type 1 cells consists of 408 cells, while the module with Type 2 cells consists of 456 cells. The number of cells or the amount of active material is an indicator for the price of the module. The more active the material, the higher the price will be.

In order to determine the suitable module connection for our wind power plant, we have to pay attention to two boundary conditions. First, we need a charging power of 100 MW.

Secondly, we need to define the voltage window. With a nominal module voltage of 48 V_{DC} or 24 V_{DC}, it is not possible to feed directly into the AC grid without a DC/DC converter. However, since the current would also have to be very high at these low voltages, a higher voltage should be selected. We already know that an inverter usually consists of two components: a DC/DC converter and the inverter. The DC/DC converter ensures that the DC link voltage is high enough so that the inverter can always feed into the grid or draw power from the grid. In the distribution grid, this required voltage is approximately 650 V_{DC}. A lower one can also be chosen, but you should have enough control reserves to be able to compensate for voltage fluctuations in the grid.

For the design of the battery, we have two possibilities: we can choose a voltage that requires a DC/DC converter, or we connect the modules so that we have a voltage that allows a DC link voltage of more than 600 V_{DC}. Since the sodium sulphur cell has a relatively narrow voltage band, we want to realize this variant and save the costs for the DC/DC converter.

Exercise 8.34 What is the needed interconnection of modules using Type 1 and Type 2 cells so that we realize a charging power of 100 MW and a minimum voltage of more than 600 V_{DC}?

What is the total capacity of these batteries?

Solution: To achieve the charging power, we need to connect strings in parallel in both cases. This leaves the question of the number of modules. Since we are interested in the minimum voltage, we have to ask how many modules have to be connected so that the minimum voltage of the modules is above 600 V_{DC}.

For module 1 with a minimum voltage of $U_{Type1} = 45.6$ V_{DC}, the following applies:

$$Y_{Type1} = \frac{600 \ V_{DC}}{45.6 \ V_{DC}} = 13.15 \approx 14.$$

We need to round up the number of modules here to ensure that we are above the voltage limit. The voltage range of the battery equals:

$$U_{Bat1} = 14 \cdot [45.6; 49.2] \ V_{DC} = [638.4; 688.8] \ V_{DC}.$$

For module 2 with a minimum voltage of 22.8 V_{DC}, the following then holds:

$$Y_{Type2} = \frac{600 \ V_{DC}}{22.8 \ V_{DC}} = 26.31 \approx 27.$$

The battery voltage is then:

$$U_{Bat2} = 27 \cdot [22.8; 24.6] \ V_{DC} = [615.6; 664.2] \ V_{DC}.$$

Since both modules have a power rate of 50 kW, we need to connect $\frac{1000 \ kW}{50 \ kW} = 20$ modules in parallel. Hence, the capacity of the battery with Type 1 cells equals:

$$\kappa_{Bat1} = 20 \cdot 14 \cdot 199.83 \ kWh = 55.82 \ MWh.$$

And for the battery with cells of Type 2:

$$\kappa_{Bat2} = 20 \cdot 27 \cdot 252.16 \text{ kWh} = 136.17 \text{ MWh}.$$

We have now defined complete battery blocks that meet the power and voltage requirements. These differ significantly in size: while the battery with Type 1 cells consists of 280 modules, the battery with Type 2 cells consists of 540 modules. The latter is therefore twice as large. We also see that the requirement of a battery voltage of 600 V_{DC} forces us to add overcapacity.

Exercise 8.35 We want to avoid overcapacity. We know from Exercise 8.32 that the low C-Rates force us to work with overcapacity. What will be the voltage window that we can achieve, if we only accept this necessary capacity?

Solution: For the battery of Type 1, we need 5 MWh capacity. Therefore, the maximum accepted number of serial modules $N_{y,Bat1}$ equals:

$$N_{y,Bat1} = \frac{\kappa_{Bat1}}{N_{x,Bat1} \cdot \kappa_{Bat1}} = \frac{5 \text{ MWh}}{20 \cdot 199.83 \text{ kWh}} = 1.25 \approx 2.$$

The resulting voltage window equals:

$$U_{Bat1} = 2 \cdot [45.6; 49.2] \text{ V}_{DC} = [91.2; 98.4] \text{ V}_{DC}.$$

We also calculate the number of serial modules for the battery with cells of Type 2:

$$N_{y,Bat2} = \frac{4.5 \text{ MWh}}{20 \cdot 252.16 \text{ kWh}} = 0.892 \approx 1.$$

In this case, the voltage is the same as for the battery. However, we know that to boost the DC power, we should have a ratio above one to four. Therefore we have to increase the total number of modules and we have to live with overcapacity.

We stick to our design from Exercise 8.34 and accept the overcapacity. A question now arises: which of the two batteries should we choose? In terms of the number of cells in series, the batteries differ only very slightly. The differences in capacity are relatively small, so it is to be expected that at the overall system level the costs are comparable.

But let's look at the overall system: one would want to avoid connecting these huge batteries with a capacity of 1.5 MW hard in parallel. The risk of balancing currents flowing between the units is simply too high. Such a monolithic battery block would also partially violate NAS-WP 6. Although we could separate individual battery sub-blocks, if the power electronics were to be repaired, this might no longer be possible, depending on the type of implementation.

If, on the other hand, we put together smaller units consisting of inverter and battery, we can take individual units off the grid without any problems. It is also possible to expand the power plant. In addition, studies have shown that such modular storage power plants allow more efficient operation, and thus the efficiency of the entire system can be optimized (Ried *et al.*, 2020).

However, each battery would then need its own inverter, cables, and connection technology. The wiring and maintenance effort would therefore be higher for the battery with Type 1 modules than for the battery with Type 2 modules.

We do not want to make this design decision here. The s.pdf to make this final decision consist of offsetting the investment and operating costs against the expected profits and calculating the LCOE for the different realizations. Instead, we now consider the system components involved and the RTM of this system.

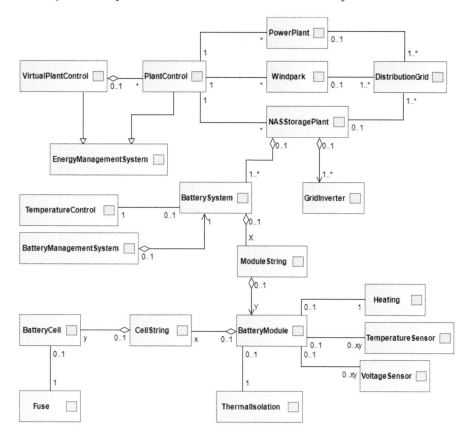

Fig. 8.49 System components of an island grid using wind and sodium sulphur battery to reduce the fossil fuel consumption of the power plant.

In Fig. 8.49, we have shown the system components of the power plant park consisting of wind turbines (`WindPark`), power plant (`PowerPlant`), and storage (`NASStoragePlant`). These three system components are interconnected via the `DistributionGrid`. All three power plants have their own `PlantControl`, whose task is to regulate and control the respective power plant. So that the power plants can be controlled in the network, the `PlantControl` of the power plants is aggregated into a `VirtualPlantControl`. This thus has responsibility for the NAS-WP 3 and NAS-WP 4 requirements, as we can also see in Tab. 8.17. `PlantControl` and

Table 8.17 Requirement traceability matrix for the island energy system.

	Lithium Ion Batteries				Application Requirements						
	ECS-NAS 1	ECS-NAS 2	ECS-NAS 3	ECS-NAS 4	WP 1	WP 2	WP 3	WP 4	WP 5	WP 6	WP 7
VirtualPlantControl							X	X			
PlantControl									X		
PowerPlant									X		
Windpark									X		
NASStoragePlant					X	X			X	X	X
DistributionGrid											
GridInverter										X	X
BatterySystem										X	X
BatteryManagementSystem		X									
TemperatureControl	X										
TemperatureSensor	X										X
Heating	X										
VoltageSensor		X									X
Fuse				X							
ThermalIsolation	X										
MechanicalConstruction			X	X							

VirtualPlantControl are realizations of an **Energy Management System**, each with a different focus.

The NASStoragePlant represents the aggregation of the batteries. Regardless of which of the discussed realizations is chosen, this system component consists of an aggregation of GridInverter and BatterySystem. The control of the temperatures of the modules, as well as the monitoring of the charging states of the modules, is the responsibility of the TemperatureControl and the BatteryManagementSystem. Both are assigned to the BatterySystem. Unlike a lithium ion battery, modules with sodium sulphur cells do not require a BatteryManagementSystem at module level. Compliance with requirement ECS-NAS 2 can be achieved at a system level if the modules are equipped with VoltageSensor and TemperatureSensor, whereas the TemperatureSensor monitors compliance with ECA-NAS 1.

The construction of the modules and the system components involved are comparable to the previous components. Compared to a lithium ion battery module, there are fewer components responsible for battery safety. Here, only the Fuse, which is to prevent a short circuit and thus responsible for ECS-NAS 3 and ECS-NAS 4, and the ThermalIsolation are included. The ThermalIsolation has a responsibility for ECS-NAS 4 on the one hand, but also for ECS-NAS 1. If we have good thermal isolation, maintaining the operating temperature becomes more efficient.

Let us now look at the responsibilities of the different system components for individual requirements. Unlike previous battery technologies, where the focus was on adding heat from the battery, a high-temperature battery adds heat. ECS-NAS 1 requires 300 °C, and the responsible system components are the Heating, the TemperatureSensor, the ThermalIsolation, and the TemperatureControl. Four system components are necessary to meet this requirement. In fact, they have different roles. The role of the `Thermal Isolation` is to ensure that the heat stays with the cells for as long as possible and is not released into the environment. `Heating` supplies the heat needed to reach 300 °C to the module. The `ThermalControl` regulates the process and uses the information from the `TemperatureSensor` to control the `Heating`. These components operate at the module level.

To prevent deep discharge (ECS-NAS 2), the `BatteryManagementSystem` uses the `VoltageSensor` to control the voltage of the cells or strings.

The `MechanicalConstruction`, on the other hand, must ensure that ECS-NAS 3 and ECS-NAS 4 are complied with, and ECS-NAS 4 also supports a `Fuse` at the cell level.

Let us now look at the requirement arising from the application. While the previous requirements were met at the sub-level of the `BatterySystem`, the application requirements are largely met by other, external system components.

NAS-WP 1 and NAS-WP 2 are fulfilled by the totality of all `BatterySystem` and `GridInverter` aggregated as `NASStoragePlant`. This follows the basic idea of combining smaller units into a larger unit that can then be viewed as a whole.

NAS-WP 3 and NAS-WP 4, utility requirements for operational control, are met by `VirtualPlantControl`. This represents an aggregation of the `PlantControl` of PowerPlant, `WindPark`, and `NASStoragePlant`. These subcomponents are also responsible for compliance with NAS-WP 5, ensuring the availability of the electric power grid.

The service has two requirements. In NAS-WP 6, it requires that individual subcomponents of the battery can be switched off in order to perform maintenance during operation. This is initially a requirement for the `NASStoragePlant`, but will be met by the `GridInverter` and the `BatterySystem`. The `GridInverter` and its subcomponents will ensure that no AC power from the grid is routed to the shutdown areas and the `BatterySystem` will ensure that no DC power is released from the modules in an uncontrolled manner.

The requirement for sensor data and information for preventive maintenance (NAS-WP 7) must again primarily be met by the main component `NASStoragePlant`. However, a number of subcomponents are involved and one could add more here.

8.6.4 Conclusion

In this section, we have learned about sodium sulphur cell chemistry. Unlike the lead acid battery or the lithium ion battery, it uses a melted active material. The previously known construction of anode, separator, and cathode is realized here by a molten anode and cathode material, which are removed from each other by a separator. For this realization, the sodium and sulphur must be heated to 300 °C, which is a disadvantage for various applications.

Sodium sulphur cells are energy cells. The E-Rate at which they can be operated is below 1/3. However, we have seen in our application example that there are use cases where this limitation is not a problem.

In the next section, we will learn about a battery technology that implements the idea of using liquid anode and cathode material at normal temperatures. Redox flow cell chemistries use liquid active material. We will see that this technology allows further degrees of freedom in the realization of storage systems and then apply this technology to the example introduced here.

8.7 Redox flow batteries

The classical structure of an electrochemical cell is that we insert two electrodes into a container with an electrolyte. The two electrodes consist of or are coated with an active material that is needed for the reaction. To avoid a short circuit, a separator is inserted between the two electrodes, which can be passed by ions but not by electrons. This structure creates a conflict in the design of an electrochemical cell, which we have to resolve in the development of a battery cell.

To realize a battery cell with high energy content, we have to increase the amount of active material. For example, the storage capacity of a lithium ion cell depends on the available mobile lithium ions. The more lithium ions that can be reversibly stored in the cathode and anode material, the greater the capacity of the cell.

If, on the other hand, we want to realize a battery cell that allows a high charging and discharging power, we have to increase the surface area of the active material. This is because after the charge carriers have separated in the active material, they have to exit via the surface of the material into the electrolyte, migrate through the separator to the other side and re-enter the active material via the surface. In order to realize a battery cell that allows a high C-Rate, we must therefore increase the surface area of the active material. We can achieve this, for example, by reducing the thickness of the active material coated on the electrodes or by increasing the surface of the electrode. This allows the charge carriers to escape more quickly from the surface and, in addition, the heat generated by the electric current is dissipated better.

When designing an electrochemical cell, we have the task of judging between the amount of active material we want to use and the size of the surface. This decision, plus the influences of the electrochemistry itself, ensures that different cell chemistries allow different C-Rates.

With the redox flow cell, we can solve this problem. Similar to the sodium sulphur battery, we use liquid active material. It is no longer bound to the surface of an electrode. This allows us to circumvent the link between volume and surface restriction. We transport the active material to a surface that is large enough for the required power and we store the active material in a volume that is large enough for the required capacity.

8.7.1 Chemistry and main reactions

The basic structure of a redox flow battery is shown in Fig. 8.50. A redox flow battery consists of two tanks and a reaction cell where the reaction takes place. A system of pipes and pumps ensures that the active material is transported from the tanks to the cell.

The reaction cell again follows the structure of an electrochemical storage cell. We have two electrodes, an anode and a cathode, which are separated from each other by a separator. The liquid active materials, called anolyte and catholyte, is pumped into the active cell, where the redox reaction takes place. In this process, the separator, analogous to the electrochemical cell, separates the electrons from each other, but the ions can pass through the separator.

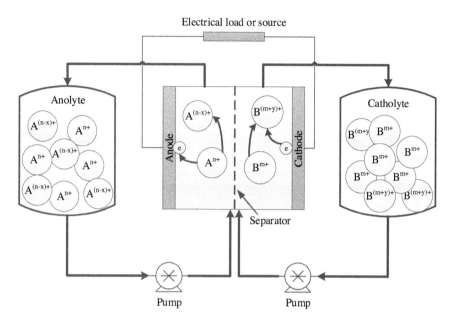

Fig. 8.50 Structure of a redox flow battery: the electrochemical cell works with liquid active materials, the anolyte and the catholyte. Both materials are stored in tanks. Pumps transport the active material to the reaction cell, where the redox reaction takes place.

We see that in both tanks the active material is present in two forms, A^{n+} and $A^{(n-x)+}$. The tank with the catholyte contains B^{m+} and $B^{(m+y)+}$. These are therefore different ions of the respective active material.

Let us look at the charging process. Anolyte and catholyte are pumped between the anode and cathode on both sides. The following reaction now takes place at the anode:

$$A^{n+} + x\,e \rightleftharpoons A^{(n-x)+}. \tag{8.78}$$

The A^{n+} donates x electrons across the anode and becomes an $A^{(n-x)+}$ ion. It is $n > x$. After the reaction takes place, the $A^{(n-2)+}$ ions are transported back into the tank. Via the electric circuit, the electrons reach the cathode and react there with the catholyte:

$$B^{m+} - y\,e \rightleftharpoons B^{(m+y)+}. \tag{8.79}$$

The reaction product is also transported back to the tank after the reaction. Both anolyte and catholyte are present in the tanks as a mixture of $A^{n+}, A^{(n-x)+}$ and $B^{m+}, B^{(m+y)+}$ respectively.

Table 8.18 Overall reactions of different redox flow battery chemistries.

Name	Reaction	Voltage
Vanadium oxide	$V^{2+} + VO_2{}^+ + 2\,H^+ \rightleftharpoons V^{2+} + VO^{2+} + H_2O$	1.26 V
Iron chrome	$Cr^{2+} + Fe^{3+} \rightleftharpoons Cr^{3+} + Fe^{2+}$	1.2 V
Polysulphide bromide	$S_4{}^{2-} + Br_2 \rightleftharpoons 2\,S_2{}^{2-} + 2\,Br$	1.36 V
Zinc cerium	$Zn + 2\,Ce^{4+} \rightleftharpoons Zn^{2+} + 2\,Ce^{3+}$	2.6 V
Zinc chlorine	$Zn + Cl_2 \rightleftharpoons ZnCl_2$	2.12 V
Vanadium bromine	$ClBr_2{}^2 + Vcl_2 + 2\,Cl^- \rightleftharpoons Cl^- + 2\,Br_2{}^- + 2\,VCl_3$	1.1 V

There are a number of substances that can be used for a redox flow cell. In Tab. 8.18 some known reactions are listed (Weber *et al.*, 2011; Sterner and Stadler, 2017). Of these substances shown, the vanadium redox flow battery is the most common. Its great advantage is that the same substances are used in both the anolyte and the catholyte. This means that there is no problem of cross-contamination of the active material via the separator (Tang, Bao, and Skyllas-Kazacos, 2014).

8.7.2 Requirements and behaviour of redox flow batteries

By separating the active material from the electrical reaction cell, the storage capacity is determined solely by the size of the tank. We can make the tank as large as the storage capacity requirements request. But we also have to deal with the question of how we ensure that enough active material is transported to the reaction cell.

ECS-RF 1: AS A design engineer, I WANT to design the tanks and transport system to the required capacity and power demand SO THAT enough active material can be stored and transported to the reaction cell for power delivery.

To realize the required charging power, we need to consider fluid transport as well as electrochemistry.

In both reactions, eqns (8.78) and (8.79), an electron exchange takes place. The voltage at the electrode is composed of three influencing variables:

$$U = U_a + R_c \cdot I + U_{con}. \tag{8.80}$$

The first voltage term in eqn (8.80) is the activation voltage U_a. This is the minimum voltage that must be present for a reaction to take place. The second term results from the internal resistance of the reaction cell R_c and the current I. The last voltage term U_{con} results from the amount of active material on the surface of the electrode. Depending on whether we charge or discharge, the voltage rises or falls sharply. If c_S represents the concentration of active material at the surface and c_B represents the concentration far from the surface, then the voltage U_{con} is given by

$$U_{con} = \frac{RT}{nF} \ln \frac{c_S}{c_B}. \tag{8.81}$$

If the concentration of the required material is greater at the surface than inside the cell, U_{con} is positive; if there is less material at the surface than inside the cell,

U_{con} is negative. To better understand this concentration potential, let's do a simple thought experiment.

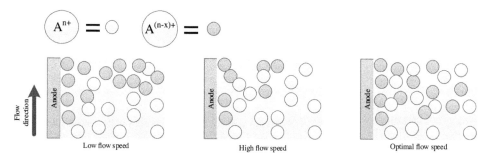

Fig. 8.51 Influence of the flow speed on the amount of active material reacting in the reaction cell.

In Fig. 8.51, we have shown the anode side of a redox flow cell three times. The active material flows at different speeds in the three illustrations. The battery is completely discharged and should now be charged. If we pump the active material through the cell at a very slow speed, as in the first diagram, all A^{n+} can change into $A^{(n-x)+}$ during the transport through the cell. However, this happens much too soon at a very slow transport speed. The concentration of $A^{(n-x)+}$ is identical everywhere in the cell, i.e. $U_{\text{con}} \approx 0$.

But if U_{con} is small, then no electrical forces act on the ions. The remaining $A^{(n+}$ can only reach the electrode by diffusion. We are therefore limited in the charging power.

If we now take the middle case, where the flow speed is much higher, we see that there is not enough time to convert enough A^{n+} into $A^{(n-x)+}$. The concentration of A^{n+} is higher at the surface than inside the cell, so the voltage is greater than zero: $U_{\text{con}} > 0$.

In the third case, we have optimized the flow speed. There is an exchange of charge carriers over the entire course of the transport. The concentration is now balanced even at the top of the cell, so there are still forces acting on the active material that is still available.

To ensure that there is always enough active material in the cell, one should work with a high flow rate. However, this means that the pumping rate and thus the pumping losses are high. This will reduce the overall efficiency of the battery. On the other hand, we only need this high flow rate when the battery is very full or almost empty, because in these two cases the amount of active material is unbalanced. In a wide range of applications, the concentration of the active material has a more balanced distribution and we can work with a lower flow rate. This results in the second requirement for redox flow cells:

`ECS-RF 2: AS A design engineer, I WANT to select the pumping power so that at the set point, the active material can exchange the charge carrier` in sufficient quantity at the electrode SO THAT the charging and discharging requirements are met at the set point.

In Fig. 8.52, different charging and discharging processes with a vanadium redox flow cell are shown (Blanc and Rufer, 2010). The charging and discharging currents were varied in each case.

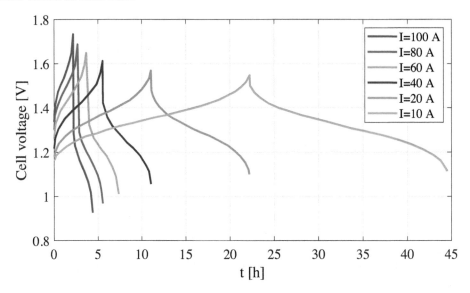

Fig. 8.52 Charge and discharge voltage in a redox flow cell. A vanadium redox flow cell was used, which is charged with different charging and discharging currents.

Exercise 8.36 From previous investigations with electrochemical storage units, we know that the storage capacity also depends in part on the charging current. Using the data from Fig. 8.52, we want to determine the stored energy for $I = 10$ A and $I = 100$ A. What is this in Ah?

Solution: To determine the capacitance, we have to identify how much current is transferred into the cell and withdrawn from it in what time interval. At $I = 100$ A, the charging and discharging process take about 2.2 h. The capacity is therefore:

$$\kappa = 2.2 \text{ h} \cdot 100 \text{ A} = 220 \text{ Ah}.$$

At $I = 10$ A, the charging and discharging process takes about 22.2 h. Thus, the capacity is:

$$\kappa = 22.2 \text{ h} \cdot 10 \text{ A} = 222 \text{ Ah}.$$

The results from Exercise 8.36 show that the capacity is independent of the charging and discharging current. This is where the redox flow cell differs from other cell chemistries. This behaviour can be easily explained. The limiting factor in all previous cell chemistries was that the active material first had to reach the surface of the cell at high currents. If the pump capacity of a redox flow cell is set correctly, this problem no longer occurs. There is always enough active material available, regardless of the desired charging or discharging current.

However, what is noticeable in Fig. 8.52 is the course of the voltage window at different charging currents. As we have already seen with the sodium sulphur battery, cell voltage is higher during charging than during discharging due to the internal resistance. This means that as a function of the charging current, the terminal voltage increases linearly. If we double the charging current, the share of internal resistance in the terminal voltage doubles. (The terminal voltage is the voltage that we measure directly at the contacts of the battery.)

However, we do not observe a linear increase in the redox cell. Instead, we can see in Fig. 8.52 that the voltage peaks rise parabolically. Here, U_{con} still has an amplifying effect on the cell voltage (Weber *et al.*, 2011; Pan and Wang, 2015). This is due to the fact that in a redox flow cell we generate additional voltage by transporting the active material during charging and discharging because there are either not enough or too many charge carriers at the electrode. As we can see, this effect increases the terminal voltage from 1.5 V_{DC} to 1.7 V_{DC}. This does not sound like much for a single cell, but it corresponds to a voltage increase of about 13%. If we connect several cells to a high-voltage system, this effect can cause damage to the components due to overvoltage. This results in the following requirement:

ECS-RF 3: AS A design engineer, I WANT to regulate the supply of the active material and the charging and discharging current so that the difference in concentration does not produce an overvoltage SO THAT there is no damage to the components.

We have discussed that we have three physical mechanisms which have an influence on the performance of the cell: the transportation mechanism, which is relevant to ensure that the active material gets in contact with the electrode, the transport of the charge carrier, and the electrochemical interaction between substances with different potentials. These three effects dominate the losses of the cell in dependence of the state of charge.

If we look at the shape of the curve in Fig. 8.53, we can divide it into three areas (Shigematsu *et al.*, 2011). The first region of the curve is that generated by the activation losses of the active material. We already know this area from the other cell chemistries. This is followed by the area in which the losses are determined by the electrical resistance of the components. In this charging phase, the transport of charge carriers and the transport of the active material are balanced. Towards the end of the charging process, less and less active material is available. As a result, U_{con} gains more influence, and the little active material still available has to find its way to the electrode. Here, the losses due to the diffusion process dominate. In the plot in Fig. 8.53, we see that the voltage rises far above 2 V_{DC} for this cell. Even with a redox flow cell we can overcharge the cell; in this case, as the active material is completely depleted, U_{con} will still continue to rise, resulting in damage to the cell and the active material.

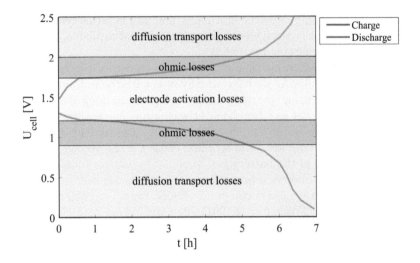

Fig. 8.53 Distribution of the different losses during the charging and discharging process. The losses are determined by the dominant physical process in each specific phase.

ECS-RF 4: AS AN operature, I WANT to ensure that there is no overcharging of the battery during the charging process SO THAT the voltage rise of U_{con} does not damage the active material or the cell.

While for storage capacity we can very easily create the desired capacity by scaling the tanks, power delivery is harder. It depends, as we have seen, on the transport of the active material. We can only increase the power to a limited extent by pumping active material through the cell faster. Instead, we need to increase the surface area. We could do this by increasing the height of the single reaction cell. But then the required pumping power increases because the fluid has a longer path along the cell. Alternatively, we can connect several cells with the optimal geometry in parallel. This is shown in Fig. 8.54. The active material is pumped through several cells in parallel, which allows a higher output.

The cells are lined up one behind the other. Instead of realizing the cathode, insulator, and anode as three individual system components, we use a bipolar plate— that is, a metal plate that is conductive on both surfaces but insulating on the inside. In this way, we combine three components into one component.

We also connect the cells electrically in parallel so that we achieve an average cell voltage of 24 V_{DC} or 48 V_{DC}. This quantity corresponds in its voltage to that of battery modules, which can then in turn be interconnected to form battery systems.

To determine the state of charge of a redox flow battery, we have three possibilities. First, as with any other storage technology, we can determine the state of charge through coulomb counting. Since this measurement method is subject to errors, the second possibility would be to estimate the state of charge via the terminal voltage. However, with this type of measurement, we face the problem of determining the influence of U_{con}. Unlike the internal resistance, which we have to determine once and which is independent of the state of the cell and the tanks, U_{con} depends on

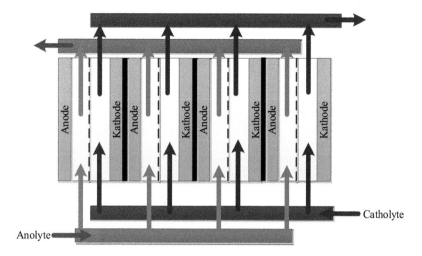

Fig. 8.54 Representation of a redox flow cell stack. Parallel reaction cells can be used to scale the charging power of a redox flow battery.

the concentration of the active material and its temperature inside the cell. Both are influenced by the charging current and the pump power.

The third alternative is to measure directly the concentration of A^{n+}, $A^{(n-x)+}$ on the anode side and B^{m+} and $B^{(m+y)+}$ on the cathode side. The ratio of both substances in the respective tanks represents the state of charge of the system very well and can be measured at any time.

Let us look at the ageing behaviour of redox flow batteries. The cycle ageing that we observe with previous cell technologies has a negligible share in the ageing. The active material undergoes a reversible reaction in which charge carriers are exchanged. A chemical reaction with possible side reactions does not take place. Due to the permanent movement of the active material, spatial anisotropies—spatial differences in concentration at the electrode or in the active material—are prevented.

The situation is different with calendar ageing. Here, too, the active material is hardly affected. However, the substances are dissolved in the anolyte and catholyte, and they can settle out again over time. Since a redox flow battery also contains moving parts, these are subject to wear and tear. Pumps and valves, for example, have to be maintained and replaced regularly. Leaks can occur, leading to a loss of active material. All this limits the life of the system and increases maintenance costs.

Unlike previous technologies, however, we can replace the active material at any time, and this also applies to the mechanically stressed parts. This means that there is always the possibility of extending the service life of the redox flow battery in the long term. We do not have this possibility with the other technologies. Since the active material is permanently embedded in the battery, replacing it always means replacing the battery.

Let's summarize the system components of a redox flow battery. In Fig. 8.55, the system components are shown. A `RedoxFlowBattery` has at least two `Tanks` in which the `Electrolyte`, either `Anolyte` or `Catholyte`, is stored. Since these chemicals are

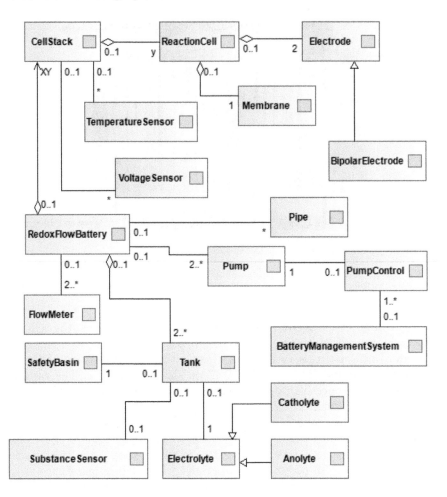

Fig. 8.55 Overview of the system components of a redox flow battery.

not to be released into the environment, the `Tanks` are provided with a `SafetyBasin`. In the event of a leak, the leaked fluid is collected there before it can escape into the environment.

With the previous battery cells, we had a design principle that kept recurring. y cells were connected in series to form a string, and x strings were connected in parallel to form a module. In this way, the voltage and maximum current of a cell could be adjusted. Here the wiring principle is a little different. We have already seen that within a stack the y `ReactionCell` components are connected in series, and the active material then flows through the cells in parallel. However, within a stack, the power is determined by the size of the `ReactionCell` area. In a `CellStack`, therefore, no additional parallel connection of the cells takes place. For the `RedoxFlowBattery`, the `CellStack` components are then connected in parallel and in series, depending on which current and voltage values are to be achieved.

The `BatteryManagementSystem` also monitors the functions of the entire battery in this system. Voltage monitoring is carried out for each cell stack, for which the BMS measures the voltage of the cell with the aid of a voltage sensor. A `TemperatureSensor` measures the cell temperature. To determine the state of charge, a `SubstanceSensor` is inserted into the tanks, which measures the concentration of the anolyte and the catholyte.

8.7.3 Application example: Integrating a redox flow battery to a wind farm in an island grid

Let us now implement the application example presented in Section 8.6.3 with a redox flow battery. The operational management and equations do not change.

With a redox flow battery, we can configure power and capacity separately. Therefore, we first determine the size of the tanks we will need for a redox flow system with vanadium as the active material. The energy density of redox flow cells is between 25 and 75 Wh/l, depending on the technology. Compared to fossil fuels with approximately 10 kWh/l or other electrochemical storage technologies, for example lithium ion batteries with approximately 1 kWh/l, this is a significantly lower energy density. This becomes clear in the volume requirement.

Exercise 8.37 For our island system, we need a storage capacity of 5 MWh. What is the volume of active anolyte and catholyte if the energy density is 50 Wh/l? How high would a tank have to be if it had a diameter of 4 m?
Solution: The volume is given by:

$$V = \frac{5 \text{ MWh}}{50 \text{ Wh/l}} = \frac{5{,}000 \text{ kWh}}{50 \text{ kWh/m}^3} = 100 \text{ m}^3.$$

Since anolyte and catholyte are used in equal quantities and in a vanadium redox flow cell anolyte and catholyte are made of the same material, we need a total of 200 m³ volume.
The required height of the tank is calculated as:

$$2\pi r^2 h = 100 \text{ m}^3$$
$$h = \frac{100 \text{ m}^3}{2\pi r^2}$$
$$= \frac{100 \text{ m}^3}{2\pi 2^2 \text{ m}^2}$$
$$= 3.97 \text{ m}.$$

We see that due to the low energy density of a redox flow system, large tank systems and transport systems may be needed. Nevertheless, we want to pursue the design further and lay out the cell stack. In our example, we assume that the cell stack is to generate a voltage of 48 V. The cells are built to carry a maximum current of 100 A.

Exercise 8.38 As we know from Fig. 8.52, the average voltage of a redox flow cell is 1.5 V_{DC}. How many reaction cells must be connected together to maintain a nominal voltage of 48 V_{DC}?

What power could this cell reach with a maximum current of 100 A? What would be the maximum and minimum voltage if the operation management is careful to always stay within the range of ohmic losses only?

For our application, a power of 150 MW is required. How many stacks are needed?

Solution: The number of cells is given by:

$$N = \frac{48 \text{ V}_{DC}}{1.5 \text{ V}_{DC}} = 32.$$

According to Fig. 8.53, the diffusion transport losses occur at about 2 V_{DC}. Therefore, the maximum voltage of the stack is:

$$U_{max} = 32 \cdot 2 \text{ V}_{DC} = 64 \text{ V}_{DC}.$$

For discharging, the voltage equals 1 V_{DC}. The minimum voltage is therefore:

$$U_{min} = 32 \cdot 1 \text{ V}_{DC} = 32 \text{ V}_{DC}.$$

The nominal power we can achieve with the stack at a voltage of 1.5 V_{DC} is given by:

$$P = 48 \text{ V}_{DC} \cdot 100 \text{ A} = 4.8 \text{ kW}.$$

The stack number is therefore:

$$N_{stacks} = \frac{150 \text{ kW}}{4.8 \text{ kW}} = 31.25 \approx 32.$$

The stack power of 4.8 kW is quite high compared to other technologies, so just 32 stacks would be perfectly sufficient to deliver the required power. This is because the liquid active material is removed after its reaction and new material is added. To deliver the required power, we only need enough surface area at the electrode and enough active material.

To build the storage power plant, we can essentially follow the same ideas as we did for the high-temperature battery. What is new here is that we have to pay attention to the transport of the active material. Since the pipe and pump system has to be maintained, it makes more sense to divide the storage tanks into smaller, separate groups.

8.7.4 Conclusion

We have learned about redox flow technology. Unlike previous electrochemical storage technologies, the active material is pumped to the electrode. It is therefore no longer a fixed component of the electrode or the electrolyte. This has the advantage that capacity and performance can be considered separately when designing a storage system. However, the lower energy density of materials that can be used in a redox flow is a disadvantage and limits the applications.

When designing and operating a redox flow battery, it is important to note that the transport of the active material has an influence on the charging and discharging performance and the voltage of the reaction cell. We had already seen with the lead acid battery that a high charging current can lead to the effect that active material is not available quickly enough at the electrode. With a redox flow battery, this effect is stronger and can lead to damage of the electronics.

8.8   Conclusion

In this chapter, we have learned about different types of electrochemical storage technologies: the lead acid battery, the lithium ion battery, the sodium sulphur battery and the redox flow batteries. Although the individual technologies differ in their characteristics, the basic mechanism and a number of requirements are identical. A reaction takes place that reduces the number of electrons in a substance and a second reaction takes place that simultaneously recaptures these electrons. If these reactions are reversible, then we have a secondary battery that we can charge and discharge.

In the next chapter, we turn to the last major group of storage technologies, the chemical storages. Whereas electrochemical storages really only exchange electrons, chemical storages store energy by transforming substances.

Chapter summary

- An electrochemical cell consists of two electrodes which are placed in an electrolyte. Between these two electrodes, a separator may be placed to ensure that no short circuit will happen.
- The basic reaction is that on the anode side, b parts of B change into d parts of D, freeing n electrons, while on the cathode side a parts of the substance A receive these electrons and turn into d parts of D.
- As in every reaction, heat will be generated, which results in losses. Each cell chemistry has its own roundtrip efficiency.
- The open-circuit voltage of an electrochemical cell depends on the state of charge. Its characteristics depend on the cell chemistry.
- We can aggregate battery cells to larger modules and modules to larger systems by connecting cells or modules in series to strings and connecting these strings in parallel. The voltage window is defined by the number of cells or modules in series and the maximum charge and discharge current can be adjusted by connected the cell or module strings in parallel.
- Two different ageing mechanism are known: calendaric ageing and cycle life.
- Calendaric ageing is caused by degradation of the active material, which can be forced by increasing the temperature of the cell.
- Cycle life is caused by the degradation of active material during the charge and discharge process. It's mainly influenced by the DoD.
- In a battery string, the components with the smallest capacity dominate the behaviour of the complete string. To reduce this effect, balancing systems are used. We distinguish between active and passive balancing methods.

- Active balancing methods distribute energy from cells with a higher state of charge to cells with a lower state of charge. This can be done with capacitors, transformers, or inductances.

- Passive balancing methods equalize the state of charge of the cells by discharging the cells with a higher state of charge using a resistor.

- Four cell chemistries are very common: lead acid, lithium ion, high-temperature, and redox flow batteries.

- Lead acid batteries consist of two lead electrodes. One is coated with lead oxide. Both electrodes are planced into an electrolyte consisting of water and sulphuric acid.

- In the case of lead acid batteries, the substances in the electrolyte are part of the reaction.

- Lead acid batteries have an open-circuit voltage between 1.8 V_{DC} and 2.2 V_{DC}. They are able to provide very high C-Rates, but show a quite small capacitance in this case. Lead acid batteries prefer to be used with C-Rates below 1C.

- Lithium ion batteries are a family of different combinations of anode and cathode materials. The anode materials are carbon, silicon, or lithium titanium oxide. The cathode materials are lithium metal oxides.

- In the case of lithium ion batteries, a number of lithium ions are transported from the anode to the cathode material and vice versa. In parallel, the electronics are transported. The electrolyte is not part of the reaction.

- In comparison to other cell chemistries, the open-circuit voltage of lithium ion batteries is larger. It can go up to 4.2 V_{DC}. The voltage window depends on the combination of anode and cathode material.

- Lithium ion batteries can reach a state, where the cell chemistry becomes unstable. Therefore, lithium ion batteries shall be used in a defined temperature and voltage window.

- If lithium ion batteries become unstable a thermal runaway may occur, which can cause danger for man and mashine. Therefore dedicated safety measures needs to be implemented.

- In high temperature batteries like the natrium sulphur batteries the active material are melted. Here, we do not have an electrolyte. The active material is flowing through a separator.

- Natrium sulphur batteries have a quite high cycle life because the active material is melted and therefore no mechanical degradation can occur.

- Redox flow batteries only uses fluid active material. They have the advantage that the capacity and the power rate can be varied simply by increasing the size of the fluid tank and the size or number of the reaction cells.

- To have a good performance of the redox flow cell, it needs to be assured that enough active material is transported into the cell.

9

Chemical storage systems

9.1 Introduction

In this chapter, we will look at technologies that use chemical conversions to store energy. This type of energy storage is very familiar to us. Plants use sunlight to convert carbon dioxide and water into oxygen and sugars and store these in the form of biomass. The plants use this to grow. We harvest the plants, store them, and use the stored energy at a later time. The human body is also a chemical storage. We eat breakfast in the morning to have energy for the way to school or work. Once we get there, we end up discharging our bodies by working or studying. We get hungry when our storage threatens to run out.

With electrical storage systems, we store energy by changing the distribution of charge carriers or by permanently setting charge carriers in motion. In electrochemical storage systems, charge carriers are exchanged in the active material. The chemical reactions that accompany the main and secondary reactions usually serve to enable the exchange of charge carriers. In chemical storage technologies, we use chemical reactions to be able to store the supplied energy. In doing so, chemical transformation processes take place. These are usually not reversible—that is, the process with which we store energy is not the process with which we extract energy. Let's take fossil fuels as an example. In a complicated process, biomass became oil and coal over a very long period of time. But this energy is received back through combustion. The heat is used to generate kinetic energy and the kinetic energy is used to generate electrical energy. Using fossil fuels as an example, we also see that chemical storage enables us to separate storage and consumption on a much larger space and time scale.

Chemical storage makes use of the three elements hydrogen (H_2), oxygen (O_2) and carbon (C). The different technologies used in chemical storage can be represented and related to each other in the CHO triangular diagram (Cairns and Tevebaugh, 1964; Damiani and Trucco, 2009; Sterner and Stadler, 2017). The three sides of the diagram represent the concentration of the substances C, H, and O. The pure substances are located on the vertices of the triangle of Fig. 9.1. The lines show the relative molar fraction of the substances. Take water H_2O, for example. Water consists of one third O and two thirds H_2. Carbon is not present at all. We therefore enter water on the base of the triangle, as this line describes all those substances that do not contain carbon. The base describes the oxygen content, which is one third for water. We read off the hydrogen content by going diagonally to the hydrogen axis and reading off what percentage we have there. We should be at two thirds, so 66%, which is the case.

Energy Storage Systems. Armin U. Schmiegel, Oxford University Press. © Armin U. Schmiegel (2023).
DOI: 10.1093/oso/9780192858009.003.0009

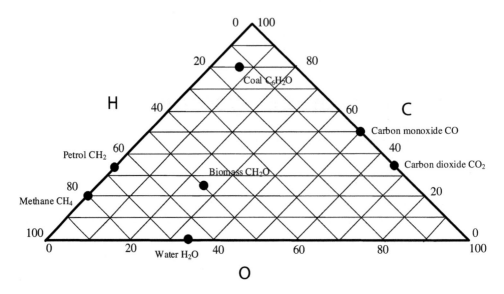

Fig. 9.1 The different technologies used in chemical storage can be shown in the CHO triangular diagram.

Let's look at methane, CH_4. Since methane contains no oxygen, methane is on the left side of the triangle. Methane is 80% hydrogen and 20% carbon. We can read the hydrogen content directly from the left side of the triangle. We obtain the carbon content by walking along a line parallel to the base to the carbon axis, the right side of the triangle, where we arrive at 20%, as expected.

In this diagram we have plotted the most important substances for chemical storage: methane (CH_4CH_2), coal (C_6H_2O), biomass (CH_2O), water (H_2O), carbon monoxide (CO) and carbon dioxide (CO_2). We can now describe the main chemical storage techniques as steps within the diagram.

Let's start with hydrogen technology. Here, water H_2O is split into O and H_2. This corresponds to a change from the point corresponding to water to the right corner point of the triangle.

Next, let us consider the use of fossil fuels. At the beginning, there is the biomass CH_2O. This is produced from $CO_2 + H_2O$ through photosynthesis. The biomass becomes coal C_6H_2O through a carbonization process. If we burn the coal, we produce $CO_2 + H_2O$ again. We actually have a cycle process here. But since the process of turning biomass into coal takes a very long time, we use energy that has been stored for millions of years.

The process of creating biofuels or fossil oil is similar. In this case, $CO_2 + H_2O$ are taken from biomass, which then becomes CH_2 through an oiling process. Combustion then releases $CO_2 + H_2O$ again.

Another process is methanization. Here, methane is produced from $CO_2 + H_2O$ methane, which is then used through combustion. Since the methanization process can be driven, for example, by biogas plants or by methanization plants powered by solar or wind energy. This process is also a circular process.

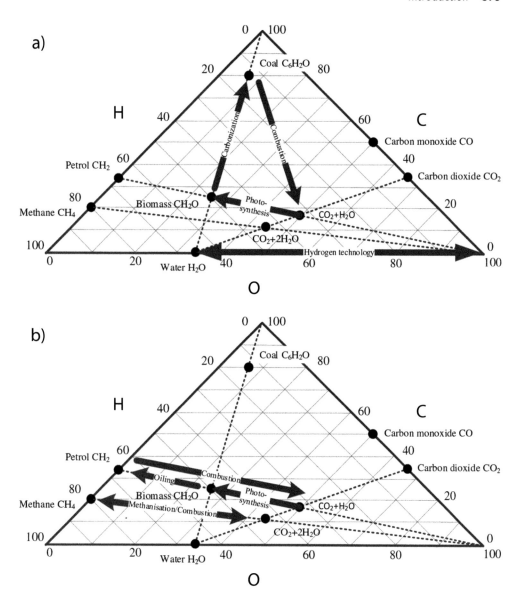

Fig. 9.2 The main chemical storage technologies can be represented as steps within the CHO triangle diagram. a) hydrogen technology and carbonisation. b) fossil oil or bio oil and methanisation.

We have simplified very complicated processes and material compositions here. Biomass consists of many different substances (for example starch $(C_6H_{10}O_5)_n$ or glycose $C_6H_{12}O_6$). Fuels contain very different hydrocarbon chains: petrol, for example, consists of light flammable hydrocarbons (pentane (C_5H_{12}), hexane (C_6H_{14}) and heptane (C_7H_{16})). An example of biofuel, on the other hand, is bioethanol (C_2H_6O).

But using the CHO triangle diagram, we were able to get an overview of the different chemical storage technologies. In the next section, we will look at the general requirements for chemical storage. Then we will take a closer look at two important storage technologies: hydrogen technology and methanization. We will also get to know the fuel cell. This is a technology that allows us to produce electricity from hydrogen, for example.

9.2 General function and requirements

Unlike electrochemical storage, chemical storage does not separate charge carriers. No electric current flows. The basic idea of a chemical storage device is to use mechanical or electrical power to drive a certain chemical process. In this process, a reaction takes place whose products are stable enough to be used later. For example, we can use electrical power to split water into its components, hydrogen (H_2) and oxygen (O). We can use the energy stored in this process by burning the H_2 with O, forming water (H_2O) later again.

As in Chapter 8, we can describe the basic equation of chemical storage as follows:

$$aA + bB \rightleftharpoons cC + dD. \tag{9.1}$$

a parts of substance A and b parts of substance B are converted into c parts of substance C and d parts of substance D. In this reaction, we observe two effects. The substances combine with each other. This process consumes energy, which is described by the reaction enthalpy ΔH_R. ΔH_R thus represents the energy stored in the substance in the storage process from aA + bB to cC + dD.

However, we also observe that in eqn (9.1) either heat is required for the reaction to occur at all, or heat is released. This is the heat of reaction $T\Delta S$, where T is the reaction temperature and ΔS is the change in entropy of the system from Aa, bB, cC, and dD. If the heat of reaction is zero, all the stored energy is available for reuse. For a reaction to be suitable for storage, the heat input $T\Delta S$ should be as low as possible for the reaction:

$$\Delta G_R = \Delta H_R - T\Delta S. \tag{9.2}$$

The term $T\Delta S$ is responsible for the efficiency of the storage process, and the term ΔH_R for the amount of energy that is stored. If $\Delta G_R = \Delta H_R$, we would have an ideal reaction.

Not every combination of substances is suitable. A very obvious requirement is that we can control the storage process and the energy extraction.

CS 1: AS AN operator, I WANT the chemical storage process to have an industrially controllable reaction SO THAT I can safely charge and discharge the storage on an industrial scale.

The CS 1 requirement means that we do not want to use reactions that only store energy on a laboratory scale. Our goal is to be able to store energy on a large, industrial scale. Chemical storages are also more commonly used to store energy on a megawatt-hour scale. This is also due to the fact that chemical storage systems are complex in their construction. It is important that the reaction can be realized in a reproducible, cheap, and controlled manner.

The application of chemical storage is such that electrical energy and, if necessary, thermal energy are used to start a chemical reaction. The reaction product is then physically stored. Since the reaction products are usually gaseous or liquid or liquefied, the substances can be stored and transported in tanks or pumped into a pipe system, where they are then transported to different storage sites or directly to the consumer.

Discharge takes place by physically removing the reaction product—that is, by refuelling (a vehicle) or by removing it from a transport system, for example a natural gas pipeline. This results in the requirement that the reaction product must not be short-lived—that is, decompose after a certain time. It should be possible to fill the reaction product into a container and be able to extract it again later. In summary, the following must apply:

CS 2: AS AN operator, I WANT to have a chemically stable reaction product SO THAT I can store it over a long period of time.

It is not enough that CS 1 and CS 2 are satisfied. There are a number of chemical processes that are industrially feasible and whose end products can also be stored. What is produced here as a reaction product should also be usable—that is, with the help of the reaction product one should be able to generate energy.

CS 3: AS AN operator, I WANT to use a reaction product from which I can extract stored energy SO THAT I can do more than simply store energy.

As we saw in Fig. 9.2, basically anything that can be burned satisfies CS 3. The energy stored in biomass, petrol, coal, and hydrogen is usually recovered by combustion. The thermal energy recovered is then converted back into kinetic energy and then into electrical energy. The efficiency of this process is not high. If we recall, of the 10 kWh of energy stored in a litre of fuel, we use just 3–4 kWh in an internal combustion engine. The advantage of chemical storage, however, is that it often has a higher energy density. For example, a litre of lithium ion battery contains roughly a half kilowatt hour and a redox flow battery only $25 - 75$ Wh. Another advantage is that the products of the chemical storage process can be easily transported. It takes three to four minutes to refuel a car with a 50 litre tank.

We can increase the efficiency a little by also using the thermal energy that is released during the combustion process. This is done in combined heat and power systems. Here, on the one hand, the kinetic energy is converted into electrical energy, and the heat is additionally fed into a heat cycle. And there are also applications in which the combustion gases are fed directly into a greenhouse, where they support the formation of new biomass.

Another alternative is the fuel cell, which we will learn about in Section 9.3.2. Here, the recovery of the stored energy takes place via a different oxidation process.

9.3   Hydrogen as a storage technology

We begin our consideration of chemical storage with hydrogen technology. The basic equation is:

$$2\,H_2O \rightleftharpoons 2\,H_2 + O_2. \tag{9.3}$$

We store energy by splitting water into its components, hydrogen and oxygen. The reverse reaction, which takes place through combustion, releases energy and corresponds to the discharge process. We have already learned about this reaction in Fig. 9.2. For the lead acid battery in Section 8.3, eqn (9.3) appeared as a side reaction. As soon as the battery was fully charged, the water in the electrolyte could be split by a further sustained charging current.

For the application, the discharge process via the combustion of hydrogen has a great advantage. Since 21% of the atmosphere consists of oxygen, we only have to transport the hydrogen. We can take the oxygen from the atmosphere. However, to be able to transport more hydrogen, we will transport it in compressed or in liquid form.

If we look at the energy content of liquid hydrogen in relation to its weight, we find that it has an energy density of $33.3\,\frac{kWh}{kg}$, which is three times higher than petrol. $(11.1\,\frac{kWh}{kg} - 12.1\,\frac{kWh}{kg})$, or diesel $(11.8\,\frac{kWh}{kg} - 11.9\,\frac{kWh}{kg})$, and an energy density 300 times higher than that of a lithium ion battery $(0.1\,\frac{kWh}{kg})$. The latter is not so surprising, because we know that in a lithium ion battery, the energy is stored in the active material on the surface of the electrodes. Electrolyte, electrode, and separator add to the weight, even though they don't contain energy. With petrol and diesel, the energy is stored chemically in long hydrocarbon chains, whereas with hydrogen, the energy is stored in the separation process.

In terms of volume, however, the ratio is unfortunately reversed. Liquid hydrogen has an energy density of $(2.1\,\frac{kWh}{l} - 2.4\,\frac{kWh}{l})$. Here, petrol $(8.2\,\frac{kWh}{l} - 8.8\,\frac{kWh}{l})$ and diesel $(\approx 9.8\,\frac{kWh}{l})$ have a significantly higher energy density. Compared to the energy density of a lithium ion battery $(0.25\,\frac{kWh}{l} - 0.67\,\frac{kWh}{l})$, hydrogen proves to be more energy dense.

In order to use hydrogen, we must consider the issue of safety. The oxygen in the atmosphere and the hydrogen together form the highly flammable oxyhydrogen gas. In order to comply with general requirement G 5, we must therefore derive two additional requirements:

`CS-H 1: AS A Design Engineer, I WANT to make sure that the storage tank does not form cracks SO THAT no hydrogen can leak and then form oxyhydrogen.`

Since a violation of CS-H 1 is dangerous, it makes sense to formulate another safety requirement for the unlikely event as well:

`CS-H 2: AS A Design Engineer, I WANT to implement measures that will detect a rupture or leakage of hydrogen SO THAT I can initiate appropriate countermeasures.`

There are three possible methods for storing hydrogen. The simplest and most obvious method is to compress the gas in a container. In doing so, we increase the density of the gas $\rho = \frac{kg}{V}$ by reducing the volume. From Section 6.4, we know that there is a relationship between pressure, volume, and temperature for a gas. Assuming we compress isothermally—that is, we ensure that the temperature of the gas remains constant—the pressure after compression is given by:

$$p_2 = \frac{p_1 \cdot V_1}{V_2}, \tag{9.4}$$

where p_1, V_1 are pressure and volume before compression and p_2, V_2 are pressure and volume after compression. We see that a decrease in volume $V_1 > V_2$ causes the pressure to increase.

Exercise 9.1 We want to remove the petrol tank from our car and switch from burning fossil fuels to burning hydrogen. How much does the volume of gaseous hydrogen ($\rho_{H,m} = 0.082658 \frac{kg}{m^3}$) theoretically need to be compressed so that we get the same energy density of petrol ($\rho_{B,V} = 8.5 \frac{kWh}{l}$)? The energy density of hydrogen is $33.3 \frac{kWh}{kg}$.

Solution: We must first determine what the volumetric energy density $\rho_{H,V}$ of gaseous hydrogen is:

$$\rho_{H,V} = \frac{0.082658}{33.3} \frac{kg}{m^3} \frac{kWh}{kg}$$

$$= 2.48 \cdot 10^{-3} \frac{kWh}{m^3}$$

$$= 2.48 \frac{kWh}{l}.$$

We see that in the gaseous state, the volumetric energy density is much worse at normal pressure and temperature conditions. This is not surprising, since we are comparing the energy density of a gas with that of a liquid. In order to achieve the same energy density in purely mathematical terms, we have to determine the ratio between the volumetric energy density of petrol and the volumetric energy density of hydrogen gas:

$$N = \frac{8.5 \frac{kWh}{l}}{2.48 \frac{kWh}{l}} = 3.427.$$

To be able to store the same amount of energy in a hydrogen tank with a volume of 1 litre, we need to compress 3.427 l of hydrogen!

The considerations from Exercise 9.1 show that storing hydrogen by compression certainly provides an improvement, but we do not come close to the energy density that fossil fuels provide. This is not surprising, of course, since fossil fuels are made up of complex hydrocarbon chains that contain within them a high chemical binding energy, whereas hydrogen consists of only one proton. But since we have to work with very high pressures, the approximations of the ideal gas law are also no longer so applicable. We therefore have to work with a somewhat more precise law, the Van der Waals model, to determine the relationship between pressure and volume.

The Van der Waals law is (Blundell and Blundell, 2010):

$$\left(p + a \left(\frac{n}{V}\right)^2\right)(V - nb) = n\,R\,T. \tag{9.5}$$

Here a represents the additional repulsive force between the gas particles and b the volume occupied by n mol gas particles. For hydrogen, $a = 2.476 \cdot 10^{-2} \mathrm{m}^6$ Pa mol^{-1} and $b = 2.661 \cdot 10^{-5}$ m^3 mol^{-1} (Züttel, 2004).

Transforming eqn (9.5) gives us an expression for the pressure:

$$p = \frac{n\,R\,T}{V - nb} - a\left(\frac{n}{V}\right)^2. \tag{9.6}$$

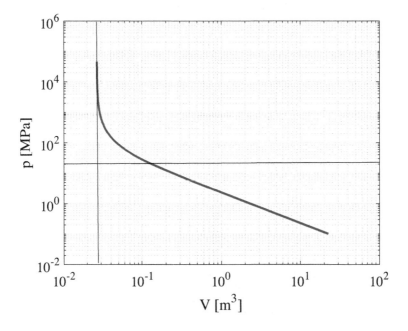

Fig. 9.3 Pressure of the hydrogen as a function of the volume. In this case, hydrogen is described as a Van der Waals gas, with $a = 2.476 \cdot 10^{-2} \mathrm{m}^6$ Pa mol^{-1} and $b = 2.661 \cdot 10^{-5}$ m^3 mol^{-1}.

The first term is proportional to $\frac{1}{V}$ and its course corresponds to that of the ideal gas law. But the second term goes with $\frac{1}{V^2}$. This becomes relevant, especially in the case of strong compressions. We now want to compress one kilogram of gaseous hydrogen. The temperature of the gas is to remain at $T = 273$ K and normal conditions apply to the pressure—that is, $p_1 = 101.325$ Pa. With these general conditions, a kilogramme occupies a volume of 22.4 m^3. We now assume that we compress this quantity step by step and measure the pressure in the process. In Fig. 9.3, the pressure curve is shown. High-pressure vessels can withstand a pressure of 20 MPa. In Fig. 9.3, this limit is plotted as a horizontal line. This limit is reached at a volume of about 0.15 m^3. We would thus have an energy density of 22.2 $\frac{\mathrm{kWh}}{\mathrm{m}^3}$. If we reduce the volume further, we find that the pressure increases faster and faster until we reach a value of about 0.027 m^3. Now the hydrogen begins to liquefy. In the process, we reach a density of 36 $\frac{\mathrm{kg}}{\mathrm{m}^3}$. This represents the maximum density of gaseous hydrogen at $T = 273$ K.

Exercise 9.2 The maximum density of gaseous hydrogen at $T = 273$ K is $36 \frac{\text{kg}}{\text{m}^3}$. What is thus the theoretically achievable volumetric energy density?

Solution: We know that the gravimetric energy density of hydrogen is $33.3 \frac{\text{kWh}}{\text{kg}}$. Thus, the volumetric energy density is given by:

$$\rho_{\text{H,V}} = 36 \, \frac{\text{kg}}{\text{m}^3} \cdot 33.3 \, \frac{\text{kWh}}{\text{kg}} = 1.198.8 \, \frac{\text{kWh}}{\text{m}^3} \approx 1.2 \frac{\text{kWh}}{1}.$$

We see that even close to the phase transition, we can bring the energy density of gaseous hydrogen to just under 1/8 of the volumetric density of petrol.

However, we must also take into account that work is done in the compression process. The work that has to be done in an isothermal compression process is calculated for the Van der Waals gas as follows:

$$W = R \, T \, \ln \frac{V_2 - b}{V_1 - b} + a \left(\frac{1}{V_2} - \frac{1}{V_1} \right). \tag{9.7}$$

Exercise 9.3 Since our pressure vessel can only withstand a pressure of 20 MPa, we can only compress the 22.4 m^3 to 0.15 m^3. We keep the temperature fixed during the compression process. How much energy does this compression process cost us?

Solution: We use eqn (9.3) and insert the initial, as well as the final volume:

$$W = R \, T \, \ln \frac{V_2 - b}{V_1 - b} + a \left(\frac{1}{V_2} - \frac{1}{V_1} \right)$$

$$= R \, T \, \ln \frac{0.15 \text{ m}^3 - b}{22.4 \text{ m}^3 - b} + a \left(\frac{1}{0.15 \text{ m}^3} - \frac{1}{22.4 \text{ m}^3} \right).$$

$$= -3.156 \text{ kWh}.$$

The negative sign shows that we have to apply energy here. This corresponds to about 30% of the energy we want to store.

Transporting hydrogen in gas containers is an established technology. Pressures of 35–70 MPa can be achieved with carbon-fibre pressure vessels. During this storage process, safety measures must be taken; should a rupture occur, it must be ensured that the container 'only' opens and does not tear into pieces. Since the compression process itself also costs energy, it should be designed efficiently.

Let us now look at the second possibility of storing hydrogen. Instead of 'only' compressing it, we can liquefy the hydrogen. We make the phase transition and hope to obtain a significantly higher volumetric energy density. The boiling temperature of hydrogen, the temperature at which liquid hydrogen becomes gaseous, is about 20 K under a normal pressure of 101.3 kPa. If we look at the evaporation process more closely, we find that part of the hydrogen evaporates at 20.39 K and another part of the hydrogen evaporates at 20.26 K. The two hydrogens differ in the total spin of the nucleus. In the first case, the spins of the fermions are anti-parallel—one therefore speaks of para-hydrogen—in the second case, the spins of the fermions are parallel—one therefore speaks of ortho-hydrogen. Apart from the different boiling temperatures, both types of hydrogen have different physical properties. At room temperature, about 25% of hydrogen is para-hydrogen and 75% is ortho-hydrogen. During the conversion of ortho-hydrogen into para-hydrogen, energy is released. As a result, the gas heats up again during the cooling process due to the conversion from ortho-hydrogen to para-hydrogen. To prevent this, the proportion of ortho-hydrogen in the gas to be liquefied is deliberately reduced by means of suitable catalysts which accelerate this conversion.

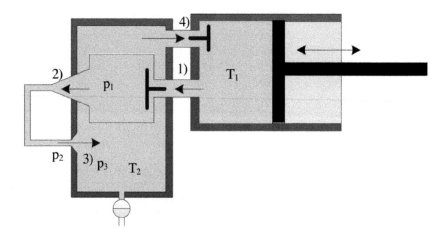

Fig. 9.4 Liquefaction of a gas using Linde's method or Claude's process. 1) A piston presses the gas to be cooled into a chamber; 2) there it is first compressed to the pressure p_2 with the pressure p_1; 3) and then expanded into a second chamber. In the process, the pressure drops to p_3 and the temperature is also reduced. The cooled gas returns to the piston chamber.

For liquefaction, it is exploited that gases cool down during expansion. Linde's method or Claude's process can be used for this. To liquefy a gas, it is first compressed and then expanded. The temperature drops from T_1 to T_2. This process is illustrated in Fig. 9.4. The process is driven by a piston. The gas with its initial temperature T_1 is first introduced into a cooling chamber. The gas is then compressed from pressure p_1 to pressure p_2 and then brought to p_3 by an adiabatic expansion. In the process, the gas cools down to T_2. The cooling chamber is surrounded by the cooled gas and pre-cooling already takes place here. The now colder gas is sucked back into the flask and the process is repeated. As soon as the boiling temperature is reached, the gas begins to liquefy and collects at the bottom of the cooling chamber. To be able to use the process for hydrogen, the hydrogen is first precooled to below 203 K.

For the gaseous hydrogen, we had determined a maximum density of 36 $\frac{kg}{m^3}$. Liquefaction increases this density to 71 $\frac{kg}{m^3}$.

Exercise 9.4 We have not yet given up hope and want to put a tank of liquid hydrogen in our vehicle. The previous tank held 70 litres of diesel ($\rho_{D,V} = 9.7 \, \frac{kWh}{l}$). How big would the tank with liquid hydrogen have to be if the density of hydrogen is $\rho_H = 71 \, \frac{kg}{m^3}$ and the energy content is $\rho_{H,G} = 33.3 \, \frac{kWh}{kg}$?

Solution: We first calculate the volumetric energy density of hydrogen:

$$\rho_V = \rho \cdot \rho_{H,G} = 71 \, \frac{kg}{m^3} \cdot 33.3 \, \frac{kWh}{kg} = 2.364 \, \frac{kWh}{m^3} = 2.365 \, \frac{kWh}{l}.$$

We see that the volumetric energy density of liquid hydrogen is also smaller than that of diesel:

$$N = \frac{\rho_{D,V}}{\rho_{H,V}} = \frac{9.7 \, \frac{kWh}{l}}{2.365 \, \frac{kWh}{l}} = 4.101.$$

To transport the same energy, we have to put a hydrogen tank in the vehicle, which has a size of $4.101 \cdot 70 \, l = 287.01 \, l$!

As we saw in Exercise 9.4, the energy density for liquid hydrogen is still not high enough. We still need four times more volume to transport the same amount of energy in the form of fossil fuels. But this is not the relevant quantity for our application! What is relevant is how much usable energy we can extract from a storage process. We know that an internal combustion engine has an efficiency of approx 30–40%. The factor four in the volume is only relevant if we also burn the hydrogen. However, in Section 9.3.2, we will learn about a technology that allows us to use hydrogen much more efficiently.

The third way to store hydrogen is to chemically or physically bond the hydrogen to another material. One form of bonding is to physically bond the hydrogen to the surface of another material. Carbon is often used here because it has a large surface area in relation to its mass. Another variant is metals that form metal hybrids when hydrogen is added.

The basic principle in this type of storage is that the hydrogen flows over the surface of the storing material. Since we are interested in high energy density, this is a porous material through which the gas flows. The gas can be bound to the surface of the metal by two mechanisms. A chemical reaction can occur:

$$M + xH_2 \rightleftharpoons MH_x + \text{heat}. \tag{9.8}$$

Here M is the metal used and x is the number of atoms bonded per metal compound. In this method, heat is needed either in the charge or in the discharge process.

The second variation is an electrochemical reaction; here additional charge carriers are used to start the reaction:

$$M + xH_2O + xe^- \rightleftharpoons MH_x + xOH^-. \tag{9.9}$$

Whereas with pressure vessels and with liquid hydrogen we had the problem that hydrogen could escape, this type of storage is safe. The hydrogen is chemically bonded, and as long as no energy is added to the metal hybrid storage, no detachment of the hydrogen takes place.

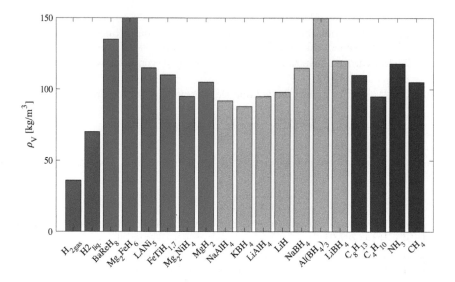

Fig. 9.5 Volumetric density of various hybrid materials suitable for storing hydrogen.

Fig. 9.5 shows various metal hybrids suitable for storage (Züttel, 2004; IEA, 2006). The largest gravimetric density achieved so far with these technologies is $150 \; \frac{kg}{m^3}$, which is a doubling of density compared to liquefaction.

Exercise 9.5 We are still looking for a storage option that allows us to take the same tank volume for our combustion car with hydrogen. No matter what storage method we use, the energy density of hydrogen remains $\rho_{H,G} = 33.3 \; \frac{kWh}{kg}$. To what gravimetric density would we need to compress hydrogen to match the volumetric density of gasoline ($\rho_{B,V} = 8.5 \; \frac{kWh}{l}$ or diesel $\rho_{D,V} = 9.7 \; \frac{kWh}{l}$)?

Solution: From Exercise 9.4, we know that gravimetric energy density is the product of density and volumetric energy density:

$$\rho_V = \rho \cdot \rho_{H,G}.$$

To find the required density we need to compensate for gasoline, we just need to substitute the volumetric energy density of gasoline and find the required density:

$$\rho \cdot \rho_{H,V} = \rho_{B,V}$$
$$\rho = \frac{8,5 \; \frac{kWh}{l}}{\rho_{H,V}}$$
$$= \frac{8500 \; \frac{kWh}{m^3}}{33,3 \; \frac{kWh}{kg}}$$
$$= 255,25 \; \frac{kg}{m^3}.$$

For diesel, the target density is:

$$\rho \cdot \rho_{H,V} = \rho_{B,V}$$

$$\rho = \frac{9,7 \ \frac{kWh}{l}}{\rho_{H,V}}$$

$$= \frac{9700 \ \frac{kWh}{m^3}}{33,3 \ \frac{kWh}{kg}}$$

$$= 291,29 \ \frac{kg}{m^3}.$$

The energy densities that stored hydrogen has compared to fossil fuels are small, but still very high compared to the storage technologies presented so far. We will see that if a fuel cell is used, the lower volumetric energy density can be compensated. In the next section, however, we will first look at the production of hydrogen.

9.3.1 The extraction of hydrogen

Hydrogen only occurs in nature in a chemically bonded form. If we want to use hydrogen, we first have to produce it by releasing it from its chemically bonded form. There are three sources here: first, we can obtain it from organic hydrocarbon compounds by reformation. Another method is the partial oxidation of hydrocarbon chains. Alternatively, we can obtain it by splitting water. All methods can be implemented industrially—that is, we can produce hydrogen in large quantities in one place so that it can be used in another place, at another time.

In steam reformation, a hydrocarbon chain is produced by adding heat and steam to produce hydrogen and carbon monoxide:

$$C_n H_m + n\,H_2O \longrightarrow \frac{n+m}{2} H_2 + nCO. \tag{9.10}$$

Methane steam reformation is a special form:

$$CH_4 + H_2O \longrightarrow CO + 3\,H_2. \tag{9.11}$$

This reaction requires an enthalpy of $\Delta H^0 = 206 \ \frac{kJ}{mol} = 57.2 \ \frac{Wh}{mol}$. This type of reformation is implemented in large industrial plants. The reaction temperature is $T = 800\text{--}900\,°C$ and is carried out at a pressure of $20\text{--}40$ bar. This reaction releases carbon monoxide as well as hydrogen. This can be used in a further step to produce additional hydrogen:

$$CO + H_2O \longrightarrow CO_2 + H_2. \tag{9.12}$$

In this reaction, the enthalpy of reaction is negative: it is $\Delta H^0 = -41.2 \ \frac{kJ}{mol} = 11.14 \ \frac{Wh}{mol}$—in other words, energy is released in this reaction. The reaction temperature is $T = 250\text{--}450\,°C$.

For this process to take place, methane (CH_4) and water (H_2O) must be present. Methane has a gravimetric energy density of $19.3 \ \frac{kWh}{kg}$ and a volumetric energy density of $5.9 \ \frac{kWh}{l}$. The gravimetric energy density is thus smaller, but the volumetric energy density is twice that of hydrogen. The reformation of methane to hydrogen therefore only makes sense in applications where the higher energy density of hydrogen is required. The methane can, for example, come from a biogas plant or be taken from the gas grid in the form of natural gas. In the latter case, of course, a fossil fuel, natural gas, would be used. Reformation is not suitable for using short-term energy surpluses to produce hydrogen. Although it would be possible to regulate the supply of biogas, natural gas, and water, temperature and pressure requirements need a continuous supply of energy and reaction material.

The same applies to partial oxidation. The general formula is:

$$C_nH_m + \frac{n}{2}O_2 \longrightarrow \frac{m}{2}H_2 + nCO. \tag{9.13}$$

If we use oil, the reaction is:

$$C_{12}H_{26} + 6O_2 \longrightarrow 13H_2 + 12CO. \tag{9.14}$$

The released carbon monoxide can be additionally converted via the hydrogen shift reaction in eqn (9.12). But here, too, the question arises as to why an already existing energy carrier should be converted into another energy carrier at great expense.

The two previous methods required heat energy for the reaction in addition to a substance from which hydrogen was to be separated. We basically store chemical binding energy by adding heat energy into a new form of chemical binding energy. We have not yet encountered the use of electrical energy. This results from the splitting of water by electrolysis.

The setup is shown in Fig. 9.6. Basically, it is similar to the electrochemical cells we have already described in Chapter 8. Again, we have two conductors immersed in an electrolyte. The electrolyte is water, and to speed up the reaction, salts are added to the water. A voltage is applied to the two electrodes. At the cathode we observe the following reaction:

$$2H_2O + 2e^- \longrightarrow H_2 + 2OH^-. \tag{9.15}$$

The hydrogen rises at the cathode and can be captured. The hydroxide OH^- moves to the anode, where the following reaction takes place:

$$2OH^- \longrightarrow \frac{1}{2}O_2 + H_2O + 2e^-. \tag{9.16}$$

The addition of electrons splits the hydroxide, giving off one oxygen atom, while the other is bonded with two hydrogen atoms to form water. The oxygen rises at the anode and can also be captured. So with this method, we succeed in splitting the water.

If we consider the two reactions (eqns (9.15) and (9.16)) together, we find that the enthalpy of reaction is $\Delta H = 285.83 \frac{kJ}{mol} = 0.079 \frac{kWh}{mol}$. This is the total amount of energy we need if we want to carry out this reaction. The free energy is only $\Delta G = 237.13 \frac{kJ}{mol} = 0.065 \frac{kWh}{mol}$.

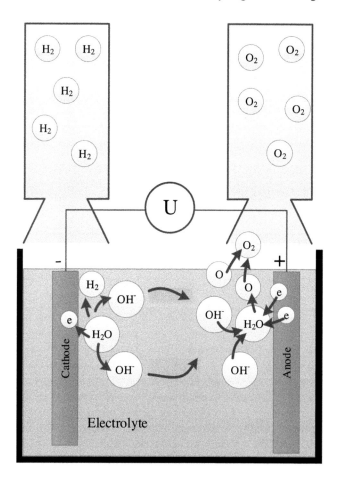

Fig. 9.6 Construction of an electrolysis device for splitting water. Two electrodes are immersed in an electrolyte consisting of water and salt. On the cathode side, water is split to form hydrogen, which is collected in a container. On the anode side, water is formed, releasing oxygen, which is collected in the reservoir.

Exercise 9.6 The free energy is the amount of energy that would have to be used for the reaction alone. The enthalpy of reaction is the amount of energy we need for electrolysis. How high are the thermal losses? And what efficiency is thus theoretically achievable for electrolysis?

Solution: We calculate the difference between the enthalpy of reaction and the free energy and thus obtain the thermal losses:

$$T\Delta S = \Delta H - \Delta G = 285.83 \, \frac{\text{kJ}}{\text{mol}} - 237.13 \, \frac{\text{kJ}}{\text{mol}} = 48.7 \, \frac{\text{kJ}}{\text{mol}} = 0.13 \, \frac{\text{kWh}}{\text{mol}}.$$

The efficiency is given by the ratio of the enthalpy of reaction to the free energy:

$$\eta = \frac{\Delta G}{\Delta H} = \frac{237.13 \, \frac{\text{kJ}}{\text{mol}}}{285.83 \, \frac{\text{kJ}}{\text{mol}}} = 82.96\%.$$

Table 9.1 Operating parameters of a PEMEL cell.

Operating parameters	Value range
Operating temperature	$20–100\,°C$
Pressure	$30–50\,\text{bar}$
Anode reaction	$H_2O \longrightarrow 2\,H^+ + \frac{1}{2}O_2 + 2\,e^-$
Cathode reaction	$2\,H^+ + 2\,e^- \longrightarrow H_2$
Efficiency	$67–82\,\%$
Cell area	$10–750\,\text{cm}^2$
Current density	$< 2.5\frac{A}{\text{cm}^2}$
Cell voltage	$2.2\,V_{DC}$
Cells per stack	<120
Power	$0.5–160\,\text{kW}$

The needed voltage for splitting the water is given by:

$$U_{\text{rev}} = \frac{\Delta H}{n\,z\,F} = \frac{285.83\frac{\text{kJ}}{\text{mol}}}{1\,\text{mol} \cdot 2 \cdot 96{,}496\frac{\text{As}}{\text{mol}}} = 1.48\ V_{DC}. \tag{9.17}$$

In order for water to split, we need to create a voltage between the two electrodes that is greater than 1.48 V_{DC}. By varying the current, electrolysis basically has the possibility of responding to variable excess currents. We could therefore use an electrolyser to use temporally fluctuating surpluses of a solar power system or a wind power system to produce hydrogen. The hydrogen produced in this way can then be stored and used when needed. Unlike a battery, however, we cannot run rapid charging and discharging cycles here. We are talking about an application that only stores energy.

Unfortunately, during high current fluctuations, the chemical stress on the electrodes is very high. This affects the lifetime of such an electrolyser. But there is another process that can split water with the help of electrical energy: proton exchange membrane electrolysis, or PEMEL for short. Tab. 9.1 shows the operating parameters of such a cell.

Fig. 9.7 shows the structure of a PEMEL cell. The cell has an inlet for water. Hydrogen and oxygen exit at two other openings on separate sides. Inside the cell there is a reaction chamber. The electrodes are attached to the wall side. The water that enters the reaction chamber through the lower opening remains on the anode side. A membrane separates this area of the reaction cell from the area of the cathode. Similar to the separator in an electrochemical cell, the membrane only allows protons to pass it (Vidas and Castro, 2021).

A voltage is applied between the anode and the cathode. On the anode side, the applied voltage causes the water to split:

$$H_2O \longrightarrow 2\,H^+ + \frac{1}{2}O_2 + 2\,e^-. \tag{9.18}$$

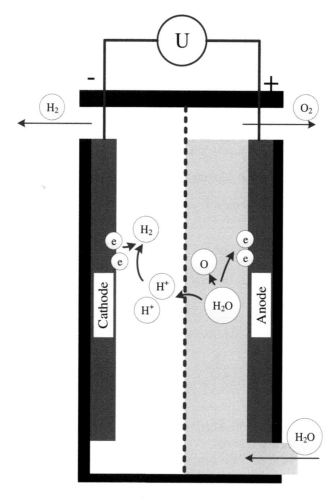

Fig. 9.7 Structure of a PEMEL cell for the splitting of water. The construction is similar to an electrochemical cell. Water is pumped into the cell. The voltage forces the water to split into hydrogen H^+ and oxygen O. The hydrogen passes the separator and reaches the cathode side, while the oxygen stays on the anode side.

The oxygen forms O_2 molecules. Due to the electric field between the anode and cathode, the hydrogen ions diffuse to the cathode side, where the hydrogen is then formed:

$$2\,H^+ + 2\,e^- \longrightarrow H_2. \tag{9.19}$$

The two gases rise into the upper part of the cell and are separated there.

The advantage of the PEMEL cell is that it has a high efficiency and can better handle power fluctuations. Since the reaction is electrically driven, little waste heat is produced. By using catalysers, the reaction in the reaction chamber can even be accelerated.

Exercise 9.7 A PEMEL electrolyser is to be designed for a wind turbine with a capacity of 2 MW, using 10% of the peak power. Stacks are used which have an active cell area of $A = 750$ cm². The voltage is $U = 2$ V_{DC}. The maximum current density of a reaction cell is $\rho_A = 2.5$ $\frac{A}{cm^2}$. How many stacks are needed?

Solution: We first determine the power we need to take in total:

$$P = 2{,}000 \text{ kW} \cdot 10\% = 200 \text{ kW}.$$

Now we determine the maximum power that a cell can provide. To do this, we first determine the maximum cell current:

$$I_{max} = A \cdot \rho_A = 750 \text{ cm}^2 \cdot 2.5 \, \frac{A}{cm^2} = 1{,}875 \text{ A}.$$

The maximum power of a cell is then given by:

$$P_{max} = U \cdot I_{max} = 2 \text{ V}_{DC} \cdot 1{,}875 \text{ A} = 3{,}750 \text{ W}.$$

Thus, we can determine how many stacks are needed for power consumption:

$$N = \frac{P}{P_{max}} = \frac{200 \text{ kW}}{3.75 \text{ kW}} = 53.3 \approx 54.$$

Photolytic processes are another form of hydrogen production: one possibility is the use of microorganisms that split water with the help of sunlight. Another variant is photocatalysis, in which sunlight generates a charge carrier in a semiconductor, which is then used directly to split water. These processes are not yet used on an industrial scale.

9.3.2 The fuel cell: How we get electricity from stored hydrogen

With the processes presented here, the production and storage of hydrogen is feasible. We have seen that hydrogen can be stored by adding electrical or thermal energy. So far, however, only combustion has been described as a method for recovering energy from stored hydrogen. The combustion of hydrogen to water is a natural application of the stored energy. However, it only uses 17 % of it, since part of the available energy is used to generate the heat of reaction. Generating electrical power by combustion in a combustion engine with a downstream electrical generator would further reduce the efficiency.

An ideal use of hydrogen would be to combine hydrogen and oxygen to form water with the release of electricity without the diversions of combustion driving a mechanism that drives a generator. In essence, we have just learned about such a possibility. Let us consider the PEMEL cell in Fig. 9.7. In electrolysis, a voltage is applied, water is introduced into the cell, and oxygen, as well as hydrogen, flows out of the cell. If we could reverse this process, we would have a process where oxygen and hydrogen flow into the cell and water flows out. Whereas before we were using electrical energy, now electrical energy would be released instead.

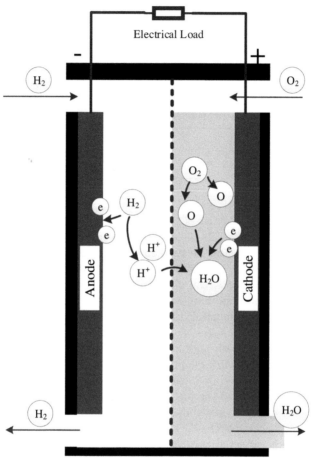

Fig. 9.8 Structure of a PEMEL fuel cell. Hydrogen H_2 enters the anode side of the cell, while oxygen O_2 enters the cell on the cathode side. The hydrogen passes the separator and, together with the oxygen, forms water, H_2O.

A PEMEL fuel cell allows this approach. Fig. 9.8 shows its basic structure. The structure and reaction are very similar to those of a PEMEL cell (Fig. 9.7). The main difference is the direction in which the reaction takes place. In a PEMEL cell, water is split to obtain oxygen and hydrogen. In a fuel cell, the reverse reaction is used. On both the anode and cathode sides, the active material is introduced. On the anode side, the fuel is ionized in the fuel cell:

$$H_2 \longrightarrow 2\,H^+ + 2\,e. \tag{9.20}$$

Table 9.2 Examples of different reactions suitable for a fuel cell.

Fuel	Reaction	ΔH_0 $\left(\frac{kJ}{mol}\right)$	Efficiency(%)
Hydrogen	$H_2 + \frac{1}{2} O_2 \longrightarrow H_2O$	286.0	82.97
	$H_2 + Cl_2 \longrightarrow 2\,HCl$	335.5	78.33
	$H_2 + Br \longrightarrow 2\,HBr$	242	85.01
Methane	$CH_4 + 2\,O_2 \longrightarrow CO_2 + 2\,H_2O$	890.8	91.87
Propane	$C_3H_8 + 5\,O_2 \longrightarrow 3\,CO_2 + 4\,H_{20}$	2,221.1	94.96
Decan	$C_{10}H_{22} + 15{,}5\,O_2 \longrightarrow 10\,CO_2 + 11\,H_2O$	6,832.9	96.45
Carbon monoxide	$CO + \frac{1}{2} O_2 \longrightarrow CO_2$	283.1	90.86

The ions reach the cathode side via the electrolyte, while the electrons use the electrodes as a transport path. There, the electron and ion combine with the added oxygen to form water:

$$O_2 + 4\,H^+ + 4\,e \longrightarrow 2\,H_2O. \tag{9.21}$$

The reaction represents a reversal of the basic reaction for water splitting described in eqn (9.3). The efficiency of this technology is correspondingly high.

Fuel cell technology can also be used for other substances. Tab. 9.2 shows some of these with their reactions and thermal efficiency. Methane, which is produced in so-called power-to-gas concepts, can also be used in a fuel cell. The efficiencies listed in Tab. 9.2 represent the maximum values. However, there are further losses when using the fuel cell. On the one hand, there are the purely ohmic losses that occur due to contacting and conduction losses. Furthermore, at low currents, there are losses due to electrode activation. At very high currents, the inertia of the ions is added. These ion diffusion losses dominate at high currents. Therefore, the optimal performance of a fuel cell—that is, the preferred operating point—is 60−80% of the maximum current.

If we compare the design of the fuel cell with the design of a redox flow cell, we see similarities. In both cases, we have a liquid or gaseous active material that is fed into a reaction chamber to react there. The physical effects that we observed in the redox flow cell are also to be expected here. In a fuel cell, for example, we have to make sure that there is enough oxygen and hydrogen available to extract enough power from the cell. The voltage of the cell depends on various factors: the concentration of the active material, the diffusion of the ions from the anode to the cathode side, the time scale of the reaction dynamics, and the charge carrier transport. In general, the following applies to the voltage of a fuel cell:

$$U = U_0 - b \log \frac{I}{I_0} - R\,I - m\,e^{nI}, \tag{9.22}$$

where U_0 is the electric potential, I_0 is the expected current of eqns (9.18) and (9.19), and R is the internal resistance of the fuel cell. $m\,e^{nI}$ describes the drop in voltage at high currents due to diffusion processes and counterpotential (Srinivasan, 2006). m, n represent parameters that depend on the chemistry and realization of the cell.

Let us look at the course of the cell voltage as a function of the current density. To do this, we carry out the following experiment. We take a fuel cell and ensure

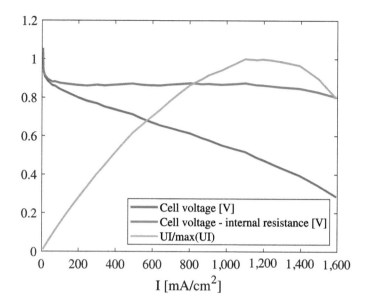

Fig. 9.9 Cell voltage of a fuel cell as a function of current density.

a constant mass flow of oxygen and hydrogen. Then we close the circuit and vary the current flowing between the anode and the cathode. In doing so, we measure the observed cell voltage. This experiment is shown in Fig. 9.9 (Srinivasan, 2006; Sazali *et al.*, 2020). Not only the cell voltage is shown, but also the internal resistance. So we can identify here the three ranges that we already know from the redox flow cell. At very low voltages, the potential differences that result from concentration differences of the active material dominate. These decrease exponentially, so that already at a current density of a few milliamperes we reach the range where the ohmic resistances dominate the voltage. If the current density becomes too large, diffusion effects become more and more apparent. We remove charge carriers from the system faster than they can be renewed.

Fig. 9.9 also shows the extracted power, normalized to the maximum achievable power. At low current densities, the power extraction is proportional to the current density. As long as we are still in the ohmic range, this still applies. Only when the diffusion effects begin to become apparent does the power that can be extracted reduce, as there are now no longer enough charge carriers available.

Compared to combustion, the fuel cell is a much better choice. The efficiency is higher than that of electrochemical storage technologies and the energy density of stored hydrogen is considerably higher. However, the electrical voltage of a fuel cell is low, at 1.23 V_{DC}, which makes it necessary to integrate power-electronic components into the overall system and also limits the power. In addition, fuel cells are not capable of recuperation. In the next section, we will deal with the question of how we can design a truck with a combination of fuel cell and lithium ion battery.

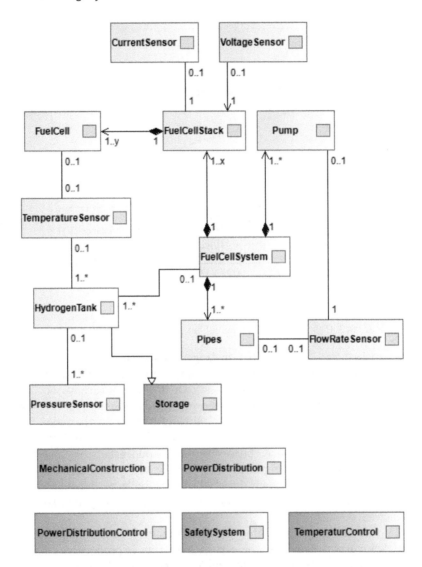

Fig. 9.10 System components of a fuel cell.

We conclude our consideration of the fuel cell with a diagram of the system components involved. In Fig. 9.10, the system components are shown. The FuelCellSystem,
like what we have already learned about electrochemical storage, consists of a $xpys$
connection of x parallel-connected FuelCellStack consisting of y series-connected Fu-
elCell. Each FuelCellStack is equipped with a CurrentSensor and a VoltageSensor, which are used by both the PowerDistributionControl and the SafetySystem.

The actual storage is the HydrogenTank, which is provided with at least one Tem-
perateSensor and at least one PressureSensor. Both are used by the PowerDistributionControl and the SafetySystem.

Since we need to transport hydrogen within the system, the `FuelCellSystem` is also equipped with at least one `Pump` and a number of `Pipes`. For the `PowerDistributionControl` and the `SafetySystem`, these components are also equipped with `FlowRateSensor`, which determines the flow rate so that monitoring of the mass transport is possible.

We can see that a fuel cell has a similarly high level of system components to a redox low cell. This is due to the fact that we not only have to perform and monitor electrical power, but also material transport. Since hydrogen is also a hazardous substance, as it becomes oxyhydrogen with the ambient air. Additional safety measures have to be fulfilled.

9.3.3 Application example: Hybrid powertrain for a commercial vehicle

In this section, we look at the design of a powertrain for a commercial vehicle: a truck equipped with a fuel cell and a lithium ion battery. Before we start with the question of the technical implementation and requirements, we will first look at the question of what configuration makes sense for such a vehicle.

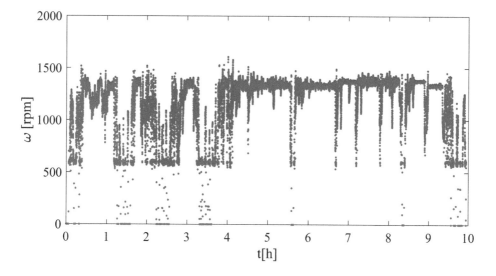

Fig. 9.11 Speed of a long-haul transporter during a 10 h tour.

As in other application examples, we first consider the load profile. To do this, we first look at the speed data of a long-haul transporter over the course of a representative trip. It is clear that this example will not be sufficient in practice to make technical and commercial decisions.

For the design of a powertrain, it is necessary to collect data from a very large number of journeys in different countries and to determine suitable vehicle classes from these data. Here we only want to look at the data from a journey over 10 hours. We are not so much interested in the speed of the vehicle, but in the power called up by the drive train. In Fig. 9.11, we have presented the speed data (NREL, 2022).

Speed data alone are not sufficient for powertrain design, especially if we don't have vehicle data available either. Much more crucial than the vehicle speed is information about the number of revolutions of the engine ω and the torque T that the engine produces. These data also depend on the vehicle and the transmission used. If we wanted to design a drive train in general, we would have to analyse the entire chain of action from the engine to the gearbox, the axle, and the wheel. In this example, we limit ourselves to replacing the existing drive motor with an electric motor.

The rotational power is given by:

$$P_{rot} = \omega \cdot T.$$

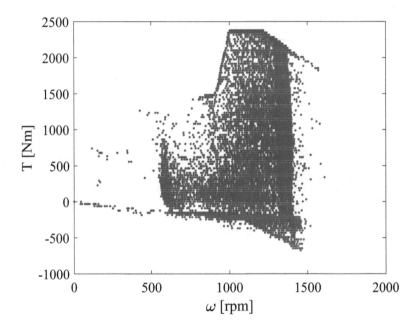

Fig. 9.12 Torque and speed set points during the 10 h trip.

We must therefore find the product of ω and T at each data point. In Fig. 9.12, these pairs are shown for the speed profile shown in Fig. 9.11. ω is always positive—that is, the motor always turns in one direction only. The torque, on the other hand, varies and tells us whether a braking process or an acceleration process is taking place. With the help of these data, we can determine a histogram of the observed motor power. A braking process then has a negative power and an acceleration process a positive power. These data are shown in Fig. 9.13.

We see that the power required for an acceleration process is up to 300 kW. The distribution is relatively even; there is no distinct structure. The maximum braking power is 100 kW. Larger braking power is not called up, which is due to the fact that

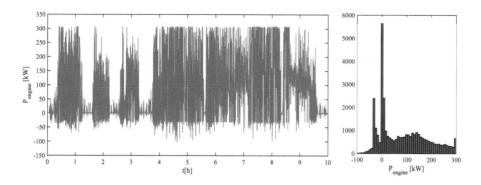

Fig. 9.13 Histogram of engine power. These data were determined using the data shown in Fig. 9.12. Negative data correspond to deceleration, positive data to acceleration.

the mechanical brake is used more in braking manoeuvres that require higher power. From this, we can derive two requirements:

FCT 1: AS A Manufacturer, I WANT my drivetrain to have an engine power of at least 300 kW SO THAT the driver does not notice any loss of power by using the new drivetrain.

FCT 2: AS A Manufacturer, I WANT my storage to be able to absorb a braking power of at least 100 kW SO THAT my storage can also absorb the braking power during recuperation.

These are requirements for the charging and discharging capacity of the storage. However, we also need requirements for the size of the storage. Using the data from Fig. 9.13, we can estimate how much energy the vehicle needs in total for the measured trips. It can be seen that a total of 913 kWh is required for acceleration. An energy of 42 kWh can be obtained from the recorded braking processes. The total energy requirement is therefore 871 kWh. Based on these data, we can make a requirement for the size of the storage tank.

FCT 3: AS A Manufacturer, I WANT my powertrain to have a storage capacity of at least 871 kWh SO THAT the range of the vehicle remains comparable.

Requirement FCT 3 is not sufficient, however, because the 871 kWh was based on the assumption that we can also use recuperation energy. If we were to use only a fuel cell, this would not be the case. Since fuel cells do not allow bidirectional power flow, we therefore still need to explicitly point out the need for recuperation energy to be stored.

FCT 4: AS A User, I WANT to temporarily store the energy released during deceleration in the form of electrical energy SO THAT I can use this energy to accelerate.

This gives rise to a requirement for the realization of the hybrid drive train. We see that we need to carry a lot of stored energy to realize the range requirement. This argues for using a fuel cell, as hydrogen has a higher energy density than electrochemical storage technologies. However, we cannot turn the fuel cell into an electrolyser by reversing the voltage: for recuperation, we need an additional technology that enables this bidirectional energy flow. We want to realize this with the help of a lithium ion battery.

FCT 5: AS A developer, I WANT to combine a fuel cell technology with a lithium ion battery SO THAT I can meet the range requirement and at the same time be capable of recuperation.

Exercise 9.8 In addition to the technical requirements, there are other requirements that are set by the different stakeholders. In this task, four requirements are to be created from the product manager's point of view.

Solution: Since the product manager has to look at his product portfolio, he will be interested in the hybrid powertrain being usable for different vehicles. He knows that the previous design is based on the evaluation of a time series. Against this background, in performance and in energy content.

FCT 6: AS A product manager, I WANT to be able to scale the fuel cell power and the battery power SO THAT I can also equip vehicles that have a higher or lower power requirement.

FCT 7: AS A product manager, I WANT the battery to be able to be integrated into the overall system in a scalable way SO THAT I can also realize applications that require a higher battery share.

The FCT 7 requirement also enables the realization of a fully electric vehicle.

Due to the combination of two energy sources, the question arises of on which basis the mix of fuel cell energy and battery energy can be determined. Here the product manager demands the following:

FCT 8: AS A product manager, I WANT the mix of fuel cell energy and battery energy to be determined by efficiency-controlled operation SO THAT the range of the vehicle can be optimized.

It would also be possible to define the mix according to economic aspects. However, a suitable economic model would have to be available for this. Since we cannot feed electricity into the grid and thus sell it, we lack a profit. However, we could use the operating and maintenance costs here.

As a final requirement, the product manager demands that the entire system can be integrated into an existing vehicle.

FCT 9: AS A product manager, I WANT the installation space of the powertrain to be comparable to the installation space of the previous powertrain SO THAT existing vehicles can be converted.

As understandable as this requirement is from the product manager's point of view, it will be difficult to meet requirement FCT 9 entirely. We know that the energy density of hydrogen is not as high as that of fossil fuels. The same is true for the lithium ion battery. The higher efficiency of an electric drive and a fuel cell can certainly help here.

Having gathered the requirements, let us look at the technical implementation and the power flow diagram. In Fig. 9.14, we have shown a conceivable realization. Here we work with a DC bus that is fed by the battery and the fuel cell. The drive inverter

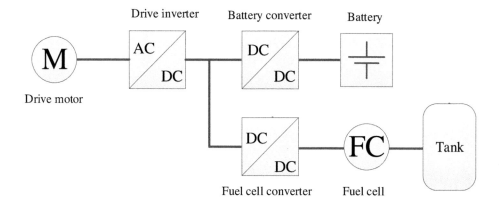

Fig. 9.14 Power-electronic components of a hybrid powertrain equipped with a fuel cell and a battery as energy supply.

draws the required power for the drive motor from this DC bus. If we recuperate, the drive inverter feeds power to the DC bus and the battery converter charges the battery.

Exercise 9.9 The design shown in Fig. 9.14 represents only one of several alternatives. What other possibilities would there be? What are their advantages and disadvantages?

Solution: We present here three alternative designs. In Fig. 9.15, we have presented these variations.

a) The design shown in Fig. 9.14 uses one engine, which is used for both motor and generator. Alternatively, we can introduce a second motor into the system that is used for braking—that is, regenerative operation. Since we know that the recuperation power is considerably smaller than the propulsion power, this motor would be smaller. If the generator inverter and generator are well matched, the losses of both could be optimized for the expected operating points.

The disadvantage of this design is that we have an additional motor and another power-electronic component in the system. Since both drive components have an influence on the driving behaviour of the vehicle, control effort is to be expected.

b) In the original design, the fuel cell feeds its power to the DC bus via the DC/DC controller. When the engine is not calling for this power, the battery converter stores this power in the battery. Two power-electronic components were used for this path. Alternatively, we could use the fuel cell directly to charge the battery.

The advantage is an increase in efficiency and a reduction in costs. The disadvantage is that this means that the voltage levels of the fuel cell and the battery are connected to each other. The battery cells have to be wired up exactly right and the fuel cell stacks as well.

c) In this version, we are not using a DC bus. Battery and fuel cell feed AC directly into the motor and regulate it together. The advantage here is a compact implementation. The disadvantage is that such a split motor control is relatively complex. However, it can be realized with a master–follower concept.

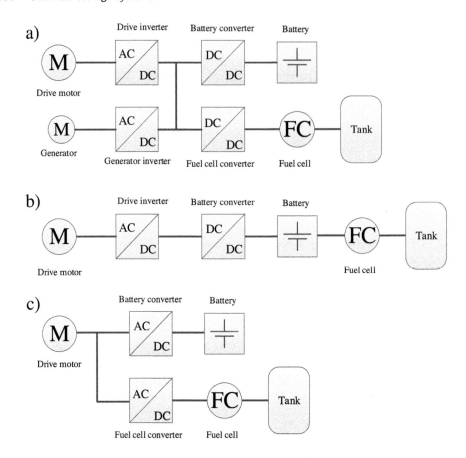

Fig. 9.15 Three alternative designs for the realization of a hybrid truck.

The power flow diagram is shown in Fig. 9.16. It contains three nodes: the two storage systems fuel cell F and lithium ion battery B, as well as the load L. We see that the possible technical realizations from Exercise 9.15 are generally found in the same power flow diagram.

We start by formulating the equations for the fuel cell node; this includes the hydrogen tank and the fuel cell stack. The fuel cell can only be operated unidirectionally, the exception being time shifts of power. The fuel cell can be used to directly cover the load F_L or to charge the lithium ion battery F_B. The refuelling process is not described here. Since the fuel cell can only be discharged, it is sufficient to specify a tank size and a starting value for the tank content for the calculation. Thus the balance equation results in:

$$0 = \eta_{FF} F_F(-\Delta) - (F_B + F_L + F_F). \tag{9.23}$$

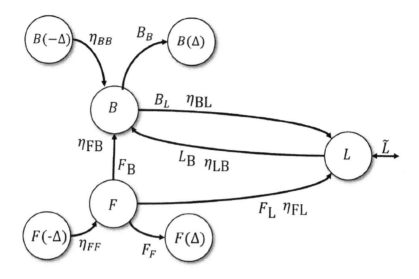

Fig. 9.16 Power flow diagram for a vehicle with a hybrid powertrain. The power node B represents the lithium ion battery, the power node F the fuel cell. The consumers are represented by L with the load profile $\tilde{L}$.

The lithium ion battery allows a bidirectional power flow. It can be charged by the fuel cell, but also by the load in case of recuperation. The equations are:

$$0 = \eta_{FB}F_B + \eta_{LB}L_B + \eta_{BB}B_B(-\Delta) - (B_B + B_L + B_B(\Delta)). \qquad (9.24)$$

The load is supplied by both the battery and the fuel cell and can deliver surpluses to the battery. This gives the following power flow equation:

$$\tilde{L} = \eta_{BL}B_L + \eta_{FL}F_L - L_B. \qquad (9.25)$$

With eqns (9.23), (9.24) and (9.25), we have described the system.

Exercise 9.10 From FCT 2, we know that $L_B <= 100$ kW. We have the choice between the cells presented in Tab. 9.3. Since we are using a battery converter, we are free to choose the voltage window. It must only not be larger than 800 V_{DC}. Which cells should we choose with which configuration if only the number of cells and the compliance with the boundary conditions are relevant?

Table 9.3 Technical data of the lithium ion battery cells used in this application example.

	LFP	NMC
Voltage range	[2.6 V_{DC}; 3.65 V_{DC}]	[3 V_{DC}; 4.2 V_{DC}]
Nominal capacity	3.3 Ah	4.4 Ah
Maximum charge power	2 C	1 C

Solution: We choose the maximum possible voltage of $U_{max} = 800$ V$_{DC}$. For such a voltage, we need for the LFP cell a total of:

$$N_y = \frac{800 \text{ V}_{DC}}{3.65 \text{ V}_{DC}} = 219.17 \approx 220$$

LFP cells.

When these 220 cells are discharged, their voltage is $220 \cdot 2.8$ V$_{DC} = 572$ V$_{DC}$. We now need to calculate what current we need in this case to use the maximum power to charge the battery. The following is valid:

$$I = \frac{P}{U} = \frac{100,000 \text{ W}}{572 \text{ V}_{DC}} = 174.825 \text{ A}.$$

The maximum current we are allowed to charge the LFP cells with is 6.6 A, so we need to connect several strings in parallel. The number is then given by:

$$N_x = \frac{174.825 \text{ A}}{6.6 \text{ A}} = 26.48 \approx 27.$$

So for the LFP variant we need a 28p220s circuit. In total, that is 6,160 battery cells.

We now perform the same calculation for nickel, manganese, cobalt (NMC) cells. First, the number of cells in a string is:

$$N_y = \frac{800 \text{ V}_{DC}}{4.2 \text{ V}_{DC}} = 190.76 \approx 191.$$

When fully charged, such a string has a voltage of $191 \cdot 3.65$ V$_{DC} = 697.15$ V$_{DC}$. The maximum current then required for recuperation is:

$$I = \frac{P}{U} = \frac{100,000 \text{ W}}{697.15 \text{ V}_{DC}} = 143.44 \text{ A}.$$

The NMC cells can drive a maximum current of 4.4 A; again, we need to connect multiple strings in parallel. The number is then calculated as:

$$N_x = \frac{143.44 \text{ A}}{4.4 \text{ A}} = 32.6 \approx 37.$$

So for the NMC cells, we need a 37p191s circuit. In total, this then amounts to 7,067 cells.

With 6,160 or 7,067 lithium ion cells, we have a very large battery. The size is because we want to recuperate the full braking power as well. If we were to forgo this requirement, or, for example, did not want to use the batteries in the full voltage window, we could reduce the number of cells. Alternatively, we could use the storage capacity of the cells.

The design of the fuel cell is analogous to that of the redox flow cell, through the fact that storage capacity and power provision are separated from each other. For the power supply, the physical boundary conditions of the stacks are decisive. So we have to see how many stacks we can connect in series at most, and how many stacks we then want to connect in parallel to provide the necessary power.

Exercise 9.11 From Fig. 9.9, we know that due to the internal resistance, the cell voltage is between 0.5 and 1 V_{DC}. At the optimal operating point, the current density is about 1.2 $\frac{A}{cm^2}$ and the voltage is 0.5 V_{DC}. Our stack has a surface area of $10 \times 20 \, cm^2$. How many stacks must be connected in parallel and series to provide the power of 300 kW? No more than 150 stacks may be connected in series.

Solution: We first calculate how much current such a stack can carry:

$$I = 1,2 \, \frac{A}{cm^2} \cdot 10 \times 20 cm^2 = 240 \, A.$$

The voltage we reach at 150 stacks is:

$$U = 150 \cdot 0.5 \, V_{DC} = 75 \, V_{DC}.$$

A single stack thus delivers a power of

$$P = UI = 240 \, A \, 75 \, V_{DC} = 18,000 \, W = 18 \, kW.$$

So to achieve the target power of 300 kW, we need to connect

$$N_x = \frac{300 \, kW}{18 \, kW} = 16.6 \approx 17$$

stacks in parallel.

The previous derivation was based on the fact that we pursued the approach of replacing an existing drive train with the fuel cell and the lithium ion battery. Since we also wanted to recuperate the entire braking power, we designed a relatively large battery. We now want to take a different approach. With the load profile, we have the information about how the user uses the vehicle. We now want to ask ourselves how we have to configure the lithium ion battery and the hydrogen in order to achieve the range desired by the customer.

We define the operation in the following way. When the vehicle brakes, as much power as the battery allows to charge is used to charge the battery. When the vehicle needs power, the battery is used first, and if this power or the energy content is not enough, the fuel cell is used. The algorithm is shown in 9.3.3. To keep the algorithm simple, we have not considered efficiency.

In Fig. 9.17, we have shown the power flows for a hybrid truck. The battery had a capacity of $\kappa_{B,max} = 20$ kWh with a maximum charging power of 40 kW. The fuel cell tank had a capacity of $\kappa_{F,max} = 870$ kWh and a maximum power of 350 kW. We see that the discharge power B_L of the battery very often corresponds to the maximum power. The charging power of the battery L_B also often reaches the maximum. However, the capacity of the battery κ_B usually does not even reach one kilowatt hour. So we see that the lithium ion battery is used here less as an energy source and more as a power storage device.

As we know, lithium ion batteries can operate at an E-Rate of up to $4E$ depending on the technology—that is, a battery with 10 kWh capacity allows a maximum charging power of 40 kW. The E-Rate depends on the cell chemistry. We want to see what the optimal storage capacity looks like at different E-Rates. For this purpose, we define the recuperation fraction as the ratio between the actually stored recuperation energy and the theoretically available recuperation energy. In Fig. 9.18, we have plotted this fraction as a function of the capacity of the lithium ion battery for an E-Rate of $1E$, $2E$, and $4E$. We see in all curves an initially almost linear increase of the recuperation fraction. This means that any increase in capacity, and thus any increase in power, has a direct effect on the recuperation rate. After a certain point, the slope decreases. An increase in capacity no longer causes the recuperation rate to increase at the same rate as before. We interpret this to mean that the extra power is sometimes used, but not as often as before. An increase in available power no longer leads to a strong increase in the recuperation rate.

Algorithm 8 Operating mode for the hybrid truck. κ_B and κ_F corresponds to the storage capacity, and $\kappa_{B,\max}$ and $\kappa_{F,\max}$ is the maximum storage capacity.

$\kappa_B \leftarrow 0,5\kappa_{B,\max}$
$\kappa_F \leftarrow \kappa_{F,\max}$
while Truck is driving **do**
    if $\tilde{L} \geq 0$ **then**
        $L_B \leftarrow -\tilde{L}$
        if $L_B \geq L_{B,\max}$ **then**
            $L_B \leftarrow L_{B,\max}$
        end if
    else
        if $\kappa_B \geq 0$ **then**
            $B_L \leftarrow \tilde{L}$
            $F_L \leftarrow \tilde{L} - B_L$
            if $F_L < 0$ **then**
                $F_L \leftarrow 0$
            end if
        else
            $F_L \leftarrow \tilde{L}$
        end if
    end if
    $\kappa_B \leftarrow \kappa_B + \Delta t \left(L_B + B_L \right)$
    $\kappa_F \leftarrow \kappa_L - \Delta t F_L$
end while

If we look at these points, we find that for an E-Rate of $4E$ this point is about 8 kWh, for $2E$ it is 16 kWh, and for $1E$ it is 32 kWh. Thus, if we convert the corresponding powers, the point is always at a power of 32 kW. At this power, the recuperation rate is about 90%. If we want to increase the recuperation rate to approximately 98%, we have to increase the power to approximately 56 kW.

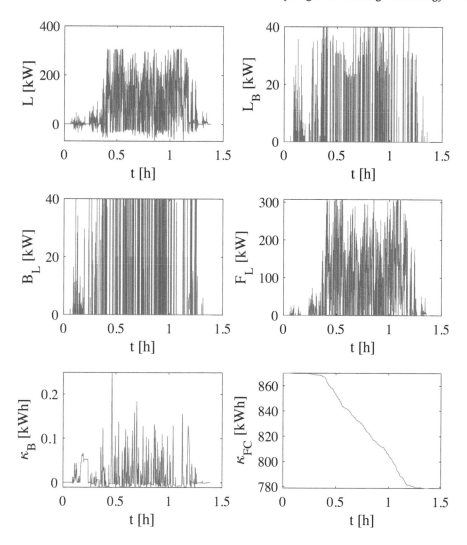

Fig. 9.17 Simulation of the power flows of a hybrid truck. The battery had a capacity of $\kappa_{B,max} = 20$ kWh with a maximum charging power of 40 kW. The fuel cell tank had a capacity of $\kappa_{F,max} = 870$ kWh and a maximum power of 350 kW.

For the design of the lithium ion battery, we are therefore faced with the challenge of finding cells that enable the highest possible E-Rate. Otherwise, we would have to work with excess capacities that we hardly use.

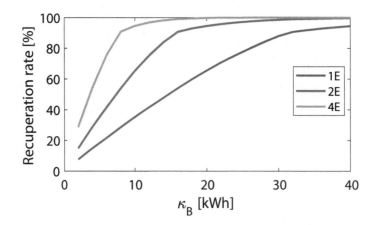

Fig. 9.18 Recuperation rate—that is, the fraction between used recuperation power and possible recuperation power as a function of the lithium ion battery capacity with different E-Rates.

Exercise 9.12 We want to configure the lithium ion battery so that we use as few cells as possible. For this, we take the lithium iron phosphate battery presented in Tab. 9.3, which has a charging power of 2C. The target power shall be 56 kW.

Solution: We first calculate the power P_{LFP} provided by a cell. Here we take the lower voltage, since the current limitation has a stronger effect here and we usually want to charge empty cells:

$$P_{cell} = 2.2 \text{ V}_{DC} 3.3 \text{ Ah } 2C = 17.16 \text{ W}.$$

The total number of cells we need to reach the target power is given by:

$$N = \frac{56{,}000 \text{ W}}{17.16 \text{ W}} = 3{,}263.4 \approx 3{,}264.$$

So we need at least 3,264 cells to absorb the power. We need to see how to divide these cells evenly between strings without the voltage rising above 800 V_{DC} when fully charged. The figure 3,264 can be easily divided by 2, 4, 8, and 16. If we go to 16 strings, we have a voltage per string of:

$$U = \frac{3{,}264}{16} \text{ 3.65 V}_{DC} = 744.6 \text{ V}_{DC}.$$

We therefore choose a 16p204s connection.

The capacity of this battery is:

$$\kappa = \frac{2.6 \text{ V}_{DC} + 3.65 \text{ V}_{DC}}{2} \text{ 3.3 Ah } 3{,}264 = 33{,}660 \text{ Wh} = 33.6 \text{ kWh}.$$

We must realize that this battery design is not in accordance with requirement FCT 2. This required that we take a braking power of 100 kW. Of course, we cannot do this without consulting the stakeholder. Additional consultation is required here.

Since we have carried out a cost–benefit assessment in our analysis, it can usually be assumed that these necessary discussions will lead to an amicable solution.

The fuel cell is responsible for most of the drive energy. We need about 900 kWh to maintain operation for 10 hours if we do not want to schedule a stopover. We can also use the 33 kWh of the lithium ion battery, as recuperation provides little energy. This allows us to reduce the fuel cell tank to 870 kWh. With an energy density of 2.4 $\frac{kWh}{l}$, we need a tank with a 334 l volume, which is about the size of a truck tank. These have a capacity of between 200 l and 500 l.

We must point out, however, that in order to make the algorithm clearer, we have not taken efficiency losses into account in this simulation. Efficiency losses can still increase the amount of hydrogen considerably.

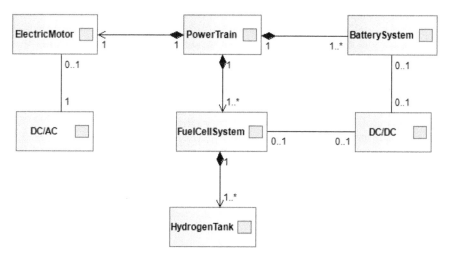

Fig. 9.19 System components of the powertrain for a hybrid commercial vehicle.

Let us conclude this section with a consideration of the system components and the requirement traceability matrix (RTM). In Fig. 9.19, the system components of the powertrain are shown. The `PowerTrain` consists of at least one `ElectricMotor` which has a `DC/AC`. This ensures that the `ElectricMotor` is able to convert the electrical energy into kinetic energy. We assume that each `ElectricMotor` has exactly one `DC/AC` associated with it, although theoretically it would also be possible for one `DC/AC` to drive several motors.

The `PowerTrain` still needs the two energy sources, the `BatterySystem` and the `FuelCellSystem`. Both system components can be connected to a `DC/DC`, which sets the necessary voltage level. While the `BatterySystem` stores the energy in its electro-chemical cells, the `FuelCellSystem` needs an external storage, the `HydrogenTank`.

In Tab. 9.4 the RTM is shown. We refer here only to the requirements FCT 1 to FCT 2 and assume that the general and technology-specific requirements are met by the system components `BatterySystem` and `FuelCellSystem`. FCT 1 dealt with the power requirements of the PowerTrain. Here all those system components must

Table 9.4 Requirement traceability matrix for the system components of the drive train of the hybrid commercial vehicle.

	Application requirements								
	FCT 1	FCT 2	FCT 3	FCT 4	FCT 5	FCT 6	FCT 7	FCT 8	FCT 9
ElectricMotor	X			X					
DC/AC	X			X					
FuelCellSystem	X				X	X			
HydrogenTank			X						
BatterySystem	X	X		X	X	X	X		
DC/AC	X	X		X	X	X	X		
PowerDistributionControl								X	
MechanicalConstruction									X

be involved that are involved in the power transport and its generation—that is, the Electric Motor and its DC/AC must be able to convert the motor power, and the energy sources Battery System and Fuel Cell Stack must be able to provide the energy.

FCT 2 looked at the recuperation system. Based on our analysis, we had found that the originally specified power was too high, or would have meant very high costs with little benefit. Therefore, we reduced the braking power to 52 kW. The system components required for this are the BatterySystem and the DC/DC.

FCT 3 is a clear requirement for the HydrogenTank. Here the responsibility is clearly defined. FCT 4, the recuperation function, requires four system components: the ElectricMotor, the DC/AC, the BatterySystem, and the DC/DC. This is because all four components are also involved in the power flow that occurs during recuperation.

FCT 5 is fulfilled by the FuelCellSystem, the BatterySystem, and the DC/DC. However, we must note here that the system component DC/DC can occur in two versions: once to connect the BatterySystem to the DC bus and once to connect the FuelCellSystem to the DC bus.

FCT 6 is a direct requirement of the FuelCellSystem, the BatterySystem, and the DC/DC, since the requirement here is that all three components can be combined in a scalable manner. The more specific requirement, FCT 7 only goes to the BatterySystem and its DC/DC.

So far, we have only had requirements for the specific system components. FCT 8 is a general requirement for the type of operation control implemented by the PowerDistributionControl. Likewise, FCT 9 is a requirement for the general system component MechanicalConstruction.

In this example, we have designed a hybrid powertrain for a vehicle. These considerations can be deepened considerably, because the aspects that apply to the layout and design of a hybrid powertrain are relatively complex, especially when driving dynamics and requirements are also considered in detail (Ehsani *et al.*, 2018). In the next section, we investigate another chemical storage technology: methanization.

9.4 Methanization: Power to gas or power to liquid

We have seen how we can use electrical energy to split water to obtain hydrogen as an energy carrier. With this we have a process which allows us to store energy, for example regeneratively produced energy for a long time and used again by direct combustion or by a fuel cell. However, we have also had to recognize that the volumetric energy density of hydrogen is smaller than that of fossil fuels. This can be partly compensated by the way we use hydrogen, but it is a disadvantage, especially in transport applications. Another disadvantage is that the world's energy consumption and energy infrastructure has been constructed to burn fossil fuels. In the mobility sector as well as in the housing and energy sector, the infrastructure to transport, store, and burn fossil fuels is already existing. Small parts can be taken over in a conversion to hydrogen, but the majority would have to be replaced.

If we had a technology to convert electrical energy into methane (CH_4), we would have found a solution. Methane is the main component of natural gas, which has been used for many decades in the residential and energy sectors, but also in mobility. With methane we would have a storage medium for which we have transport, storage, and utilization infrastructure.

Methanization gives us such a technology. It is labelled 'power to gas' or 'power to liquid' (Götz *et al.*, 2016). Electrical energy is used to produce methane. This is then stored and used later. Let us first familiarize ourselves with the basic reaction; then we will see how we can add a methanization plant to the island grid we described in Section 8.6.3.

9.4.1 The basic reaction for power to gas

To produce methane, we need a hydrogen source and a carbon source. We can produce the hydrogen by splitting water. We have already learned about the technologies, and they represent a large part of the electrical demand for methanization. Carbon dioxide from the atmosphere can serve as our carbon source.

In the first step, we decompose the carbon dioxide and form carbon monoxide and water from hydrogen and carbon dioxide:

$$H_2 + CO_2 \longrightarrow CO + H_2O. \tag{9.26}$$

This reaction needs $41 \frac{kJ}{mol}$ energy.

In the second step, further hydrogen is added to the carbon monoxide and methane and water are formed:

$$3\,H_2 + CO \longrightarrow CH_4 + H_2O. \tag{9.27}$$

This reaction is exothermic and releases $206 \frac{kJ}{mol}$.

Both overall reactions can be combined into one reaction equation:

$$4\,H_2 + CO_2 \longrightarrow CH_4 + 2\,H_2O. \tag{9.28}$$

The reaction takes place at a temperature of $250-300\,°C$ and pressures of $20-30\,bar$, and is exothermic. Thus, we only need to initially realize the appropriate environmental conditions for the reaction and can use the released heat energy to sustain it.

In both eqns (9.26) and (9.27) water is released. We can feed this water in to an electrolyser to recover hydrogen there; the whole process is shown in Fig. 9.20. Water and carbon dioxide are fed into the methanizer. The reaction products are water and methane. The methanization process converts eight hydrogen atoms into one part methane, which binds four of the hydrogen atoms, and two parts water, which binds

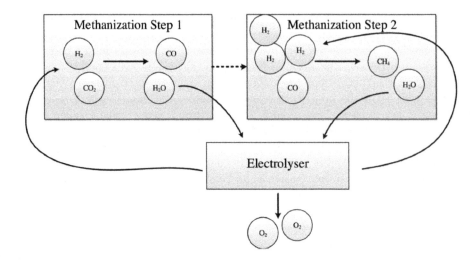

Fig. 9.20 Illustration of the methanization process. This is done in two separate steps, which both need hydrogen produced by an electrolyser.

the other four hydrogen atoms. The formed water is returned to the electrolyser so that the hydrogen atoms can again be separated from the oxygen and returned to the methanization process. We can therefore estimate that the efficiency of a methanization step is about 50%, since only half of the hydrogen becomes the targeted end product. If we combine this with the 80% efficiency of an electrolyser, we see that this process has an efficiency of only 40% (Ghaib and Ben-Fares, 2018; Götz *et al.*, 2016).

Compared to other storage technologies, methanization has a low efficiency. However, it also has advantages. We can store large amounts of methane for a very long time in an existing infrastructure. The energy density is high, which is an advantage for mobile applications. Methane can not only be used to generate energy, but also serves as a raw material supplier in the chemical industry.

In addition to the production of methane via the industrial process described in Fig. 9.20, it is also possible to feed the hydrogen together with the carbon dioxide into a bioreactor, where the reaction takes place in a biological process. In this process, biomass is decomposed. Methane and carbon dioxide are produced, which in turn are converted into water and methane.

The high temperatures that are generated during catalytic methanization can be used for additional interconnection of the sectors. Together with carbon dioxide, which is taken from the atmosphere or a biogas plant, we produce heat and methane. The heat is fed into the heating network, and methane is stored for later usage. In the process, methane can either be utilized through combustion or used in suitable fuel cells to generate electrical power. Hence, the power-to-gas process has two advantages: we use an infrastructure which already exists and we combine three sectors—the electricity sector, the mobility sector, and the heat sector of our energy system. In the next section, we investigate power to gas a little bit more, but restrict our investigation only to the electricity sector.

9.4.2   Application example: Integrating power to gas into island grid

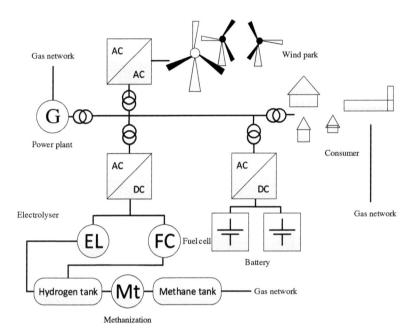

Fig. 9.21 Autonomously supplied island grid. A wind farm and a power plant powered by natural gas serve as the main energy sources. Two storage units are integrated into the system: a battery and a power-to-gas plant. The methane produced is stored and fed into the gas network.

In Section 8.6.3, we equipped an island grid with a wind turbine and a high temperature battery. In addition, the island grid was equipped with a gas-fired power plant that uses fossil fuels for power generation. In this example, we want to add a power-to-gas plant to the system. Fig. 9.21 shows the setup. The power-to-gas system consists of a bidirectional inverter. This allows us to take power from the grid or feed it into the grid. The electrical energy that we take from the grid is fed to an electric generator that produces hydrogen. This hydrogen is stored temporarily. Using a fuel cell, we can convert this hydrogen into electrical energy and feed it into the grid. Alternatively, we can convert it into methane in a methanization process. The methane is temporarily stored in a tank and fed into the gas network.

To create the power flow diagram, we need to add two power nodes to the diagram shown in Fig. 8.45. On the one hand, we have node H, which represents the hydrogen tank, and on the other we have node M, which represents the methane tank. If we were to integrate all power paths and the two new nodes in Fig. 8.45, we would get a confusing picture, so we will limit ourselves to looking at the power flows of the new nodes for the time being and supplement the representation of Fig. 8.45 with additional inflows and outflows.

The hydrogen tank receives its power via the wind plant W, the power plant G, the solar power plant P, or the storage S. The power output is either to the load L, the storage S, or the methanization plant M.

The methane tank has only one inlet, and that is from the hydrogen tank. On the other hand, it can deliver its power to the power plant G. Since we have so far only considered households as electrical loads, and we assume that the loads in the household do not have their own generators that produce electricity from fossil fuels, in our current representation we cannot cover the loads directly through the methane tank. Alternatively, there is the possibility of describing the gas demand of the households as a separate load profile $\tilde{L}_G$; if this is done, then another node L_G will exist here.

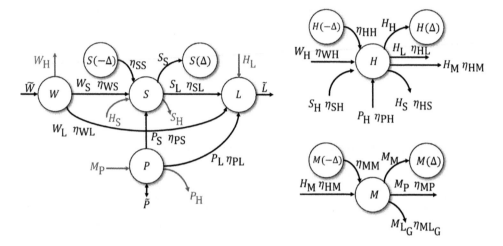

Fig. 9.22 Extended power flow diagram for a self-sufficient island grid that is equipped with a wind farm and a battery as well as a power-to-gas plant. For a better overview, the two additional power nodes H and M are shown separately with their inflows and outflows. In the overall overview, the new power flows are entered in grey.

In Fig. 9.21, the new parts of the power flow diagram are shown. We have first entered the additional power flows in the power flow diagram from Fig. 8.45. Furthermore, we have shown the power nodes H and M with their inflows and outflows separately.

We now want to set up the power flow equation for H and M. We start with the hydrogen tank H:

$$0 = \eta_{\mathrm{WH}}W_{\mathrm{H}} + \eta_{\mathrm{SH}}S_{\mathrm{H}} + \eta_{\mathrm{PH}}P_{\mathrm{H}} + \eta_{\mathrm{HH}}H_{\mathrm{H}}(-\Delta) - (H_{\mathrm{H}}(\Delta) + H_{\mathrm{L}} + H_{\mathrm{M}} + H_{\mathrm{S}}). \quad (9.29)$$

The hydrogen gets power from the wind farm W_{H}, the storage S_{H}, and the powerplant P_{H}. It can transfer its power to the load, the methane tank with its methanization device, and the storage. Next we look at the methane tank:

$$0 = H_{\mathrm{M}}\eta_{\mathrm{HM}} + M_{\mathrm{M}}(-\Delta)\eta_{\mathrm{MM}} - (M_{\mathrm{M}}(\Delta) + M_{\mathrm{P}} + M_{\mathrm{L}_{\mathrm{G}}}). \quad (9.30)$$

Here we use the power flow $M_{\mathrm{L_G}}$ only if we also have a load profile $\tilde{L}_\mathrm{G}$ describing the household gas demand. In this case, there is an additional power flow equation to consider:

$$\tilde{L}_\mathrm{G} = M_{\mathrm{L_G}}\eta_{\mathrm{ML}_G} + \tilde{G}_\mathrm{C}. \tag{9.31}$$

We had to introduce in eqn (9.31) the quantity $\tilde{G}_\mathrm{C}$, which describes the demand for fossil gas that cannot be met by the methanization plant. In this way, we have the possibility of adjusting the dimensioning of the methanization plant and of varying the quantities.

Exercise 9.13 We have formulated the equations for the hydrogen tank and the methane tank in eqns (9.29) and (9.30). We need to look at the power flow equations for the wind turbine W, the battery S, the gas power plant P, and the loads L. What are the extended powerflow equations for these nodes? (Hint: To calculate them, we need to extend the equations from Section 8.6.3.)

Solution: We start with the power flow equations for the wind farm. As can be seen in Fig. 9.22, this gets an additional power outflow W_H—that is, part of the power is used to produce hydrogen. The equation is therefore:

$$\tilde{W} \geq W_\mathrm{S} + W_\mathrm{L} + W_\mathrm{H}.$$

The battery can receive power from the hydrogen tank via the fuel cell. It has an additional inflow H_S. Conversely, it can output power to fill the hydrogen tank. So there is still a power flow S_H:

$$0 = \eta_{\mathrm{WS}}W_\mathrm{S} + \eta_{\mathrm{SS}}S_\mathrm{S}(\Delta) + \eta_{\mathrm{PS}}P_\mathrm{S} + \eta_{\mathrm{HS}}H_\mathrm{S} - (S_\mathrm{L} + S_\mathrm{S}(\Delta) + S_\mathrm{H}).$$

The gas power plant can draw additional power from the methane tank. So we have an additional power flow M_P. But we can also use the power from the power plant to supply power to the electrolyser to fill the hydrogen tank. So we still have a power drain P_H:

$$0 = \eta_{\mathrm{MP}}M_\mathrm{P} + \tilde{P} - (P_\mathrm{S} + P_\mathrm{L} + P_\mathrm{H}).$$

This formulation is slightly different from that presented in Section 8.6.3. If we interpret $\tilde{P}$ as a load profile, or consumption profile, then it is clear why we put $\tilde{P}$ here as a power inflow. $\tilde{P}$ corresponds in size to the power demand of storage and loads corrected for the power inflow from the methane tank.

The consumption node L, which only describes the electrical loads, now receives an additional inflow via the hydrogen tank H_L. The power flow equation is therefore:

$$\tilde{L} = \eta_{\mathrm{WL}}W_\mathrm{L} + \eta_{\mathrm{SL}}S_\mathrm{L} + \eta_{\mathrm{PL}}P_\mathrm{L} + \eta_{\mathrm{HL}}H_\mathrm{L}.$$

Using eqns (9.29), (9.30), and (9.31), as well as the extended equations from Exercise 9.13, we have assembled the power flow equations. The extension significantly increases the complexity of the overall system. We have to specify a total of 10 additional power flows in the operation management and observe compliance with additional boundary conditions and balance equations.

By introducing the hydrogen and methane tank, we have created the possibility of introducing a seasonal storage. The self-discharge of both storage facilities is extremely low. We can store methane for months and years, and the same applies to hydrogen, although the losses will be measurable here, since hydrogen is a volatile gas. This gives us an additional problem in determining how to operate the energy system in a reasonable manner. The operation management presented in Section 8.6.3 worked without forecasting. That is, to determine a power flow at time t, all information from time $t - \Delta t$ was collected and combined with information at time t. The decision regarding which power flows to evaluate, and how, depended on this information. This approach does not apply to seasonal storage, because the information that a power flow is needed in a few days or months which will not be covered by wind or the battery is not available, and has not been included in the previous power flow equations. There are various approaches to solving this problem.

The hydrogen tank and the methane tank are given a minimum fill level $\kappa_{H,min}$, $\kappa_{M,min}$ as a boundary condition. This fill level can be time-dependent. This boundary condition adds a seasonal component to the existing equations. For example, if we want to fill the reservoir in the summer and autumn months so that discharging can take place in the winter and spring months, we can specify this filling process through this boundary condition.

We reward or penalize the charging and discharging of the hydrogen and methane tanks using a time-dependent tariff. Since the efficiencies of the power flows from the hydrogen and methane tank are significantly lower than those of the other power flows—recall that methanization had an efficiency of 50%—we can increase the power flows through financial assets. We can incorporate seasonal effects into the tariffs. In this case, we would adjust the revenue function that looks at the costs for the whole system.

The first two approaches have the disadvantage that we are, so to speak, introducing specifications to the system 'from outside'. Such an approach always bears the risk that the one who gives the specifications also determines the system's behaviour. When analysing the solution, a question always arises: is an optimal system configuration with an optimal system behaviour really an optimal solution or are there better solutions? Another possibility is that the operational management determines the power flows not for one point in time alone, but for several points in time simultaneously. This corresponds to a solution using a perfect forecast. In such cases, we do not need to introduce assumptions about tariffs or minimum capacities. Such calculations are demanding, but technically solvable.

9.5   Conclusion

In this chapter, we have learned about chemical storage technologies. These are essentially the production of hydrogen and methane. Both storage media have the advantage that they have a high energy density and can store energy over a very long period of time. If appropriate measures are taken during storage, then the self-discharge is negligible compared to the self-discharge of other storage technologies. Both storage media can be used in two ways. One is through combustion of the storage medium. Both methane and hydrogen can be used in modified combustion engines. In the home, both media can be burned with suitable equipment to produce heat or electricity. Alternatively, both media can be used via cold combustion, which takes place in a fuel cell. Here, a chemical reaction takes place, producing electricity and waste heat. The efficiency is considerably higher than that of a combustion engine, which is why fuel cells powered by hydrogen are becoming more and more common in mobility.

The biggest disadvantage of these technologies is their lower efficiency compared with other storage technologies. This applies above all to methanization, which can only achieve 50% efficiency due to the nature of the process. If the power-electronic conversion stages are added to this, efficiency levels of 30–40% are reached. On the other hand, many countries already have a gas infrastructure that uses fossil gas. This storage, transport, and consumption infrastructure can then be used directly.

Chapter summary

- Chemical storage uses energy to chemically convert substances. The most common reactions are hydrogen, oxygen, and carbon.
- The splitting of water allows us to store excess electrical energy. In the process, water becomes oxygen and hydrogen.
- Hydrogen can be burned together with oxygen. An alternative and energetically better solution is to use a fuel cell. Here, electrical energy is generated from hydrogen and oxygen.
- Hydrogen has a low volumetric energy density. Therefore, the hydrogen must be strongly compressed or liquefied.
- Power to gas or power to liquid are methanization processes. Here, methane is produced from hydrogen, carbon, and oxygen with the help of electrical energy.
- Methane has the advantage that it can be integrated into our existing energy system.

10
Demand side management

We explained in the Introduction to Chapter 2 that storage is used to store surplus energy so that we can use it at a later time or in another place. In this chapter, we look at another technique that can help us make better use of it: instead of storing energy, we can also postpone the consumption of energy to another time.

The storage technologies presented so far have used different physical and chemical principles to store energy. According to requirement G1, they stored surplus energy to make it available for a later point in time. To store, we therefore needed a positive residual load—that is, excess production. The demand side management (DSM) strategy is to manipulate the residual load through appropriate measures. Whenever more energy is produced than can be consumed at the moment, consumption is increased, and at times when less energy is produced, we reduce our consumption.

This approach is closely related to the terms 'smart grid' and 'smart home' (Jabir *et al.*, 2018; Mariano-Hernández *et al.*, 2021). As renewable energy sources feed in energy at different levels of the power grid, the question of how to use the surplus energy in times of overproduction, when there is not enough storage capacity, arises more and more. One approach is to use time-variable energy tariffs to give private households incentives to consume more energy in times of surplus and to reduce energy consumption in times of lower production. While some experiments worked with visual signals or printed tariff tables, in the smart home, consumption control was automated.

The smart grid extends the basic idea of the smart home, in that the energy grid is monitored in its entirety and consumers are adjusted according to production. In addition, however, there are functions that serve to comply with grid standards. For example, if increased production of solar power in the distribution grid causes local voltage increases, increased consumption or local adjustment of feed-in can ensure that the grid voltage remains within the normative specifications.

10.1 Basic functions and requirements

As we have already described, the basic idea of DSM is to adjust the residual load. In DSM, this is done by manipulating the loads. We therefore need an energy management system (EMS) that knows the current consumption as well as the current production. Furthermore, the EMS must be able to influence the consumption of the loads in the house or grid.

In order for energy to be stored via DSM, there must be loads in the system whose consumption can be controlled and/or shifted over time.

Energy Storage Systems. Armin U. Schmiegel, Oxford University Press. © Armin U. Schmiegel (2023).
DOI: 10.1093/oso/9780192858009.003.0010

DSM 1: AS AN Energy Management System, I WANT to be able to change the consumption of loads SO THAT I can adjust consumption according to production.

Essentially, there are two ways to adjust the consumption of a load. We can switch individual loads on or off, or we can change their consumption patterns. Suppose we have N_α loads in a household whose consumption we can scale and N_β loads that we can simply switch on and off. This certainly leaves a number of N loads that we cannot manipulate. If we want to determine the total consumption, we have to sum up the different loads:

$$L^t = \sum_{i=1}^{N_\alpha} \alpha_i L_i + \sum_{j=1}^{N_\beta} \beta_j L_j + \sum_{k=1}^{N} L_k. \tag{10.1}$$

β_j describes whether a load is connected or disconnected—that is, $\beta_j \in \{0, 1\}$. α_i indicates the relative consumption for controllable loads. α_i can take on continuous but also discrete values, which depends on the respective load and its controllability.

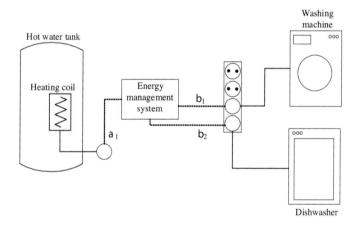

Fig. 10.1 Example of load management: the energy management system (centre) controls a washing machine (β_1) and a dishwasher (β_2) via a power strip. Furthermore, the energy management system can control the heating coil of a hot water tank (α_1).

To illustrate how to interpret eqn (10.1), consider the smart home shown in Fig. 10.1. In this illustration, we show only the controllable loads. First, we have a hot water tank which is used for heating and domestic water. Inside this hot water tank is a heating coil which heats the water electrically. Normally, this heating coil is controlled by the controller of the heating system, but it is possible to ignore this control and control the heating coil directly. In this case, we use α_1 to specify the power that the heating coil can draw.

Furthermore, we have a washing machine and a dishwasher in the household. Here, we cannot set the intensity with which these two appliances consume electricity, but we can switch them on via β_1 and β_2. The control of α_1, β_1, and β_2 is done by the

EMS, which provides DSM. We can switch on two loads with β_1 and β_2 and we can increase the consumption with α_1.

Exercise 10.1 In this exercise, we consider a industrial application. We have two motors whose consumption can be controlled between 5 kW and 15 kW. In addition, 20 consumers with a consumption of 100 W are connected. These consumers can be switched on or off. These are cooling units that can only be switched off for a limited period of time. Furthermore, there are non-controllable consumers that have a total consumption of 18 kW. What is the minimum and maximum consumption?

Solution: To find the minimum consumption, we need to switch off all the controllable loads. Only the non-controllable loads with 18 kW and the two motors that work with a minimum load of 5 kW remain as consumers. The minimum consumption thus results in:

$$L_{\min} = 2 \cdot 5 \text{ kW} + 18 \text{ kW} = 28 \text{ kW}.$$

To find the maximum consumption, we need to switch on all loads—that is, $\alpha_i = 10$ kW and $\beta_j = 1$:

$$L_{\max} = \sum_{i=1}^{2} (\alpha_i 10 \text{ kW} + 5 \text{ kW}) + \sum_{j=1}^{20} \beta_j 0.1 \text{ kW} + 18 \text{ kW}$$

$$= \sum_{i=1}^{2} (1 \cdot 10 \text{ kW} + 5 \text{ kW}) + \sum_{j=1}^{20} 1 \cdot 0.1 \text{ kW} + 18 \text{ kW} = 50 \text{ kW}.$$

Load management means controlling demand. In previous storage systems, the available power from sources and storage was distributed to the available loads. The distribution was performed by an EMS that controlled the power flows to optimize a target function. In the case of the solar power storage system, for example, the target function was the energy costs. If these were to be minimized, this would mean that solar power was consumed directly first, then energy from the storage was used. Surpluses were stored.

If DSM is added to such a system, further degrees of freedom are available for control, because now consumption can be increased or reduced. However, this has the consequence that the target function must be adjusted. Whereas previously a reduction in energy costs as a target function led to reasonable system behaviour, with load management this target function would result in all loads being consistently switched off, which would, of course, reduce energy costs considerably, but would not be a reasonable solution for the user. We therefore need an additional requirement that rules out such a solution.

DSM 2: AS A User, I WANT the total consumption of my loads to not be changed by the DSM SO THAT I can be sure that the DSM redistributes my consumption in time.

In formulating DSM 2, we have assumed that there is no waste of energy in our system and we do not want to use DSM to save energy. If we had unnecessary loads in our system, then switching off or reducing consumption would, of course, be the better option. DSM 2 can also be softened by allowing a reduction in total consumption of $x\%$. This would allow DSM to shift loads so that they operate in a more efficient mode.

Another alternative formulation of DSM 2 would involve not formulating the total consumption of the loads as the conservation variable, but the work to be done by the loads. This would be a very system-specific formulation in which we would have to incorporate a lot of knowledge about the concrete application.

Based on DSM 2, there are two basic functions that can be realized with DSM: first, consumption can be shifted in time—that is, schedules are created for the individual loads that distribute the consumption in a suitable manner throughout the day (Mahmood *et al.*, 2016)

Exercise 10.2 A solar power system is installed on a plantation, as well as a number of electrically operated pumps to irrigate the fields. The 20 pumps each have a consumption of 2 kW. The solar power plant has a forecast for daily production. Tab. 10.1 shows this forecast.

The pumps must run at least four times a day for one hour. What does a reasonable schedule look like?

Solution: To find the total energy produced as per the forecast, we need to sum up the power of the day's forecast and find the energy:

$$E_{\text{solar}} = 168 \text{ kW} \cdot 1 \text{ h} = 168 \text{ kWh}.$$

Table 10.1 Solar production of the solar plantation.

Time	9:00	10:00	11:00	12:00	13:00	14:00	15:00	16:00	17:00	18:00	19:00
Power [kW]	3	4	11	30	31	24	17	24	9	8	7

For the required pump energy, we have to consider that each pump consumes 2 kW four times a day. So we need a total energy of $4 \cdot 20 \cdot 2 \text{ kW} = 160 \text{ kWh}$. Throughout the day, the solar system generates enough energy to run the pumps. We want to create a schedule that ensures that we only run the pumps using the solar system. We can implement this in such a way that at each hour only as many pumps are switched on as there is solar power. In doing so, we rotate the pumps so that each pump is also switched on four times a day. Such a schedule is shown in Tab. 10.2.

Table 10.2 Pump schedule for the plantation supplied with solar energy.

Time	09:00	10:00	11:00	12:00	13:00	14:00	15:00	16:00	17:00	18:00	19:00
Number of pumps	1	2	5	15	15	12	8	12	4	4	3
Power[kW]	2	4	10	30	30	24	16	24	8	8	6

The example in Exercise 10.2 shows how a schedule is used to distribute loads so that the load is always high when solar production is also high. This requires a forecast or model for the expected surplus.

The second function for DSM is that the existing consumption is dynamically adjusted to the surplus. In times of surplus, consumption is increased, and when less power is available, it is reduced.

Exercise 10.3 We consider a household that has a hot water tank. As shown in Fig. 10.1, this hot water tank has a heating coil that heats the water. During the day, electric current is used to heat the water to its target temperature. This requires 3 kWh of electrical energy.

The household has a solar power system. This produces an excess power of 2 kW for two hours around midday, 1 kW for three hours in the morning and 500 W for four hours in the afternoon.

We can dynamically adjust the heating power of the heating coil to increase the solar power consumption for heating. What is a suitable usage strategy?

Solution: We want to avoid the self-discharge of the heat storage tank being too high. Therefore, the storage tank is charged with variable power as late in the day as possible. We need 3 kWh to have the right temperature in the hot water tank at the end of the day. We therefore calculate backwards and determine the time from which we have stored the 3 kWh:

$$4\,\text{h} \cdot 0.5\,\text{kW} + \frac{1}{2}\,\text{h} \cdot 2\,\text{kW} = 3\,\text{kWh}.$$

So if we charge from midday, we have transferred enough energy to the hot water tank so that we have reached the desired target temperature. The power in the morning and some of the power from noon can be fed in or used in other ways.

We can distribute the load over time, or vary its intensity. Both mechanisms lead to the same control task: minimize the residual load L_R under the constraint that the total consumption L_0 is maintained:

$$\min \int_{t=0}^{T} L_R\left(\alpha_i, \beta_j\right) dt \quad \text{w.t.b.c.:} \quad \int_{t=0} L\,dt = L_0. \tag{10.2}$$

Eqn (10.2) formulates demand side management as an optimization problem, but the formulation is not yet complete. One element is missing here, which also leads to another requirement. Let us consider our example in Fig. 10.1. Suppose we have a solar power system in the household and want to make better use of solar energy with the help of DSM. It is summer, and one of those days when the sun shines intensely throughout day. The uncontrolled consumers in the household cannot use all the power; the washing machine and dishwasher also have nothing to do. So we can only use the heating coil to consume the solar energy. However, there are limits here. The water in the hot water tank is not meant to be used for cooking, but has a target temperature that should be reached. Once this is reached, we cannot increase the consumption any further.

Conversely, if we have changing weather and the washing machine and dishwasher have been filled, we cannot simply stop the washing process because the solar power system no longer produces enough surplus energy. The controllable consumers have their own boundary conditions, which DSM must be aware of and observe.

DSM 3: AS A Energy Management System, I WANT to comply with the operating parameters of the controllable loads SO THAT I do not damage the loads or restrict their operation too much.

DSM can be localized to subsystems and to individual system levels respectively. If we consider the example from Fig. 10.1, we used an EMS that is localized at a household. If this household is located in an apartment building, each household could have its own DSM. There may additionally be an EMS that looks at multiple homes and shares information with the subsystems. DSM can therefore take place across several system levels. Each EMS regulates on its own system level. Thus, an EMS can control the power flows in the distribution network of a utility consisting of hundreds of consumers and generators, while in the respective households or buildings—that is, subsystems of the overall system—separate EMSs carry out their own controls. This has the advantage that the EMS has aggregated information at the subsystem level, which can minimize the control effort.

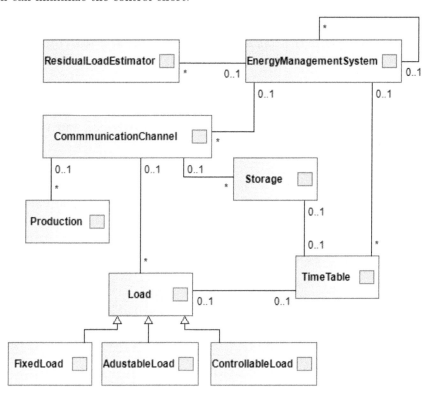

Fig. 10.2 System components of a demand side management system.

The system components are shown in Fig. 10.2. The central control unit here is the EnergyManagementSystem. This can be linked to other EMSs. It has a different CommunicationChannel, because not all Load, Production, and Storage components have the same communication interface. We have also introduced a Storage here, as DSM can of course also work in combination with the use of a storage facility.

The EnergyManagementSystem has various ResidualLoadEstimator components, which are calculation units that determine how high the residual load is. Some of these are aggregations of different forecasting units, while some of these system components work with heuristics.

The EnergyManagementSystem has a TimeTable for each load, which shows how the EnergyManagementSystem wants to use or can use this load. This also applies to the `Storage`, because here we also have to determine whether excess power should be stored or consumed, or whether we should retrieve needed power from the `Storage`.

We divide the `Load` into three groups. The `AdjustableLoads` are those `Loads` that we can control via a α. The `ControllableLoads` we can switch on or off using a β. The `FixedLoads` are those that are beyond the control of the EMS.

As simple as the idea of DSM sounds at first, its implementation is complicated. When we work in private households, load forecasting is extremely difficult. We can see a difference between workdays, public holidays, and the weekend, but quantitative statements are also not easy to come by.

Nevertheless, DSM in combination with storage is an additional component for a sustainable energy supply. In the next section, we will therefore look at an application example of DSM.

10.2   Application example: DSM for a commercial and residential complex

In this example, DSM for a commercial and residential complex is to be developed. In the previous application examples, the question was always how the respective storage technology must be designed. This is not the case with a DSM application; instead, the focus here is on identifying the control options.

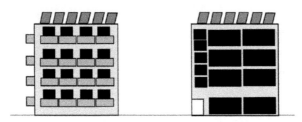

Fig. 10.3 Commercial and residential complex where demand side management is to be implemented.

The structure of the building complex is shown in Fig. 10.3. It consists of two buildings: a residential building with 20 residential units and a commercial building housing five businesses. The roofs of the two buildings are equipped with a solar power system. The installed capacity is equivalent to 800 kW$_\mathrm{p}$. This corresponds to an annual production of approximately 800,000 kWh. Individual households consume between 2,500 kWh and 4,500 kWh per year. Their maximum connected load is 15 kW. The five commercial units are a shop (4,000 $\frac{\mathrm{kWh}}{\mathrm{a}}$), a bakery (6,200 $\frac{\mathrm{kWh}}{\mathrm{a}}$), a butcher's shop (7,035 $\frac{\mathrm{kWh}}{\mathrm{a}}$), and two businesses using office space only (1,960 $\frac{\mathrm{kWh}}{\mathrm{a}}$ each).

In Tab. 10.3, the controllable loads are listed and the control range of the loads is shown. If a range is given for a load, then it is a load which can be adjusted with the help of a α. If only one value is given for a load, it is a load that we can switch on or off with a β. In Tab. 10.3, we have additionally determined how large the total control range is that we can set with these loads.

Table 10.3 Controllable loads of the different building units and the control range.

Unit	Controllable loads	Control range of the different loads		Control range
Household	Kitchen Appliances	300–700 W	500 W	300–1,200 W
	Heating			
Retail shop	Freezer cabinets, air conditioning	400–1,000 W	2,000 W	400–3,000 W
Bakery	Oven	1,000–3,000 W		1,000–3,000 W
Butchery	Refrigeration	1,000–3,000 W		1,000–3,000 W
Office	Air conditioning	1,000 W		1,000 W

DSM is a business model of the building operator. Tenants have agreed to join the overall system's load management. Profits resulting from the DSM and the feed-in of solar electricity are distributed proportionally to the tenants, with participants receiving a higher share if they have made a higher standard contribution.

From the building operator's point of view, the following requirement arises:

CRC 1: AS AN operator, I WANT to reduce the energy costs of the overall system SO THAT I can make an attractive offer to the tenants.

Of course, the building operator knows that in addition to a financially attractive offer, comfort must also be ensured. None of the tenants will want to cope with important loads simply being switched off.

CRC 2: AS AN operator, I WANT to meet the mandatory load requirements of individual customers SO THAT the customers do not suffer a loss of comfort.

CRC 2 poses a problem for the operator, who has to know the mandatory load requirement of each customer and observe it in the central DSM. She simplifies this task. She equips every rental unit with its own EMS. Each system can be configured by the tenant. There are various concepts for determining this local available power for DSM. One solution, for example, is for users to prioritize their consumers and define time slices in which they must be served.

The central DSM, which determines the production of the solar power system, analyses the respective reserves and gives the EMS of the residential and commercial units the control instructions to be realized. With this concept, every tenant can decide how they participate in the DSM and the operator gets aggregated information, which allows her to optimize the power consumption.

The power flow diagram is shown in Fig. 10.4. For clarity, the 20 residential units have been grouped as L_{r_i} nodes and the five commercial units as L_{c_i} nodes. We have a production node P representing the production of solar power.

Solar energy can be transmitted to residential units, $P_{L_{r_i}}$ or shops, $P_{L_{c_i}}$. Furthermore, there is the possibility to sell solar power to the grid, P_G:

$$\tilde{P} = \sum_{i=1}^{20} P_{L_{r_i}} + \sum_{j=5}^{5} P_{L_{c_i}} + P_G. \tag{10.3}$$

We can use some of the solar energy for the loads and sell any remaining surplus to the grid for a profit.

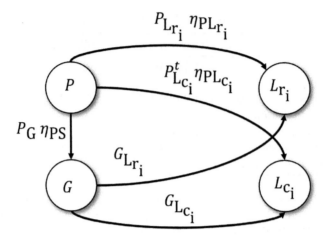

Fig. 10.4 Power flow diagram for load management of a residential and commercial complex.

For the consumption nodes, the load demand adjusted by the EMS must be covered by solar power and grid purchase:

$$\forall i: \tilde{L}_{\mathrm{r}_i}\left(\alpha_{\mathrm{r}_i}, \beta_{\mathrm{r}_i}\right) = P_{\mathrm{L}_{\mathrm{r}_i}} + G_{\mathrm{L}_{\mathrm{r}_i}} \tag{10.4}$$

$$\forall j: \tilde{L}_{\mathrm{c}_j}\left(\alpha_{\mathrm{c}_j}, \beta_{\mathrm{c}_j}\right) = P_{\mathrm{L}_{\mathrm{c}_j}} + G_{\mathrm{L}_{\mathrm{c}_j}}. \tag{10.5}$$

According to CRC 1, the objective function should be to minimize energy costs. Here, the electricity tariffs depend on the respective consumer. With c_{P_c} as the feed-in tariff for the solar electricity and $c_{G_{L_{r_i}}}$ or $c_{G_{L_{c_j}}}$ as the electricity tariffs of the consumers, the objective function is:

$$\min Y = c_{P_G} P_G - \left(\sum_{i=1}^{20} c_{G_{L_{r_i}}} G_{L_{r_i}} + \sum_{j=1}^{5} c_{G_{L_{c_j}}} G_{L_{c_j}} \right). \tag{10.6}$$

In eqn (10.6), the optimization is done at each time point—that is, at every time step t the energy costs are optimized. An alternative approach would be to optimize over a period T—that is, a longer period of time is considered for the control:

$$\min Y = \int_{t=0}^{T} Y \, dt. \tag{10.7}$$

However, such a method needs reliable load forecasts and needs to be robust in the face of mispredictions. It is very easy to see how difficult it is to forecast consumption if, for example, a resident changes his diet from one day to the next and suddenly spends a lot of time preparing elaborate meals. The consumption forecast may not be able to predict this behavioural change correctly.

With the power flow equations and the yield function, we have described the system completely. The responsibilities in terms of requirements are relatively simple. The central EMS determines the currently available solar power production and obtains the control reserves of the decentralized EMS, on which an optimization takes place in the sense of eqn (10.6) or (10.7), and with this result the EMS then retrieves the control reserves of the various consumers.

Exercise 10.4 The solar power system generates $\tilde{P} = 17{,}550$ kW of solar power. The feed-in tariff is $c_{P_G} = 0.88 \frac{\text{€ct}}{\text{kWh}}$. From the different EMSs, the central EMS receives the consumption information shown in Tab. 10.4.

Table 10.4 Consumption information of individual households and commercial units.

Building unit	Minimum consumption [kW]	Maximum consumption [kW]	Electricity tariff $\left(\frac{\text{€ct}}{\text{kWh}}\right)$
Household 1	150	300	0.24
Household 15	800	1,200	0.22
Household 17	250	900	0.23
Remaining households	550	550	0.245
Retail shop	1,500	2,500	0.13
Bakery	900	1,000	0.11
Butcher's shop	1,000	1,250	0.12
Office 1	700	800	0.18
Office 2	200	500	0.24

The minimum consumption can be completely covered by the solar system. There is still 2,700 kW of solar electricity left. Thus, it has to be prioritized to which consumer more solar power can be allocated. What could an economically sensible distribution look like?

Solution: In this example, the consumers are prioritized by the monetary value of the control energy provided. This results from the difference between the maximum and minimum consumption, integrated over a period of one hour, for example, and multiplied by the electricity tariff:

$$v = (L_{x_{\max}} - L_{x_{\min}}) \cdot c_{G_{L_x}} \cdot \Delta t.$$

For household 17, this therefore results in a value of:

$$v = (900\,\text{kW} - 250\,\text{kW}) \cdot 0.23 \frac{\text{€ct}}{\text{kWh}} \cdot 1\,\text{h} = 149.5\,\text{€}.$$

The EMS calculates these values for all households and sorts them according to monetary value. According to this order, the solar electricity is distributed to the households. This distribution is shown in Tab. 10.5.

Table 10.5 Distribution of excess solar energy to the different residential and commercial units.

Building unit	Value of control power (€)	Additional consumption [kW]	Remaining solar coverage [kW]
Household 17	149.50	650	2,050
Shop	130.00	1,000	1,050
Household 15	88.00	400	650
Office 2	72.00	300	350
Household 1	36.00	150	200
Butcher's shop	30.00	200	0
Office 1	18.00	0	0
Bakery	11.00	0	0

It should be noted that this 'distribution' is not done directly, but by switching loads on or off. In this case, households 1, 15, and 17, the shop and office 2 can cover their maximum consumption solar, while the butchery is only allowed 200 kW of additional consumption.

10.3 Conclusion

In Exercise 10.4, we calculated the optimal distribution for a point in time. In doing so, we used a relatively simple algorithm. We have monetarily ordered the possible control services, and thus obtained a prioritization. However, the question of whether DSM is a sensible investment arises. If we consider a household, the installation of controllable loads is relatively cost-intensive. In addition, the user has to enter his consumption priorities and determine which loads he will not allow to be shifted.

The use of a storage unit has clear advantages here. A storage unit is relatively easy to install in a household. The user also does not need to enter his consumption preferences, because the storage unit stores the surplus by itself and makes it available when it can be withdrawn.

The case is different when it comes to individual large consumers, such as those in commercial enterprises. For example, reducing the cooling temperature of a cold store by one degree can easily store an existing surplus, or increasing the cooling temperature is possible as long as the boundary conditions of the cold chain are not violated.

In the context of a distribution network or a city district, it therefore makes sense to integrate individual selected loads into a DSM system.

Chapter summary

- Demand side management (DSM) is based on the idea not of shifting the energy in time and space by storing it, but of adjusting the consumption to the current production.

- There are two possibilities of adjustment. Loads can be switched on or off, or loads can be adjusted in the amount of their consumption.

- DSM involves a number of constraints. For example, the users must agree to the change in consumption. Furthermore, the loads must not be damaged.

- DSM is carried out via the energy management system. This can aggregate and forward data at different system levels.

- DSM is a very specialized form of consumption optimization and requires information and intervention options. Compared to a storage system, the realization of a DSM system is associated with a greater amount of effort.

References

NREL drivecat – Chassis Dynamometer Drive Cycles (2022). National Renewable Energy Laboratory.

OEDI Data Lake, US Department of Energy.

Power Data Access Viewer.

Acone, Mariano, Romano, Roberto, Piccolo, Antonio, Siano, Pierluigi, Loia, Francesca, Ippolito, Mariano Giuseppe, and Zizzo, Gaetano (2015). Designing an Energy Management System for Smart Houses. In *2015 IEEE 15th International Conference on Environment and Electrical Engineering (EEEIC)*, pp. 1677–82. IEEE.

Akhil, Abbas A., Huff, Georgianne, Currier, Aileen B., Kaun, Benjamin C., Rastler, Dan M., Chen, Stella Bingqing, Cotter, Andrew L., Bradshaw, Dale T., and Gauntlett, William D (2013). *DOE/EPRI 2013 Electricity Storage Handbook in Collaboration with NRECA*, Volume 1, Sandia National Laboratories Albuquerque, NM.

Ardito, Luca, Procaccianti, Giuseppe, Torchiano, Marco, and Migliore, Giuseppe (2013). Profiling Power Consumption on Mobile Devices. *Energy*, 101–6.

Armstrong, Marianne M., Swinton, Mike C., Ribberink, Hajo, Beausoleil-Morrison, Ian, and Millette, Jocelyn (2009). Synthetically Derived Profiles for Representing Occupant-Driven Electric Loads in Canadian Housing. *Journal of Building Performance Simulation*, **2**(1), 15–30.

Badescu, Viorel (2007). Optimal Control of Flow in Solar Collectors for Maximum Exergy Extraction. *International Journal of Heat and Mass Transfer*, **50**(21–2), 4311–22.

Bauer, Michaela (2019). System Design and Power Flow of Stationary Energy Storage Systems. PhD thesis, ETH Zurich.

Blanc, Christian, and Rufer, Alfred (2010). Understanding the Vanadium Redox Flow Batteries. *Paths to Sustainable Energy*, **18**(2), 334–6.

Blum, Nicola U., Wakeling, Ratri Sryantoro, and Schmidt, Tobias S. (2013). Rural Electrification through Village Grids: Assessing the Cost Competitiveness of Isolated Renewable Energy Technologies in Indonesia. *Renewable and Sustainable Energy Reviews*, **22**, 482–96.

Blundell, Stephen J. and Blundell, Katherine M. (2010). *Concepts in Thermal Physics*. Oxford University Press on Demand.

Boldea, Ion, and Nasar, Syed A. (2016). *Electric Drives*. CRC Press.

Bouffard, François, and Galiana, Francisco D. (2004). An Electricity Market with a Probabilistic Spinning Reserve Criterion. *IEEE Transactions on Power Systems*, **19**(1), 300–7.

Bransden, Brian Harold, and Joachain, Charles Jean (1989). Introduction to quantum mechanics.

Breyer, C., Gerlach, A., Hlusiak, M., Peters, C., Adelmann, P., Winiecki, J., Schütze-ichel, H., Tsegaye, S., and Gashie, W. (2009). Electrifying the Poor: Highly Economic Off-Grid PV Systems in Ethiopia: A Basis for Sustainable Rural Development. In *Proceedings 24th European Photovoltaic Solar Energy Conference, Hamburg*, pp. 21–5.

Brundlinger, R. Grid Codes in Europe: Overview on the Current Requirements in European Codes and National Interconnection Standards. *NEDO/IEA PVPS Task*, **14**.

Cairns, E. J. and Tevebaugh, A. D. (1964). CHO Gas Phase Compositions in Equilibrium with Carbon, and Carbon Deposition Boundaries at One Atmosphere. *Journal of Chemical & Engineering Data*, **9**(3), 453–62.

Carroll, Aaron, Heiser, Gernot et al. (2010). An Analysis of Power Consumption in a Smartphone. In *USENIX Annual Technical Conference*, Volume 14, pp. 21–21. Boston, MA.

Chowdhury, Shahriar Ahmed, Aziz, Shakila, Groh, Sebastian, Kirchhoff, Hannes, and Leal Filho, Walter (2015). Off-Grid Rural Area Electrification through Solar-Diesel Hybrid Minigrids in Bangladesh: Resource-Efficient Design Principles in Practice. *Journal of Cleaner Production*, **95**, 194–202.

Cirrincione, Maurizio, Pucci, Marcello, and Vitale, Gianpaolo (2017). *Power Converters and AC Electrical Drives with Linear Neural Networks*. CRC Press.

Damiani, Lorenzo, and Trucco, Angela (2009). Biomass Gasification Modelling: An Equilibrium Model, Modified to Reproduce the Operation of Actual Reactors. In *Turbo Expo: Power for Land, Sea, and Air*, Volume 48821, pp. 493–502.

Daowd, Mohamed, Omar, Noshin, Van Den Bossche, Peter, and Van Mierlo, Joeri (2011). Passive and Active Battery Balancing Comparison Based on Matlab Simulation. In *2011 IEEE Vehicle Power and Propulsion Conference*, pp. 1–7. IEEE.

De Doncker, Rik W., Pulle, Duco W. J., and Veltman, André (2020). Modern Electrical Drives: An Overview. *Advanced Electrical Drives*, pp. 1–16. Springer.

Delligatti, Lenny (2013). *SysML Distilled: A Brief Guide to the Systems Modeling Language*. Addison-Wesley.

Dittmer, Manfred, Schmiegel, Armin Uwe, Cousseau, Jean-Frencois, and Lippert, Michael (2009). Demand Driven Integrated PV-system with Lithium-Ion Batteries for Storage to Boost Self Consumption. In *European Photovoltaic Solar Energy Conference*, Volume 24.

Dragoon, Ken, and Schumaker, Adam (2010). Solar PV Variability and Grid Integration. Renewable Northwest Project.

Dujic, Drazen, Kieferndorf, Frederick, Canales, Francisco, and Drofenik, Uwe (2012). Power Electronic Traction Transformer Technology. In *Proceedings of the 7th International Power Electronics and Motion Control Conference*, Volume 1, pp. 636–42.

Eckroad, S. (1999). Flywheels for Electric Utility Energy Storage. Technical Report. Electric Power Research Institute, Palo Alto, CA.

Ehsani, Mehrdad, Gao, Yimin, Longo, Stefano, and Ebrahimi, Kambiz M (2018). *Modern Electric, Hybrid Electric, and Fuel Cell Vehicles*. CRC Press.

Erdinc, Ozan (2014). Economic Impacts of Small-Scale Own Generating and Storage Units, and Electric Vehicles under Different Demand Response Strategies for Smart Households. *Applied Energy*, **126**, 142–50.

Falcones, Sixifo, Mao, Xiaolin, and Ayyanar, Raja (2010). Topology Comparison for Solid State Transformer Implementation. In *IEEE PES General Meeting*, pp. 1–8. IEEE.

Feynman, Richard P., Leighton, Robert B., and Sands, Matthew (1965). The Feynman Lectures on Physics; Vol. I. *American Journal of Physics*, **33**(9), 750–2.

Figgener, Jan, Stenzel, Peter, Kairies, Kai-Philipp, Linßen, Jochen, Haberschusz, David, Wessels, Oliver, Angenendt, Georg, Robinius, Martin, Stolten, Detlef, and Sauer, Dirk Uwe (2020). The Development of Stationary Battery Storage Systems in Germany: A Market Review. *Journal of Energy Storage*, **29**, 101153.

Fossheim, Kristian, and Sudbø, Asle (2004). *Superconductivity: Physics and Applications*. John Wiley & Sons.

Friedenthal, Sanford, Moore, Alan, and Steiner, Rick (2014). *A practical guide to SysML: the systems modeling language*. Morgan Kaufmann.

Frost, Damien F., and Howey, David A. (2017). Completely Decentralized Active Balancing Battery Management System. *IEEE Transactions on Power Electronics*, **33**(1), 729–38.

Ghaib, Karim, and Ben-Fares, Fatima-Zahrae (2018). Power-to-Methane: A State-of-the-Art Review. *Renewable and Sustainable Energy Reviews*, **81**, 433–46.

Gonzalez, Adolfo, Gallachóir, B. O., McKeogh, Eamon, and Lynch, Kevin (2004). Study of Electricity Technologies and Their Potentials to Address Wind Energy Intermittency in Ireland. Sustainable Energy Research Group, Department of Civil and Environmental Engineering, University College Cork, Cork.

Götz, Manuel, Lefebvre, Jonathan, Mörs, Friedemann, Koch, Amy McDaniel, Graf, Frank, Bajohr, Siegfried, Reimert, Rainer, and Kolb, Thomas (2016). Renewable Power-to-Gas: A Technological and Economic Review. *Renewable Energy*, **85**, 1371–90.

Govindarajan, Jaykanth (2017). Adaptive Online Estimation of State of Charge and Parameters of Li-Ion Batteries.

Häberlin, Heinrich (2007). Photovoltaik. *Strom aus Sonnenlicht für Verbundnetz und Inselanlagen*, **1**.

Haessig, Pierre, Multon, Bernard, Ahmed, Hamid Ben, Lascaud, Stéphane, and Jamy, Lionel (2013). Aging-aware NaS Battery Model in a Stochastic Wind-Storage Simulation Framework. In *IEEE PowerTech 2013 Grenoble Conference*, pp. 1–6. IEEE.

Halka, Monica, and Nordstrom, Brian (2011). *Metals and Metalloids*. Facts On File.

Halper, Marin S., and Ellenbogen, James C. (2006). Supercapacitors: A Brief Overview. *The MITRE Corporation, McLean, Virginia, USA*, **1**.

He, Zhiwei, Gao, Mingyu, Wang, Caisheng, Wang, Leyi, and Liu, Yuanyuan (2013). Adaptive State of Charge Estimation for Li-Ion Batteries Based on an Unscented Kalman Filter with an Enhanced Battery Model. *Energies*, **6**(8), 4134–51.

Hesse, Holger C., Schimpe, Michael, Kucevic, Daniel, and Jossen, Andreas (2017). Lithium-Ion Battery Storage for the Grid: A Review of Stationary Battery Storage System Design Tailored for Applications in Modern Power Grids. *Energies*, **10**(12), 2107.

Hong, G. W., Abe, N., and Baclay, M. (2011). Else an Eventual Return to Conventional Energy: Impacts and Fate of an Off-Grid Rural Electrification Project in an Island in the Philippines. In *Proc. 26th European Photovoltaic Solar Energy Conf., Hamburg, Germany*, pp. 4052–6.

Horowitz, Paul, Hill, Winfield, and Robinson, Ian (1989). *The Art of Electronics*, Volume 2, Cambridge University Press.

Ibanez, Federico Martin (2017). Analyzing the Need for a Balancing System in Super-Capacitor Energy Storage Systems. *IEEE Transactions on Power Electronics*, **33**(3), 2162–71.

IEA, Hydrogen (2006). Production and Storage: R&D Priorities and Gaps. *2006 available on: http://www. iea. org/textbase/papers/2006/hydrogen. pdf.*

Immonen, Paula et al. (2013). Energy Efficiency of a Diesel-Electric Mobileworking Machine.

Immonen, Paula, Laurila, Lasse, Rilla, Marko, and Pyrhonen, Juha (2009). Modelling and Simulation of a Parallel Hybrid Drive System for Mobile Work Machines. In *IEEE EUROCON 2009*, pp. 867–72. IEEE.

Immonen, Paula, Ponomarev, Pavel, Åman, Rafael, Ahola, Ville, Uusi-Heikkilä, Janne, Laurila, Lasse, Handroos, Heikki, Niemelä, Markku, Pyrhönen, Juha, and Huhtala, Kalevi (2016). Energy Saving in Working Hydraulics of Long Booms in Heavy Working Vehicles. *Automation in Construction*, **65**, 125–32.

Ise, Toshifumi, Kita, Masanori, and Taguchi, Akira (2005). A Hybrid Energy Storage with a SMES and Secondary Battery. *IEEE Transactions on Applied Superconductivity*, **15**(2), 1915–18.

Jabbour, Nikolaos, and Mademlis, Christos (2017). Supercapacitor-Based Energy Recovery System with Improved Power Control and Energy Management for Elevator Applications. *IEEE Transactions on Power Electronics*, **32**(12), 9389–99.

Jabir, Hussein Jumma, Teh, Jiashen, Ishak, Dahaman, and Abunima, Hamza (2018). Impacts of Demand-Side Management on Electrical Power Systems: A Review. *Energies*, **11**(5), 1050.

Jackson, John David (2009). Classical Electrodynamics. *American Institute of Physics*, **15**(11), 62–62.

Kairies, Kai-Philipp (2017). Battery Storage Technology Improvements and Cost Reductions to 2030: A Deep Dive. *International Renewable Energy Agency Work.*

Keil, Peter (2017). Aging of Lithium-Ion Batteries in Electric Vehicles. PhD thesis, Technische Universität München.

Kempener, Ruud, and Borden, Eric (2015). Battery Storage for Renewables: Market Status and Technology Outlook. International Renewable Energy Agency, Abu Dhabi, 32.

Kim, Sang-Hoon (2017). *Electric Motor Control: DC, AC, and BLDC Motors*. Elsevier.

Kondepudi, Dilip K et al. (2008). *Introduction to Modern Thermodynamics*, Volume 666, John Wiley & Sons.

Koot, Michael Wilhelmus Theodorus (2006). Energy Management for Vehicular Electric Power Systems.

Kruchinin, Sergei (2021). *Modern Aspects of Superconductivity: Theory of Superconductivity.* World Scientific.

Kubade, Priyanka, Umathe, S. K., and Tutakne, Dr D. R. (2017). Regenerative Braking in an Elevator Using Supercapacitor. *International Research Journal of Advanced Engineering and Science*, **2**(2), 62–5.

Kumar, U. Suresh, and Manoharan, P. S. (2014). Stand Alone Wind/PV/Diesel Hybrid System for Telephone Transceiver Station in TamilNadu Rural Areas. *International Journal of Applied Engineering Research*, **9**(24), 8437–45.

Kurzweil, Peter, and Dietlmeier, Otto K. (2018). *Elektrochemische Speicher: Superkondensatoren, Batterien, Elektrolyse-Wasserstoff, Rechtliche Rahmenbedingungen.* Springer-Verlag.

Kusakana, Kanzumba, and Vermaak, Herman Jacobus (2013). Hybrid Renewable Power Systems for Mobile Telephony Base Stations in Developing Countries. *Renewable Energy*, **51**, 419–25.

Li, Jianwei, Yang, Qingqing, Robinson, Francis, Liang, Fei, Zhang, Min, and Yuan, Weijia (2017). Design and Test of a New Droop Control Algorithm for a SMES/Battery Hybrid Energy Storage System. *Energy*, **118**, 1110–22.

Li, Jianwei, Zhang, Min, Zhu, Jiahui, Yang, Qingqing, Zhang, Zhenyu, and Yuan, Weijia (2015). Analysis of Superconducting Magnetic Energy Storage Used in a Submarine HVAC Cable Based Offshore Wind System. *Energy Procedia*, **75**, 691–6.

Lin, Xiao, Pan, Shuang-xia, and Wang, Dong-yun (2008). Dynamic Simulation and Optimal Control Strategy for a Parallel Hybrid Hydraulic Excavator. *Journal of Zhejiang University-SCIENCE A*, **9**(5), 624–32.

Loisel, Rodica, Mercier, Arnaud, Petric, Hrvoje, Gatzen, Christoph, and Elms, Nick (2010). *Key Elements to Valuate Large-Scale Electricity Storage.* Germany: International Renewable Energy Storage Conference, 22–4 Nov, pp. 759–74.

Ma, Hongbin, Yan, Liping, Xia, Yuanqing, and Fu, Mengyin (2020). *Kalman Filtering and Information Fusion.* Springer.

Maclay, James D., Brouwer, Jacob, and Samuelsen, G. Scott (2007). Dynamic Modeling of Hybrid Energy Storage Systems Coupled to Photovoltaic Generation in Residential Applications. *Journal of Power Sources*, **163**(2), 916–25.

Magnor, D., Soltau, N., Bragard, Michael, Schmiegel, Armin Uwe, De Doncker, R. W., and Sauer, Dirk Uwe (2010). Analysis of the Model Dynamics for the Battery and Battery Converter in a Grid-Connected 5KW Photovoltaic System. In *Proceedings of the 25th European Photovoltaic Solar Energy Conference (PVSEC), Valencia, Spain*, pp. 6–10.

Mahesri, Aqeel, and Vardhan, Vibhore (2004). Power Consumption Breakdown on a Modern Laptop. In *International Workshop on Power-Aware Computer Systems*, pp. 165–80. Springer.

Mahmood, Danish, Javaid, Nadeem, Alrajeh, Nabil, Khan, Zahoor Ali, Qasim, Umar, Ahmed, Imran, and Ilahi, Manzoor (2016). Realistic Scheduling Mechanism for Smart Homes. *Energies*, **9**(3), 202.

Mariano-Hernández, D., Hernández-Callejo, L., Zorita-Lamadrid, A., Duque-Pérez, O., and García, F. Santos (2021). A Review of Strategies for Building Energy Management System: Model Predictive Control, Demand Side Management,

Optimization, and Fault Detect & Diagnosis. *Journal of Building Engineering*, **33**, 101692.

Messenger, Roger A., and Abtahi, Amir (2018). *Photovoltaic Systems Engineering*. CRC Press.

Mokrian, Pedram, Stephen, Moff et al. (2006). A Stochastic Programming Framework for the Valuation of Electricity Storage. In *26th USAEE/IAEE North American Conference*, pp. 24–7. Citeseer.

Munzke, Nina, Schwarz, Bernhard, and Barry, James (2017a). The Impact of Control Strategies on the Performance and Profitability of Li-Ion Home Storage Systems. *Energy Procedia*, **135**, 472–81.

Munzke, Nina, Schwarz, Bernhard, and Barry, James (2017b). Performance evaluation of Household Li-Ion Battery Storage Systems. In *The 27th International Ocean and Polar Engineering Conference*. OnePetro.

Neapolitan, Richard E., and Nam, Kwang Hee (2018). *AC Motor Control and Electrical Vehicle Applications*. CRC Press.

Nge, C. L., Midtgård, O. M., Sætre, T. O., and Norum, L. (2010). Energy Management for Grid-Connected PV System with Storage Battery. In *Proceedings of the 25th European Photovoltaic and Solar Energy Conference, Valencia, Spain*.

Ookubo, Kenji, Kousaka, Masaaki, and Ikeda, Kenji (2008). Recent Power Transformer Technology. *Fuji Electric Review*, **45**, 90–5.

Pan, Feng, and Wang, Qing (2015). Redox Species of Redox Flow Batteries: A Review. *Molecules*, **20**(11), 20499–517.

Ponomarev, Pavel, Minav, Tatiana, Aman, Rafael, and Luostarinen, Lauri (2015). Integrated Electro-Hydraulic Machine with Self-Cooling Possibilities for Non-Road Mobile Machinery. *Strojniški Vestnik/Journal of Mechanical Engineering*, **61**(3).

Reddy, Thomas B. (2011). *Linden's Handbook of Batteries*. McGraw-Hill Education.

Ried, Sabrina, Schmiegel, Armin Uwe, and Munzke, Nina (2020). Efficient Operation of Modular Grid-Connected Battery Inverters for RES Integration. *Advances in Energy System Optimization: Proceedings of the 2nd International Symposium on Energy System Optimization*, pp. 165–78.

Rodrigues, E. M. G., Fernandes, C. A. S., Godina, R., Bizuayehu, A. W., and Catalão, J. P. S. (2014). NaS Battery Storage System Modeling and Sizing for Extending Wind Farms Performance in Crete. In *2014 Australasian Universities Power Engineering Conference (AUPEC)*, pp. 1–6. IEEE.

Salomaa, Ville (2017). Efficiency Study of an Electro-Hydraulic Excavator. Master's thesis.

Sazali, Norazlianie, Wan Salleh, Wan Norharyati, Jamaludin, Ahmad Shahir, and Mhd Razali, Mohd Nizar (2020). New Perspectives on Fuel Cell Technology: A Brief Review. *Membranes*, **10**(5), 99.

Schimpe, Michael, Becker, Nick, Lahlou, Taha, Hesse, Holger C., Herzog, Hans-Georg, and Jossen, Andreas (2018). Energy Efficiency Evaluation of Grid Connection Scenarios for Stationary Battery Energy Storage Systems. *Energy procedia*, **155**, 77–101.

Schmiegel, Armin Uwe (2014). Die Leiden des Alters. *PV Magazine*.

Schmiegel, Armin Uwe, and Kleine, Andreas (2013). Upper Economical Performance Limits for PV-Storage Systems. In *Proceedings of the 28th European Photovoltaic Solar Energy Conference and Exhibition*, pp. 3745–50.

Schmiegel, Armin Uwe, and Kleine, Andreas (2014). Optimized Operation Strategies for PV Storages Systems Yield Limitations, Optimized Battery Configuration and the Benefit of a Perfect Forecast. *Energy Procedia*, **46**, 104–13.

Schmiegel, Armin Uwe, Koch, Karl, Meissner, Andre, Knaup, Peter, Jehoulet, Christoph, Schuh, Holger, Landau, Markus, Braun, Martin, Bündenbender, Kathrin, Geipel, Rolf et al. (2010). The Sol-Ion System, an Integrated PV-System with Lithium-Ion Batteries: system performance. In *Proceedings of the 25th Photovoltaic and Solar Energy Conference*, pp. 5–7.

Sen, Virendra, Lal, Jaya Dipti, and Charhate, S. V. (2015). A Review: Power Consumption at Base Stations in Wireless Networks. *International Journal of Engineering and Management Research (IJEMR)*, **5**(1), 260–6.

She, Xu, Huang, Alex Q., and Burgos, Rolando (2013). Review of Solid-State Transformer Technologies and Their Application in Power Distribution Systems. *IEEE Journal of Emerging and Selected Topics in Power Electronics*, **1**(3), 186–98.

Shigematsu, Toshio et al. (2011). Redox Flow Battery for Energy Storage. *SEI Technical Review*, **73**(7), 13.

Smets, Arno H. M., Jäger, Klaus, Isabella, Olindo, Swaaij, René A. C. M. M., and Zeman, Miro (2015). *Solar Energy: The Physics and Engineering of Photovoltaic Conversion, Technologies and Systems*. UIT Cambridge.

Sopian, Kamaruzzaman, Othman, Mohd Yusof, and Rahman, Mohd Azhar Abd (2005). Performance of a Photovoltaic Diesel Hybrid System in Malaysia. *ISESCO Science and Technology Vision*, **1**, 37–9.

Srinivasan, Supramaniam (2006). *Fuel Cells: From Fundamentals to Applications*. Springer Science & Business Media.

Sterner, Michael, and Stadler, Ingo (2017). *Energiespeicher-Bedarf, Technologien, Integration*. Springer.

Stetz, Thomas (2013). Autonomous Voltage Control Strategies in Distribution Grids with Photovoltaic Systems. PhD thesis, Kassell University.

Stetz, Thomas (2014). *Autonomous Voltage Control Strategies in Distribution Grids with Photovoltaic Systems: Technical and Economic Assessment*. Volume 1. kassel university press GmbH.

Stetz, Thomas, von Appen, Jan, Niedermeyer, Fabian, Scheibner, Gunter, Sikora, Roman, and Braun, Martin (2015). Twilight of the Grids: The Impact of Distributed Solar on Germany's Energy Transition. *IEEE Power and Energy Magazine*, **13**(2), 50–61.

Strzelecki, Ryszard Michal (2008). *Power Electronics in Smart Electrical Energy Networks*. Springer Science & Business Media.

Sun, Daoming, Yu, Xiaoli, Wang, Chongming, Zhang, Cheng, Huang, Rui, Zhou, Quan, Amietszajew, Taz, and Bhagat, Rohit (2021). State of Charge Estimation for Lithium-Ion Battery Based on an Intelligent Adaptive Extended Kalman Filter with Improved Noise Estimator. *Energy*, **214**, 119025.

Tang, Ao, Bao, Jie, and Skyllas-Kazacos, Maria (2014). Studies on Pressure Losses and Flow Rate Optimization in Vanadium Redox Flow Battery. *Journal of Power Sources*, **248**, 154–62.

Tawalbeh, Mohammad, Eardley, Alan et al. (2016). Studying the Energy Consumption in Mobile Devices. *Procedia Computer Science*, **94**, 183–9.

Thiaux, Yaël, Schmerber, Louis, Seigneurbieux, Julien, Multon, Bernard, and Ahmed, Hamid Ben (2009). Comparison between Lead-Acid and Li-Ion Accumulators in Stand-alone Photovoltaic System Using the Gross Energy Requirement Criteria. In *24th European Photovoltaic Solar Energy Conference*, pp. 3982–3990.

Ulrich, Karl T. (2003). *Product Design and Development*. Tata McGraw-Hill Education.

Vanek, Francis, Albright, Louis, and Angenent, Largus (2008). *Energy Systems Engineering*. McGraw-Hill Professional Publishing.

Vidas, Leonardo, and Castro, Rui (2021). Recent Developments on Hydrogen Production Technologies: State-of-the-Art Review with a Focus on Green-Electrolysis. *Applied Sciences*, **11**(23), 11363.

Von Appen, Jan (2018). *Sizing and Operation of Residential Photovoltaic Systems in Combination with Battery Storage Systems and Heat Pumps*, Volume 5, Kassel University Press GmbH.

Von Appen, J., Schmiegel, A., and Braun, M. (2012). Impact of PV Storage Systems on Low Voltage Grids: A Study on the Influence of PV Storage Systems on the Voltage Symmetry of the Grid. In *27th European Photovoltaic Solar Energy Conference*, 3822–8.

Von Appen, Jan, Stetz, Thomas, Braun, Martin, and Schmiegel, Armin (2014). Local Voltage Control Strategies for PV Storage Systems in Distribution Grids. *IEEE Transactions on Smart Grid*, **5**(2), 1002–9.

Wang, Chao-Yang, Zhang, Guangsheng, Ge, Shanhai, Xu, Terrence, Ji, Yan, Yang, Xiao-Guang, and Leng, Yongjun (2016). Lithium-Ion Battery Structure that Self-Heats at Low Temperatures. *Nature*, **529**(7587), 515–18.

Wang, Hongmei, Wang, Qingfeng, and Hu, Baozan (2017). A Review of Developments in Energy Storage Systems for Hybrid Excavators. *Automation in Construction*, **80**, 1–10.

Wang, M., Vandermaar, A. J., and Srivastava, K. D. (2002). Review of Condition Assessment of Power Transformers in Service. *IEEE Electrical Insulation Magazine*, **18**(6), 12–25.

Weber, Adam Z., Mench, Matthew M., Meyers, Jeremy P., Ross, Philip N., Gostick, Jeffrey T., and Liu, Qinghua (2011). Redox Flow Batteries: A Review. *Journal of Applied Electrochemistry*, **41**(10), 1137–64.

Wei, Zhongbao, Tseng, King Jet, Wai, Nyunt, Lim, Tuti Mariana, and Skyllas-Kazacos, Maria (2016). Adaptive Estimation of State of Charge and Capacity with Online Identified Battery Model for Vanadium Redox Flow Battery. *Journal of Power Sources*, **332**, 389–98.

Weniger, Johannes, Tjaden, Tjarko, Bergner, Joseph, and Quaschning, Volker (2016). Sizing of Battery Converters for Residential PV Storage Systems: How Much Rated

Power Is Sufficient? In *10th International Renewable Energy Storage Conference and Energy Storage, Düsseldorf.*

Williamson, Ian, Ruth, K., Philip, T. D., David, R., Stathis, T. D., and Aristomenis, N. D. (2003). Intelligent Load Control Strategies Utilising Communication Capabilities to Improve the Power Quality of Inverter Based Renewable Island Power Systems. In *International Conference RES for Island, Tourism & Water.*

Xi, Xiaomin, Sioshansi, Ramteen, and Marano, Vincenzo (2014). A Stochastic Dynamic Programming Model for Co-optimization of Distributed Energy Storage. *Energy Systems*, **5**(3), 475–505.

Yoo, Jaeyeong, Park, Byungsung, An, Kyungsung, Al-Ammar, Essam A., Khan, Yasin, Hur, Kyeon, and Kim, Jong Hyun (2012). Look-Ahead Energy Management of a Grid-Connected Residential PV System with Energy Storage under Time-Based Rate Programs. *Energies*, **5**(4), 1116–34.

Zhang, Xiaogang, Huang, Weinan, Ge, Lei, and Quan, Long (2019). Research on Response Characteristics and Energy Efficiency of Power Unit Used for Electric Driving Mobile Machine. *IEEE Access*, **7**, 125747–53.

Züttel, Andreas (2004). Hydrogen Storage Methods. *Naturwissenschaften*, **91**(4), 157–72.

Index